A. Gvishiani · J.O. Dubois
**Artificial Intelligence and Dynamic Systems
for Geophysical Applications**

Springer
*Berlin
Heidelberg
New York
Barcelona
Hong Kong
London
Milan
Paris
Tokyo*

Alexei Gvishiani · Jacques O. Dubois

Artificial Intelligence and Dynamic Systems for Geophysical Applications

With 96 Figures and 14 Tables

 Springer

Professor Dr. Alexei Gvishiani
Schmidt United Institute of Physics of the Earth RAS
Institute of Physics of the Earth
Molodezhnaya Str. 3
117964 Moscow, Russia
E-mail: gvi@wdcb.ru

Professor Dr. Jacques Octave Dubois
Institut de Physique du Globe de Paris
Place Jussieu 4
75252 Paris Cedex 5, France
E-mail: dubois@ipgp.jussieu.fr

ISBN 3-540-43258-2 Springer-Verlag Berlin Heidelberg New York

Library of Congress Cataloging-in-Publication Data
Gvishiani, A. D. Artificial intelligence and dynamic systems for geophysical applications /
Alexej Gvishiani, Jacques O. Dubois. p. cm. Includes bibliographical references and index.
ISBN 3540432582 (alk. paper)
1. Artificial intelligence–Geophysical applications. 2. Geodynamics–Mathematical
models. 3. Differentiable dynamical systems. 4. Nonlinear theories. I. Dubois, J. O.
(Jaques Octav), 1931 - II. Title.
QE501.4.M38 G85 2002   550´.1´5118–dc21   2002020910

This work is subject to copyright. All rights are reserved, whether the whole or part of the material is concerned, specifically of translation, reprinting, reuse of illustrations, recitation, broadcasting, reproduction on microfilm or in any other way, and storage in data banks. Duplication of this publication or parts thereof is permitted only under the provisions of the German Copyright Law of September 9, 1965, in its current version, and permission for use must always be obtained from Springer-Verlag. Violations are liable for prosecution under the German Copyright Law.

Springer-Verlag Berlin Heidelberg New York
a member of BertelsmannSpringer Science+Business Media GmbH

http://www.springer.de

© Springer-Verlag Berlin Heidelberg 2002
Printed in Germany

The use of general descriptive names, registered names, trademarks, etc. in this publication does not imply, even in the absence of a specific statement, that such names are exempt from relevant protective laws and regulations and therefore free for general use.

Production: PRO EDIT GmbH, Heidelberg, Germany
Cover Design: Erich Kirchner, Heidelberg, Germany
Typesetting: Camera-Ready by Author

Printed on acid-free paper   SPIN: 10856348   30/3130/Di   5 4 3 2 1 0

Artificial Intelligence and Dynamic Systems in Geophysical Applications

A. Gvishiani, J.O. Dubois

Foreword

This volume is the second of a two-volume series written by A. Gvishiani and J.O. Dubois.
The series presents the application of new artificial intelligence and dynamic systems techniques to geophysical data acqusition, management and studies. Most of the mathematical models, algorithms and tools presented were developed by the authors. The first volume of the series, published in 1998, is entitled " Dynamical Systems and Dynamic Classification Problems in Geophysical Applications ". It is devoted to the application of dynamic systems, pattern recognition and finite vector classification with learning to a variety of geophysical problems.
This volume introduces geometrical clustering and fuzzy logic approaches to geophysical data analysis. A significant part of the volume is devoted to applying the artificial intelligence techniques introduced in volumes 1 and 2, to fields such as seismology, geodynamics, geoelectricity, geomagnetism, aeromagnetics, topography and bathymetry.
As in the first volume, this volume consists of two parts, describing complementary approaches to the analysis of natural systems. The first part, written by A. Gvishiani, deals with new ideas and methods in geometrical clustering and the fuzzy logic approach to geophysical data classification. It lays out the mathematical theory and formalized algorithms that form the basis for classification and clustering of the vector objects under consideration. It lays the foundation for the second part of this book which is the use of this classification in the study of dynamical systems.
The focus of the first theoretical part is fuzzy set mathematics and logic techniques along with the mathematical model of "geometrical illumination" for

object clustering. It leadss to numerous concrete classifications with learning and clustering algorithms such as RODIN, FLARS, etc. This approach is original and brand new: it was introduced in geophysical literature by A. Gvishiani and co-authors in late 90s and is stil being actively developed.

A few well known and widely used Internet oriented geophysical data bases are described in the first part of the book. Among them is SMDB- Strong Ground Motion Earthquake Data Base. The SMDB was originaly developed in the early 90s by J. Bonnin, A. Gvishiani, B. Mohammadioun and M. Zhizhin, as a part of a French-Russian geophysical collaboration programme. In 1995 it became an official "Key-nodal" activity of the European Mediterranean Seismological Centre, making it one of the most important geophysical data bases in Europe.

The developed mathematical techniques are applied to a wide range of important geophysical problems, including Among pattern recognition in earthquake-prone areas, syntactic classification of seismic records, recognition of magnetic anomalies along the Mid-Atlantic Ridge, clustering analysis for magnetic anomalies studies and image recognition of linear and circular structures in bathymetry data. These results were obtained by the authors in the framework of the French-Russian collaboration programme between the Institutes of Physics of the Earth in Paris and Strasbourg and the United Institute of Physics of the Earth in Moscow.

The second part, written by J.O. Dubois, is concerned with various theoretical tools and their applications to modeling of natural systems using large geophysical data sets. Fractals and dynamic systems are used to analyse geomorphological (continental and marine), hydrological, bathymetrical, gravimetrical, seismological, geomagnetical and volcanological data.

In these applications chaos theory and the concept of self-organized criticality are used to describe the evolution of dynamic systems.

This book completes the two volume series written by J.O. Dubois and A. Gvishiani. The first volume is devoted to the mathematical and algorithmical basis of the proposed artificial intelligence techniques; this volume presents a wide range of applications of those techniques to geophysical data processing and research problems. At the same time it presents a reader with another algorithmic approach based on fuzzy logic and geometrical illumination models.

Many readers will be interested in the two volumes (vol.1, J.O. Dubois, A. Gvishiani "Dynamic Systems and Dynamic Classification Problems in Geophysical Applications" and the present vol.2, A. Gvishiani, J.O. Dubois "Artificial Intelligence and Dynamic Systems in Geophysical Applications") as a package.

This book will be of interest to geophysicists, geologists, engineers, applied mathematicians, computer programmers and networking specialists, who use or are interested in using artificial intelligence methods to analyse large data sets. The authors are particularly concerned with geophysical data sets, but the techniques described have applications in many other fields. It will be of particular interest to readers concerned with the prediction of different kinds of natural processes that develop in space and time, and with the monitoring and testing of various prediction algorithms.

Professor Jacques Émile Dubois
CODATA LAST PRESIDENT
Paris, France

Acknowledgement

This second volume of the monography written by the authors in 1996-2001 presents the results of more than 20 years of intensive cooperation between French and Russian (Soviet) geophysicists, mathematicians and informaticians. The theme of the book is what is called nowadays artificial intelligence approach. It includes pattern recognition and geoinformatics applied to geophysical techniques, natural hazard, geodynamical data analysis. Working on the book, the authors maintained permanent collaboration with major international organizations dealing with global data managment matter as CODATA, ICSU Panel on World Data Centers, European-Mediterranean Seismological Centre.

Doctor John Rumble (NIST, USA) the President of CODATA and Professor Jacques Émile Dubois, the last President of CODATA encouraged the authors to orient this study towards CODATA goals and objectives what gave a real opportunity to present both volumes of this monography in the series "DATA AND KNOWLEDGE IN A CHANGING WORLD".

On different stages of the project, the French Russian working group of geophysicists, mathematician, informaticians and geologists contributed to development and evaluation of presented theories and algorithms, as well as to their applications to geophysical data analysis.

The french contributors were: Claude Jean Allègre, Jean-Louis Le Mouël, Vincent Courtillot, Raul Madariaga, Armand Galdeano, Michel Diament, Pascal Bernard (Institut de Physique du Globe de Paris). Jean Bonnin, Armando Cisternas, Michel Cara, Daniel Rouland (Institut de Physique du Globe de Strasbourg), Jean Sallantin, Hervé Philip (Université de Montpellier), Christian Weber (BRGM), Bagher Mohammadioun (CEA). The main Russian (Soviet) contributors were: Dimitri Rundkuist (Institute of Geology and Geochronology of St Petersburg), Vladimir Strakhov, Vladimir Keilis Borok, Alexander Soloviev, Vladimir Gurvitch, Vladimir Kossobokov, Michael Zhizhin, Alexander Beriozko, Valentin Mikhailov, Alexei Burov, Elena Graeva (Institute of Physics of the Earth, Moscow).

All of them are here greatly acknowledged.

We acknowledge really important contribution of the Institut de Physique du Globe de Paris that privided continuous support to this study by organizing and funding visits of Russian (Soviet) scientists in France. We are grateful to directors of IPGP, Jean-Louis Le Mouël, Vincent Courtillot and Claude Jaupart to Michel Diament director of the Laboratoire de Gravimétrie et de Géodynamique, to Guy Aubert head of foreign affairs of IPGP, as well as to IPGP officers who provided excellent administrative and coordinating job: Jeanine Mivielle (IPGP head of personal), Thérèse Chétail secretary general of IPGP, Hélène Robic, secretary. Antoine Sempere, head of CNRS office in Moscow is warmly acknowledged for his permanent efficient help of the project development.

We acknowledge as well the important contribution of Schmidt United Institute of Physics of the Earth, Russian Academy of Sciences in Moscow, headed by Vladimir Strakhov and Geophysical Centre of Russian Academy od Sciences headed by Gennadi Sobolev.

We are also grateful to Dr Wayne Crawford, who corrected English language of the manuscript, to Professor Jacques Émile Dubois who accepted to write the preface, to Hélène Robic who solved the main problems in presentation of figures and of the final form of the manuscript, and to Alexander Beriozko for preparation and computerized versions of some figures.

Contents

I Artificial Intelligence in Geophysical Data Analysis

1 Dynamic and Fuzzy Logic Clustering and Classification.   **1**
- 1.1 Syntactic Algorithms for Time Series Classification   1
  - 1.1.1 Structural representation of waveforms: time-series parametrisation   1
  - 1.1.2 Structural dissimilarity: Levenstein distance   4
  - 1.1.3 "K-mean distances" decision rule. SPARS Algorithm .   7
- 1.2 Fuzzy Logic Approach to Classification   7
  - 1.2.1 Basic Definitions   7
  - 1.2.2 Operations on Fuzzy Sets   11
  - 1.2.3 Fuzzy binary relations   14
  - 1.2.4 The fuzzy logic approach to time series classification   15
  - 1.2.5 Fuzzy version of SPARS algorithm (FSPARS)   24
- 1.3 Clustering Algorithms   34
  - 1.3.1 "Lighting" and clustering in finite metric spaces . . .   34
  - 1.3.2 Algorithm "RODIN":   44
  - 1.3.3 Fuzzy clustering   56
- 1.4 Linear and Circle Structures Recognition Algorithms   64
  - 1.4.1 Differential operations in GIS environment   64

2 Applications to Physics of the Earth, Seismology and Engineering Seismology   **71**
- 2.1 Syntactic Classification of Engineering Seismology Data . . .   71
  - 2.1.1 Strong ground motion earthquake data base   71
  - 2.1.2 Classification of strong motion records according to geotectonic regions   74
- 2.2 Seismological Data Classification   78
  - 2.2.1 Seismic observation networks and databases   78
  - 2.2.2 Syntactic classification of seismograms   81

2.3 Recognition of Magnetic Anomalies along the Mid-Atlantic Ridge 88
    2.3.1 Method 88
    2.3.2 Data, results and interpretation 94
2.4 Clustering Analysis for Magnetic Field Studies 99
    2.4.1 Euler deconvolution 99
    2.4.2 Magnetic Data Clustering 105
    2.4.3 Euler solutions by clusterization in the gulf of Saint Malo 109
2.5 Linear and Circular Clustering of Bathymetric Data in the Wharton Basin 114

3 Recognition of Earthquake-Prone Areas and Seismic Hazard Assessment   123

3.1 Earthquake-prone Areas in the Western Alps 123
    3.1.1 E.C. application to the neotectonic scheme ... 135
    3.1.2 CORA algorithm classification results 137
    3.1.3 Comparison of CORA-algorithm results with E.C.-algorithm results 142
    3.1.4 Control experiments 142
3.2 Seismically Dangerous Zones in the Pyrénées 146
3.3 Comparison between Earthquake-prone areas in the Pyrénées and the Alps 158
3.4 Strong Earthquakes Prone-areas in the Great Caucasus ... 163

II Fractals and Dynamic Systems   171

4 Fractals and Multifractals   173

4.1 A brief Review of Fractals and Multifractal Analysis 173
4.2 Geomorphology (Continental and Marine) Hydrology ... 177
    4.2.1 Continental Earth's Relief, Topography, Self-affine Fractals 177
    4.2.2 Bathymetry, Seafloor Roughness 179
    4.2.3 Fractal and Multifractal analysis applied to river basins and to river flows 184
4.3 Gravity Anomalies and Structural Inversion Modeling 192
    4.3.1 Fractal analysis of gravity anomalies 193
4.4 Geomagnetism 196

Contents XIII

	4.4.1	The Fractal Structure of the Interplanetary Magnetic Field	196
	4.4.2	Fractal dimension and power law for geomagnetic time series	197
4.5	Tectonics, Seismicity, Volcanology		199
	4.5.1	Renormalization group theory	199
	4.5.2	Fragmentation, Fracturation	202
	4.5.3	Tectonics, Fractals and Multifractals	213
	4.5.4	Tectonics. Study of Surface Faults	220
	4.5.5	Multifractals and Wavelets applied to fault fields	223
	4.5.6	Seismicity, Gutenberg and Richter Law, Multifractals	226
	4.5.7	Power Law or Poisson Law ?	229
4.6	Fragmentation, tectonics, seismicity, synthesis trial		236

5 Dynamic System Properties and Long Time Series   239
- 5.1 Geomorphology, Hydrology . . . 239
  - 5.1.1 Correlation Function and Rivers Flows . . . 239
  - 5.1.2 Wavelets Applied to Floods . . . 240
- 5.2 Seismology . . . 243
  - 5.2.1 Cantor Dust application . . . 243
  - 5.2.2 SOC applications to seismology . . . 245
- 5.3 Volcanology . . . 246
  - 5.3.1 Application to Volcanic Eruptions . . . 246
  - 5.3.2 Cantor Dust and Correlation Function applications . 248
  - 5.3.3 Volcano behaviour and Self Organized Criticality . . 253
  - 5.3.4 Multifractal analysis . . . 256
- 5.4 Geomagnetism Study at Different Time Scales . . . 258
  - 5.4.1 Geomagnetic Reversals . . . 258
  - 5.4.2 Temporal Variations of the Magnetic Field Vector . . 262
  - 5.4.3 Theoretical Modeling . . . 270
- 5.5 Others . . . 282
  - 5.5.1 Heat and Water Transport in an Underground cave . 282

6 Conclusions and Perspectives   285
- 6.1 About Part I . . . 285
- 6.2 About Part II . . . 288
  - 6.2.1 Intermittency and turbulence . . . 289
  - 6.2.2 The problem of short time series, slow and fast dynamics in coupled systems . . . 290
  - 6.2.3 Self-Organised criticality, SOC . . . 292

6.2.4	Mastering and controlling Chaos	292

III References   295

IV Index   333

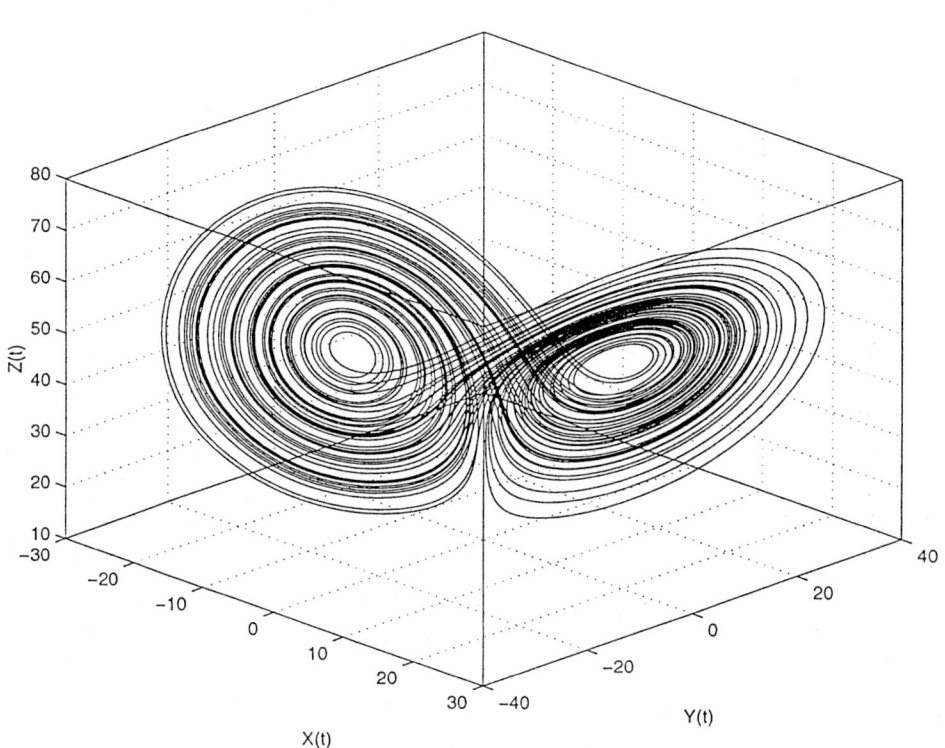

Lorenz Attractor,

$$\begin{cases} \dfrac{dX}{dt} = P_r(Y - X), \\ \dfrac{dY}{dt} = -XZ + rX - Y, \\ \dfrac{dZ}{dt} = XY - bZ, \end{cases}$$

(after Dubois and Gvishiani, 1998).

Part I

Artificial Intelligence in Geophysical Data Analysis

Chapter 1

Dynamic and Fuzzy Logic Clustering and Classification.

1.1 Syntactic Algorithms for Time Series Classification

1.1.1 Structural representation of waveforms: time-series parametrisation

The syntactic approach to classification and pattern recognition[1] is based on representation of objects of recognition as combination of letters from finite or infinite alphabet.

To apply this approach we have to obtain a structural representation of a seismic record. Here we describe a parametrisation that allows us to acquire this representation. The idea is to calculate the system of masses and springs that best fits each record and then to compare the mass-spring systems for each records to determine how the smoothed energy evolves with time.

Mathematically, we Hilbert transform a seismic record to obtain its energy envelope, we match the envelope using a set of spring-mass filters with different central frequencies and fixed damping. We use special averaging in

---

[1] The results described in this chapter have been obtained in 1991-1996 by A. Gvishiani and M. Zhizhin (Schmidt Institute of Physics of the Earth in Moscow), J. Bonnin (Institut de Physique du Globe de Strasbourg) and B. Mohammadioun (CEA, France). The project has been funded by CEA, France and European Mediterranean Seismological Centre which are friendly acknowledged.

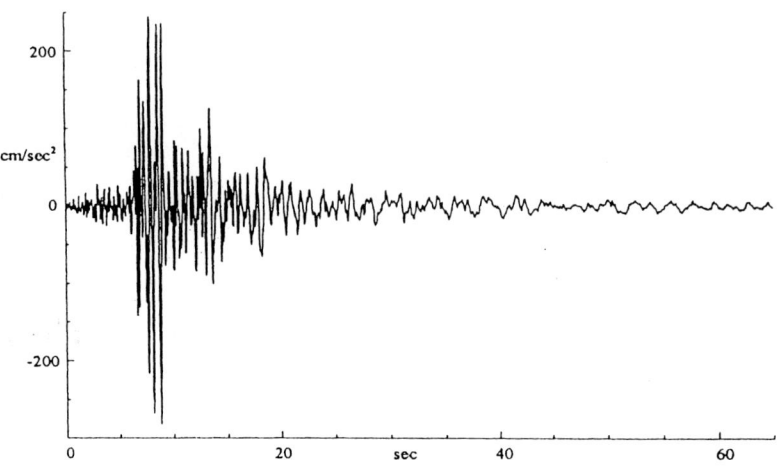

Figure 1.1: *A seismic record* This figure shows the strong motion record of the Coalinga, May 2, 1983, main shock.

---

the frequency domain to increase the sampling interval (equivalent to time-domain smoothing). As a result, we obtain a set of smoothed envelopes of spring-masse responses to the seismic signal, which we call the " discrete time-frequency diagram".

Let us assume that any time-series can be described as $a_n = a(t_0 + (n-1)\Delta t)$, $n = 1, \cdots, N_s$, where N_s is number of samples in the record and Δt is sampling interval.

Algorithm 1: Parametrisation

- *Input.* The parametrisation algorithm receives a time-series (seismic record) $a(t)$ to be parametrised (Figure 1.1), the effective frequency band of the record F_{min}, F_{max}, the number of filters to be applied N_f, the time-window duration δt, and the damping of the spring-mass filters ζ.

- *Step 1.* Fourier transform the given record and zero out negative frequencies ("Hilbert transform").

$$a \xrightarrow{H} \overline{a}.$$

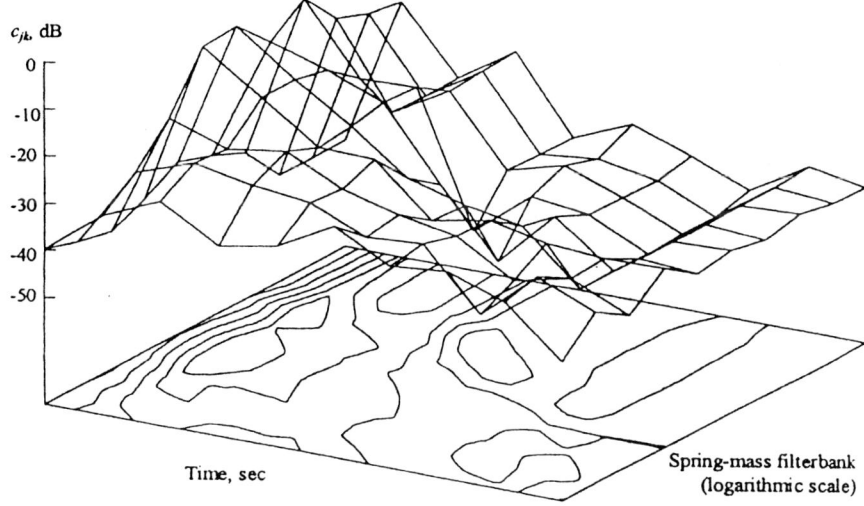

Figure 1.2: *Time-frequency diagram* Time-frequency diagram of strong motion record after narrow-band pass filtering by 10 damped spring-mass filters.

- *Step 2.* For $j = 1, \cdots, N_f$, apply spring-mass narrow-band filters

$$\widetilde{a_j}(f) = \bar{a}(f) H(f, f_j^H), \qquad (1.1)$$

where $H(f, f_j^H)$ is the j-th spring-mass filter and f_j^H is its dominant frequency.

- *Step 3.* For $j = 1, \cdots, N_f$, average/thin out the spectrum in the frequency domain, which increases the sampling interval in the time-domain: $\widetilde{a_j} \longrightarrow^T b_j$.

- *Step 4.* For $j = 1, \cdots, N_f$, reverse Fourier transform the thinned out signals: $b_j \longrightarrow^{RF} \bar{b}_j$.

- *Output.* In the case of discrete data the output of *Step 4* is a matrix $\bar{b} = (\bar{b}_{jk} exp(i\varphi_{jk}))$ with dimension $N_f \times N_t$, where N_t is the number of time-windows after smoothing in the record, $|\bar{b}_{jk}|^2$ is the amount of power energy distributed in k-th frequency band, and φ_{jk} is the phase of the smoothed response. Finally, our parametrisation out puts a peak-value normalised matrix $c = (c_{jk})$ (Figure 1.2).

Thus, we can parameterize the structural representation of the record as a sequence of instantaneous spectra calculated for fixed duration time-windows:

$$w_i(\delta t) = \{a(t) : t \in [t_i,\ t_{i+1}]\},\ i = 1,\ \cdots,\ N_t,\ t_i = t_0 + i\delta t, \quad (1.2)$$

where N_t is the number of fragments, and δt is the duration of those fragments.

1.1.2 Structural dissimilarity: Levenstein distance

Having obtained a structural representation of the records, we can measure the difference between these strutures using various methods. A standard method is to cluster the instantaneous spectra as vectors in R^{N_f}, then to assign a label to each cluster and substitute each instantaneous spectrum with the cluster label. Every seismic record is thus represented as a word consisting of letters of the finite alphabet of those labels. We can then calculate the *Levenstein distance* defined as the minimum number of editing operations necessary to transform a given word to another word.

- *Definition 1.* Let us consider a word $A \equiv a_1 \cdots a_n$, where $a_i,\ i = 1,\ \cdots,\ n$ are letters of alphabet Ω. We define three editing operations:
   - *deletion:* $del(A, i) = a_1 \cdots a_{i-1} a_{i+1} \cdots a_n$;
   - *insertion:* $ins(A, i, b) = a_1 \cdots a_{i-1} b a_i \cdots a_n$, where $b \in \Omega$;
   - *substitution:* $sub(A, i, c) = a_1 \cdots a_{i-1} c a_{i+1} \cdots a_n$, where $c \in \Omega$.

- *Definition 2.* Let us consider two words $A \equiv a_1 \cdots a_n$ (source word) and $B \equiv b_1 \cdots b_n$ (target word) where $a_i,\ i = 1,\ \cdots,\ n$ and $b_j,\ j = 1,\ \cdots,\ m$ are letters of the alphabet Ω. The *Levenstein distance* between A and B is the minimal number of editing operations (insertion, deletion and substitution) that must be applied to word A to obtain word B (Figure 1.3).

It is not difficult to show that the *Levenstein distance* introduces the structure of metric space to the set of words from the alphabet Ω.

This dissimilarity measure can be used for structural classification of seismic records or for cluster analysis. In syntactic analysis some of the letters and editing operations are more significant than others.

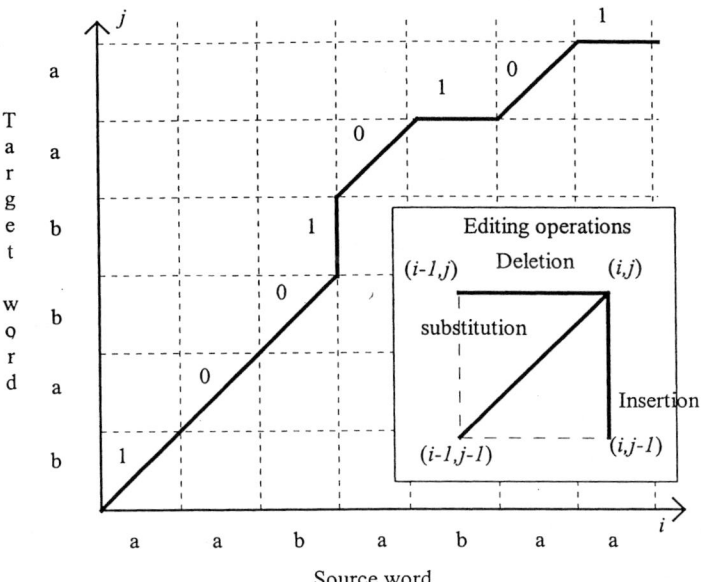

Figure 1.3: *Levenstein distance* Distance calculation between two words. (Here equal to 4).

We give more importance to these operations by introducing weighted Levenstein distances.

- *Definition 3.* To each editing operation we assign a real number which is called the weight of the operation:

$$w_{del}(A, i) \quad \text{--weight of a deletion;}$$
$$w_{ins}(A, i, b) \quad \text{--weight of an insertion;}$$
$$w_{sub}(A, i, c) \quad \text{--weight of a substitution}$$

- *Definition 4.* Let us consider the words $A \equiv a_1 \cdots a_n$ and $B \equiv b_1 \cdots b_n$ where a_i, $i = 1, \cdots, n$ and b_j, $j = 1, \cdots, m$ are letters of the alphabet Ω. The weighted Levenstein distance between A and B is the minimum total weight of the editing operations necessary to transform word A to word B.

We also introduce the notion of the empty, or *null* (" ") element, which is used to shift elements along a record to find the best fit.

Using the notion of the null element *Definition 4* is equivalent to

$$w(a_i, null) = q_1 + q_2 \| a_i \|^2 \text{ —weight of a deletion of the letter } a_i,$$
$$w(null, b_j) = q_1 + q_2 \| b_j \|^2 \text{ —weight of an insertion of the letter } b_j,$$
$$w(a_i, b_j) = \| a_i - b_j \|^2 \text{ —weight of a substitution of letter } a_i \text{ by letter } b_j$$

where $\| \cdot \|$ is the usual Euclidean norm for vectors of the cluster centroids of instantaneous spectra and q_1 and q_2 are free parameters of the algorithm. Using *null* elements the path shown in Figure 1.3 can be written as:

$$\begin{array}{llllllll}
WordA: & a & a & b & - & a & b & a & a \\
Editing: & sub & & & ins & & del & & del \\
WordB: & b & a & b & b & a & - & a & -
\end{array}$$

The distance is calculated using a dynamic programming technique [360].

Algorithm 2: Distance calculation

- *Input.* The algorithm receives two words: $A \equiv a_1 \cdots a_n$ and $B \equiv b_1 \cdots b_m$, where $a_i, i = 1, \cdots, n$, and $b_j, j = 1, \cdots, m$ are letters of the alphabet Ω, that represent weights of the editing operations (values of parameters q_1 and q_2, see above).

- *Step 1.* Initialisation: $D(0,0) = 0$ is the accumulated distance.

- *Step 2.* Recursion :

$$D(i,j) = \min \left\{ \begin{array}{ll} D(i-1) & +w(a_i, null), \\ D(i, j-1) & +w(null, b_j), \\ D(i-1, j-1) & +w(a_i, b_j) \end{array} \right\}, \quad (1.3)$$

where $1 \leq i \leq n,\ 1 \leq j \leq m$.

- *Output.* $D(n, m)$ – weighted Levenstein distance.

In our study we used an even more general way to calculate the dissimilarity measure. We considered each vector of the instantaneous spectrum as a separate cluster, (i.e. a letter of an infinite alphabet) and we calculated the weights using the above formulae. This weight calculation scheme directly estimates the importance of each fragment. It is similar to the DTW (Dynamic Time Warping) algorithms used in speech recognition and is steady with respect to the insertion/deletion weights and the weight of substitution [373].

1.1.3 "K-mean distances" decision rule. SPARS Algorithm

We have outlined how to parameterize a data set and how to calculate its dissimilarity measurement. We will now describe how to classify parametrised records based on the dissimilarity measure. This method is known as the "K-mean distances" decision rule.

Definition 4. Let us assume that we have C different classes $S_i, i = 1, \cdots, C$, and we want to discriminate an object t into them. For each class we calculate the function

$$r(t, S, K) = \frac{1}{K} \min \sum_{i=1, x_i \in S_i}^{K} \rho(t, x_i) - \text{mean distance to the}$$

K-nearest neighbours from S_i,

where $\rho(\cdot, \cdot)$ is the distance defined above. We assign t to the class S_j for which $r(t, S_j, K)$ is minimum among all $r(t, S_i, K)$, $i = 1, +\cdots, C$.
Finally, all the records $\{t\}$ under consideration are becoming classified into one of the classes S_i. However the classification changes if we change K. In other words 2, 3 and 4 mean that distance rules can give different classifications. Stability of the obtained classifications, while we change the values of K, is an important evidence in favor of the classification obtained.

1.2 Fuzzy Logic Approach to Classification

1.2.1 Basic Definitions

The main feature of a classical set is the existence of a clear boundary between elements that do and don't belong to the set. In other words, if all our elements under consideration belong to the universal set $\mathcal{U}$, then a classical subset $\mathcal{A} \subset \mathcal{U}$ can be defined by a membership function that takes only two values : "elements belonging to $\mathcal{A}$" and "elements not belonging to $\mathcal{A}$".
Fuzzy set theory extends this notion by introducing more complicated membership functions.

We start with some formal definitions. We consider a basic set $\mathcal{U}$ that we will call universal.

Defintion 1. $\mathcal{A}$ is a fuzzy subset of $\mathcal{U}$ (or $\mathcal{A}$ is a fuzzy set in $\mathcal{U}$) if there exists a map

$$\mu_\mathcal{A} : \mathcal{U} \to \mathcal{L},$$

where $\mathcal{L}$-is some ordered set.
A common example of $\mathcal{L}$ is the segment $[0, 1]$.
We call the values of the map $\mu_\mathcal{A}(x)$, $x \in \mathcal{U}$ the degree of inclusion of the element x into the fuzzy set $\mathcal{A}$
We call $\mu_\mathcal{A}$ the membership function of the fuzzy set $\mathcal{A}$.
If $\mathcal{L} = [0, 1]$, then the following terminology is used:

- " $x \in \mathcal{U}$ completely belongs to $\mathcal{A}$" if and only if $\mu_\mathcal{A}(x) = 1$

- " $x \in \mathcal{U}$ does not belong to $\mathcal{A}$" if and only if $\mu_\mathcal{A}(x) = 0$

- " $x \in \mathcal{U}$ partialy belongs to $\mathcal{A}$" if and only if $0 < \mu_\mathcal{A}(x) < 1$.

Let us introduce two special fuzzy sets:

- The empty set $\emptyset$ such that $\mu_\emptyset(x) = 0 \ \forall x \in \mathcal{U}$

- The universal set $\mathcal{U}$ such that $\mu_\mathcal{U}(x) = 1 \ \forall x \in \mathcal{U}$

Any classical subset $\mathcal{A} \subset \mathcal{U}$ can be considered as a fuzzy set in $\mathcal{U}$ with the membership function

$$\mu_\mathcal{A}(x) = \begin{cases} 1, & \text{if } x \in \mathcal{A} \\ 0, & \text{if } x \bar{\in} \mathcal{A} \end{cases}$$

If the set $\mathcal{U}$ is finite or $\mathcal{U} = \mathcal{N}$-the set of natural numbers, then, the following notation defines its fuzzy subsets $\mathcal{A}$:

$$\begin{aligned}
\mathcal{A} &= \{\mu_\mathcal{A}(x_1)/x_1, \ \mu_\mathcal{A}(x_2)/x_2, \ \cdots, \ \mu_\mathcal{A}(x_n)/x_n, \ \cdots\} \\
\mathcal{A} &= \sum_{x_i \in \mathcal{A}} \mu_\mathcal{A}(x_i)/x_i, \quad \text{or} \\
\mathcal{A} &= \mu_\mathcal{A}(x_1)/x_1 + \mu_\mathcal{A}(x_2)/x_2 + \cdots + \mu_\mathcal{A}(x_n)/x_n + \cdots
\end{aligned}$$

where $\mathcal{U} = \{x_1, x_2, \cdots, x_n, \cdots\}$, and the symbol of summation is understood as the operation of classical sets union.

Examples

1. Let $\mathcal{U} = \{x_1, x_2, x_3\}$. Then $\mathcal{A} = \{0.2/x_1; 0.5/x_2; 0.8/x_3\} = 0.2/x_1 + 0.5/x_2 + 0.8/x_3$ is a fuzzy set in $\mathcal{U}$.

2. Let $\mathcal{U} = \{0, 1, \cdots, n, \cdots\}$ be a universal set all integers. We define a fuzzy subset $\mathcal{S}$ ("Small") membership function by the formula:

$$\mu_\mathcal{S}(n) = (1 + (\frac{n}{10})^2)^{-1},$$

Then the fuzzy set $\mathcal{S}$ can be written as

$$\mathcal{S} = \sum_{n=0}^{\infty} (1 + (\frac{n}{10})^2)^{-1}/n. \qquad (1.4)$$

Definition 2 If $\mathcal{A}$ is a fuzzy set in $\mathcal{U}$, then the element $x_0 \in \mathcal{U}$ such that $\mu_\mathcal{A}(x_0) = 0.5$ is called the transfer point of the set $\mathcal{A}$. For the fuzzy set $\mathcal{S}$ ("Small") the point of transfer x_0 is equal to 10.

If the universal set $\mathcal{U}$ is equivalent to the set of real numbers $\Re$, we use the following symbolization:

$$\mathcal{A} = \int_\mathcal{U} (\mu_\mathcal{A}(x)/x) dx \qquad (1.5)$$

Definition 3 Fuzzy sets with a membership function $\mu_\mathcal{A}:\mathcal{U} \to [0, 1]$ are called fuzzy sets of the first type. If $\mu_\mathcal{A}:\mathcal{U} \to \mathcal{L}$ where $\mathcal{L}$ is a fuzzy set of the first type from the segment $[0, 1]$, then $\mathcal{A}$ is called a fuzzy set of the second type or super fuzzy set.
The following example illustrates the definition.
In this case $\mathcal{U} = \{0, 0.1, 0.2, 0.3, \cdots, 0.9, 1\}$.
Let $\mathcal{A}$ be a fuzzy set in $\mathcal{U}$ that has only three elements x_1, x_2 and x_3 for which $\mu_\mathcal{A}(x) \neq 0$. The rest of the elements $\mathcal{U} \setminus \{x_1, x_2, x_3\}$ do not belong to $\mathcal{A}$. Assume the elements x_1, x_2, x_3 have different degrees of inclusion into the fuzzy set $\mathcal{A}$: x_1 – "weakly", x_2 – "moderately", and x_3 – "strongly" belong to $\mathcal{A}$.
Then fuzzy set $\mathcal{A}$ can be written as

$$\mathcal{A} = \{\text{"weakly"}/x_1, \text{"moderately"}/x_2, \text{"strongly"}/x_3\} \qquad (1.6)$$

Now we define each of the sets "weakly", "moderately", and "strongly" as a fuzzy set of the first type in [0, 1].

$$\mathcal{W} = \text{"weakly"} = \{0.1/0,\ 0.1/0.1,\ 0.2/0.2,\ 0.8/0.3\} \qquad (1.7)$$

$$\mathcal{M} = \text{"moderately"} = \{0.5/0.3,\ 0.7/0.4,\ 1/0.5,\ 0.8/0.6\} \qquad (1.8)$$

$$\mathcal{S} = \text{"strongly"} = \{0.6/0.6,\ 0.8/0.7,\ 0.9/0.9,\ 1/1\} \qquad (1.9)$$

In other words, when we say "strongly" we mean that in formula 1.6 the fuzzy set of the first type from the segment [0, 1] defined by formula 1.9 corresponds to x_3. (Similar ideas apply to formulas 1.7 and 1.8).
Therefore $\mathcal{A}$ 1.6 is a fuzzy set of the second type or a super fuzzy set.
Based on this example we may define a fuzzy set $\mathcal{A}$ of the second type in the following way :
Suppose that $\mathcal{U}$ is finite or equal to the set of integers. Then

$$\mathcal{A} = \{\mathcal{A}_1/x_1,\ \mathcal{A}_2/x_2,\ \cdots,\ \mathcal{A}_n/x_n,\ \cdots\},$$

where $\mathcal{A}_i = 0$ or $\mathcal{A}_i$ is a fuzzy set of the first class defined by the formula:

$$\mathcal{A}_i = \{\mu_{\mathcal{A}_i}(y_1)/y_1,\ \cdots,\ \mu_{\mathcal{A}_i}(y_k)/y_k,\ \cdots\}. \qquad (1.10)$$

In 1.10 $y_j \in [0, 1]$ and $\mu_{\mathcal{A}_i}(y_j) \in [0, 1]$.

The notions of "degree of inclusion" and "probability" should not mixed. Even though we have a real characteristic from the segment [0, 1] in both case, the natures of these characteristics are fondamentally different.
The notion of degree of inclusion assumes the existence of a universal set $\mathcal{U}$, and shows how much the elements $x \in \mathcal{U}$ posess originally defined features represented by the function μ_A. In contrast, the notion of probability deals with an event, in other words, a change of the universal.
The difference can be illustrated as follows. Suppose a bus can hold 50 passegers. To define the fuzzy set "young" with the set of passengers, we must consider 50 concrete people and to define to what degree μ_A each passanger is young.
For "probability", the problem is formulated differently: what are the chances that, among an arbitrary group of 50 people, there is a young person in a particular place. Here again "young" may be formulated in terms of fuzzy sets. It is a person for which the degree of "youngness" is not equal to zero.

Let $\mathcal{A}$ be a fuzzy set in the universal set $\mathcal{U}$ with the membership function :

$$\mu_\mathcal{A} : \mathcal{U} \rightarrow [0, 1].$$

Definition 4
We define the support of $\mathcal{A}$ by the classical subset :

$$supp(\mathcal{A}) = \{x \in \mathcal{U} : \mu_\mathcal{A}(x) \neq 0\}. \quad (1.11)$$

If $supp(\mathcal{A}) = \{x_1, \cdots, x_n\}$ is a finite set, then $\mathcal{A}$ may be written as:

$$\mathcal{A} = \{\mu_\mathcal{A}(x_1)/x_1, \cdots, \mu_\mathcal{A}(x_n)/x_n\} = \sum_{i=1}^{n} \mu_\mathcal{A}(x_i)/x_i. \quad (1.12)$$

1.2.2 0perations on Fuzzy Sets

To construct algorithms using fuzzy set classifications we need to introduce principal operations on fuzzy sets analogous to those in classical set theory. Herein are the formal definitions of such operations

Definition 1.2.2.1
Let $\mathcal{A}$ and $\mathcal{B}$ be two fuzzy sets, $\mu_\mathcal{A}$ and $\mu_\mathcal{B}$ their membership functions, and $\mathcal{X}$ the universal set.
Then,

$$\mathcal{A} = \mathcal{B}, \text{ if } \forall x \in \mathcal{X} \Rightarrow \mu_\mathcal{A}(x) = \mu_\mathcal{B}(x) \text{ (equality)} \quad (1.13)$$

$$\mathcal{A} \subseteq \mathcal{B}, \text{ if } \forall x \in \mathcal{X} \Rightarrow \mu_\mathcal{A}(x) \leq \mu_\mathcal{B}(x) \text{ (inclusion)} \quad (1.14)$$

Definition 1.2.2.2
The supplement $\overline{\mathcal{A}}$ of a fuzzy set $\mathcal{A}$ is defined by the membership function

$$\begin{array}{c}\mu_{\overline{\mathcal{A}}}(x) = n(\mu_\mathcal{A}(x)) \text{ where} \\ n : [0, 1] \rightarrow [0, 1] \text{ is a negation operation}\end{array} \quad (1.15)$$

(in other words n-is a non-increasing involution function, such that $n(0) = 1$ and $n(1) = 0$)

The function $n(u)$ may have different forms. The classical negation function is:

$$n(u) = 1 - u \quad (1.16)$$

We will use this classical negation throughout this book allowing formula 1.15 to be rewritten as:

$$\mu_{\overline{A}}(x) = 1 - \mu_A(x) \qquad (1.17)$$

Analogous theories can be constructed using other negation functions. Some well known negation functions are:

$$\text{quadratic negation}: \ n_r(u) = \sqrt{1 - u^2}$$
$$\text{and}$$
$$\text{Sucheno's negation}: \ n_\lambda(u) = \frac{1-u}{1+\lambda u}$$

Definition 1.2.2.3
We define a triangle norm (t-norm) to be a function

$$T : [0, 1] \times [0, 1] \to [0, 1]$$

that satisfies the following conditions:

- $T(0, 0) = 0$;
- $T(u_1, v_1) \leq T(u_2, v_2)$,
- $T(u, v) = T(v, u)$
- $T(u, T(v, w)) = T(T(u, v), w)$

$T(u, 1) = T(1, u) = u$
if $u_1 \leq u_2, v_1 \leq v_2$

A t-norm T is called "archimedian", if T is continuous as a function of two arguments and $T(u, u) < u \ \forall u \in [0, 1]$. A t-norm T is called a "strict t-norm", if T is a strictly increasing function of both arguments.
Here are some examples of t-norms:

Minimum (Zadeh's t-norm) : $T(u, v) = \min(u, v)$
Probability t-norm : $T_p(u, v) = u \cdot v$
Lukasiewicz's t-norm : $T_m(u, v) = \max(0, x + y - 1)$
Dombi's t-norm : $T_w(u, v) = \begin{cases} \min(u, v), & \text{if } u = 1 \text{ or } v = 1 \\ 0, & \text{in all other cases} \end{cases}$

Definition 1.2.2.4
We define a triangle co-norm (t-co-norm) to be a function

$$\perp : [0, 1] \times [0, 1] \to [0, 1]$$

that satisfies to the following conditions:

$$- \perp(1, 1) = 1, \quad \perp(0, v) = \perp(v, 0) = v$$
$$- \perp(u_1, v_1) \geq \perp(u_2, v_2), \text{ if } u_1 \geq u_2, v_1 \geq v_2$$
$$- \perp(u, v) = \perp(v, u)$$
$$- \perp(u, \perp(v, w)) = \perp(\perp(u, v), w)$$

A t-co-norm $\perp$ is called archimedian if $\perp$ is continuous as a function of two arguments and $\perp(u, u) > u \; \forall u \in [0, 1]$.
A t-co-norm $\perp$ is called a strict t-co-norm, if $\perp$ is a strictly increasing function of two arguments.

The class of triangle co-norms are dual to the class of triangle norms. That means that any t-co-norm $\perp$ may be obtained from a t-norm T using the following transformation

$$\perp(u, v) = 1 - T(1 - u, 1 - v).$$

Examples of t-co-norms include:

$$\perp(u, v) = \max(u, v)$$
$$\perp_p(u, v) = u + v - u \cdot v$$
$$\perp_m(u, v) = \min(1, u + v)$$
$$\perp_w(u, v) = \begin{cases} \max(u, v), & \text{if } u = 0 \text{ or } v = 0 \\ 1, & \text{in all other cases} \end{cases}$$

The following pairs of t-norms and t-co-norms are dual:
$\min(u, v)$ and $\max(u, v)$
$T_p(u, v) = u \cdot v$ and $\perp_p(u, v) = u + v - uv$
$T_m(u, v) = \max(0, x + y - 1)$ and $\perp_m(u, v) = \min(1, u + v)$
$T_w(u, v)$ and $\perp_w(u, v)$.
The notions of t-norms and t-co-norms are used to define the intersection and union operations for fuzzy sets.

Definition 1.2.2.5
Let $\mathcal{A}$ and $\mathcal{B}$ be fuzzy sets, $\mu_\mathcal{A}$ and $\mu_\mathcal{B}$ their membership functions, and T a t-norm and $\perp$ a t-co-norm dual to T. Then the intersection and union of the fuzzy sets $\mathcal{A}$ and $\mathcal{B}$ are defined by the following membership functions :

$$\mu_{\mathcal{A} \cap \mathcal{B}} = T(\mu_\mathcal{A}, \mu_\mathcal{B})$$
$$\mu_{\mathcal{A} \cup \mathcal{B}} = \perp(\mu_\mathcal{A}, \mu_\mathcal{B})$$

Indeed, $\forall x \in \mathcal{X}$, $\mu_{\mathcal{A} \cap \mathcal{B}}(x) = T(\mu_\mathcal{A}(x), \mu_\mathcal{B}(x))$ and $0 \leq \mu_{\mathcal{A} \cap \mathcal{B}}(x) \leq 1$ by the definition of a t-norm (see definition 1.5.3.3). Analogously, $\forall x \in \mathcal{X}$, $\mu_{\mathcal{A} \cup \mathcal{B}}(x) = \perp(\mu_\mathcal{A}(x), \mu_\mathcal{B}(x))$.

1.2.3 Fuzzy binary relations

Archimedian t-norms, and therefore the intersection and union operations of fuzzy sets, can be represented by additive generators of archimedian functions. such generators are continuous monotonously decreasing functions $f : [0, 1] \to \Re^+$. ($\Re^+$-is the set of non-negative real numbers), such as:

$$T(u, v) = f^{-1}(f(u) + f(v)), \qquad (1.18)$$

where

$$f^{(-1)}(y) = \begin{cases} f^{-1}(y), & \text{if } y \in [0, f(0)] \\ 0, & \text{if } y > f(0) \end{cases} \qquad (1.19)$$

Additive generators and the duality of t-norms and t-co-norms are the base for studies of generalized intersection and union operations of fuzzy sets.

The following averaging operator also plays an important role in our constructions. The notion of "generalized mean-value" was introduced by Kolmogorov [284]. The mean-value of values $C_1, C_2, \cdots, C_N$ is defined by the formula:

$$M^F(C_1, C_2, \cdots, C_N) = F^{-1}\left(\frac{1}{N}\sum_{i=1}^{N} F(C_i)\right) \qquad (1.20)$$

where $F(u)$ is a function and F^{-1} is its reverse transformation. (1.22 is an often used particular case of the generalized mean-value:

$$M_r^F(C_1, C_2, \cdots, C_N) = F^{-1}\left(\frac{1}{N}\sum_{i=1}^{N} C_i^r\right)^{1/r} \qquad (1.21)$$

It is not complicated to show that if $C_i > 0$, $i = 1, \cdots, N$, then

$$\begin{aligned} M_{-\infty}(C_1, C_2, \cdots, C_N) &= \min_{1 \le i \le N}(C_i) \\ M_{+\infty}(C_1, C_2, \cdots, C_N) &= \max_{1 \le i \le N}(C_i) \end{aligned} \qquad (1.22)$$

$$M_0(C_1, C_2, \cdots, C_N) = \left(\prod_{i=1}^{N} C_i\right)^{1/N} \qquad (1.23)$$

$$M_1(C_1, C_2, \cdots, C_N) = \frac{1}{N}\sum_{i=1}^{N} C_i \qquad (1.24)$$

The formulas (1.23) and (1.24) obviously define geometrical and arithmetical mean-values respectively. This fact supports the use of the term "generalized mean-value" for function (1.20).

Fuzzy Sets Classification Algorithms

Let us have a mapping $\varphi\colon \mathcal{X} \to \mathcal{Y}$ and a fuzzy set $\mathcal{A} \subset \mathcal{X}$. Then the image of $\mathcal{A}$ mapped by φ is a fuzzy set $\mathcal{B} \subset \mathcal{Y}$ with the membership function

$$\mu_\mathcal{B}(y) = \sup_{x \in \varphi^{-1}(y)} \mu_\mathcal{A}(x),\ y \in \mathcal{Y} \tag{1.25}$$

where $\varphi^{-1}(y) = \{x \in \mathcal{X} | \varphi(x) = y\}$.

Definition 1.2.3.1
A binary fuzzy relation $\mathcal{R}$ between the sets $\mathcal{X}$ and $\mathcal{Y}$ is a fuzzy set in the direct product $\mathcal{X} \times \mathcal{Y}$ with the membership function

$$\mu_\mathcal{R}\colon \mathcal{X} \times \mathcal{Y} \to [0,\ 1]$$

defined by formula (1.25).

Let φ be a binary fuzzy relation between $\mathcal{X}$ and $\mathcal{Y}$ with membership function $\mu_\varphi(x,y)$, and let $\mathcal{A}$ be a fuzzy set in $\mathcal{X}$ with membership function $\mu_\mathcal{A}(x)$. Then, the image $\mathcal{B}$ of the fuzzy set $\mathcal{A}$ produced by fuzzy relation φ is the fuzzy set $\mathcal{Y}$ with the membership function

$$\mu_\mathcal{B}(y) = \sup_{x \in \mathcal{X}} \min\left(\mu_\mathcal{A}(x), \mu_\varphi(x,y)\right),\ y \in \mathcal{Y} \tag{1.26}$$

1.2.4 The fuzzy logic approach to time series classification

Herein we show one way to use fuzzy logic to classifying and recognizing geophysical data. We consider in this chapter the waveform classification problem formulated and studied in section 1.1. But, this approach can be generalized to study a wide range of geophysical problems. Furthermore, any classification problem for discrete time functions $f(t_i)$, $t_i \in [T_0, T_1]$ may be tackled using this approach.

From a formal point of view, measurements from a seismic station can be considered as time-dependent processes that define a mapping of a certain region of the Earth's surface into the set of all possible seismic signals. In other words, if $\mathcal{E}$ is a two-dimensional set of points on the surface of this region and Sgn is the set of all potential seismic signals, we have a mapping

$$St : \mathcal{E} \longrightarrow Sgn\,, \tag{1.27}$$

defined by a fixed seismic station St. The mapping 1.27 is defined as follows: $St(e) \in Sgn$ is a signal recorded by the station St from an earthquake with epicenter in the point $e \in \mathcal{E}$.

The definition domain of the function 1.27, which we symbolize by $\mathcal{D}_{St}$, satisfies the condition $\mathcal{D}_{St} \subset \mathcal{E}$ (strict inclusion). Indeed, in the finite time span when the station St is operational, the earthquake epicentre cannot be located in any point $e \in \mathcal{E}$. Therefore, the mapping St is defined only on own subset $\mathcal{D}_{St} \subset \mathcal{E}$.

At the same time 1.27 obviously depends on time and, therefore, it represents the mapping of two variables:

$$St : \mathcal{E} \times [T_0, T_1] \longrightarrow Sgn, \qquad (1.28)$$

where $[T_0, T_1]$ is the time period over which the seismic station St has been operational. Below, we will use the notation (1.27), if the time parameter is not essential.

The information given to us by (1.27) – (1.28) forms an initial database that we symbolize as $\mathcal{I}$. The initial database consists of:

1. a finite number of points $e_i \in \mathcal{E}$, where epicentres of the known earthquakes occured, $\mathcal{E}_\mathcal{I} = \{e_1, \cdots, e_n\}$,

2. a finite number of signals $Sgn_\mathcal{I} = \{s_1, \cdots, s_m\}$, and a (usually) multivariate function

$$St_\mathcal{I} : \mathcal{E}_\mathcal{I} \longrightarrow Sgn_\mathcal{I}, \qquad (1.29)$$

where $St_\mathcal{I}(e_i) = \{s_{i1}, \cdots, s_{im_i}\}$, for $i = 1, \cdots, n$ where $s_{i1}, \cdots, s_{im_i}$ are the signals registered by the station St from earthquakes with epicentres at the points e_i.

Another way to view the function St_i is as a binary relation on $\mathcal{E}_\mathcal{I} \times Sgn_\mathcal{I}$:

$$X_i St_\mathcal{I} y_j \iff y_j \in F(x_i).$$

The problem is to establish the physical nature of a new incoming signal $S^* \in Sgn \setminus Sgn_\mathcal{I}$, registered by the station St, based on information in the initial database $\mathcal{I}$.

In other words by analyzing the information in the database $\mathcal{I}$ we should be able to construct the co-image $(St)^{-1}(S^*)$. To formulate the problem, we define a mapping

$$Sp : Sgn \longrightarrow \mathcal{E}, \qquad (1.30)$$

such that $Sp(S^*) = (St)^{-1}(S^*)$.

Fuzzy Sets Classification Algorithms

The algorithm described in this chapter tackles this problem by parameterizing the incoming signal by its relation to the initial database. The relation is studied using the Levenstein distance ([305]) and indicates how to construct the " physical origin " $Sp(S^*) = (St)^{-1}(S^*)$ of the signal $S^* \in Sgn$.
In more formal terms, we model the surface $\mathcal{X}$ at the region $\mathcal{E}$ using a distance function $d(x,y)$ (for instance, the usual Euclidian distance $d(x,y) = \sqrt{(x_1-x_2)^2 + (y_1-y_2)^2}$). As a model of the set of signals Sgn we consider the space of their wave-parametrizations $\mathcal{Y}$ (see 1.1 with the Levenstein distance $\rho(S_1, S_2)(see??)$. As a model of the station St we consider every mapping $\Phi : \mathcal{X} \to \mathcal{Y}$ such that the restriction $\Phi|_{\mathcal{E}_\mathcal{I}} = St_\mathcal{I}$.
Let us introduce the following notation:

$$x_i = mod_X(e) \quad -X - \text{models of the points } e_i;\ i = 1, \cdots, n.$$
$$y_i = mod_Y(s_j) \quad -Y - \text{models of the signals } s_j;\ j = 1, \cdots, m.$$

Then the diagram

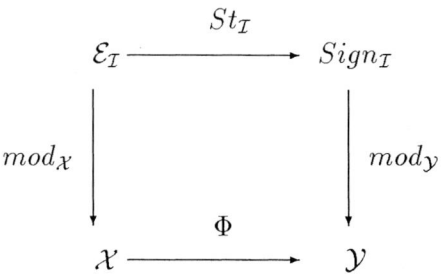

(1.31)

is commutative
We also use the notation

$$X_\mathcal{I} = \{x_i|_1^n\} = mod_X(\mathcal{E}_\mathcal{I}) \quad -X - \text{model of } \mathcal{I}$$
$$Y_\mathcal{I} = \{y_j|_1^m\} = mod_Y(Sgn_\mathcal{I}) \quad -Y - \text{model of } \mathcal{I}$$

In these terms the correspondence between Φ and St means that:
$\forall\ i = 1, \cdots, n$ the following diagram is commutative :

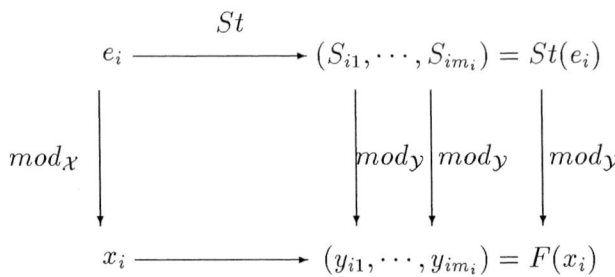

(1.32)

Reffering to 1.1 we can say that SPARS algorithms analyse $\mathcal{Y} - model\ y^* = mod_\mathcal{Y}(S^*)$ of a new signal $S^* \overline{\in} Sgn_\mathcal{I}$ in connection with the structure of the subset $\mathcal{Y}_\mathcal{I} \subset \mathcal{Y}$. As a result of such analysis SPARS algorithms defines a domain $Sp(y^*) = \Phi^{-1}(y^*) \subset \mathcal{X}$ that can be considered as a co-image of $y^* \in \mathcal{Y}$. (Here Sp is the mapping $Sp: Sgn \to \mathcal{E}$. We use the same symbol for the mapping $Sp: \mathcal{Y} \to \mathcal{X}$ between the spaces $\mathcal{X}$ and $\mathcal{Y}$ where SPARS actually operates).

The suggested solution looks as follows:

As a co-image (epicentre) $(St)^{-1}(S^*) = Sp(S^*)$ of the signal S^* the set $mod_\mathcal{X}^{-1}(Sp(y^*))$ of all points in region $\mathcal{E}$ is considered such that the $\mathcal{X}$-modules of these points are included in the set $Sp(y^*)$. This is illustrated in figure 1.4.

The described SPARS oriented decision making scheme represents a pattern recognition approach with a large fuzzy logic component. In the general case $Sp(y^*)$ is a two-dimensional domain in $\mathcal{E}$ in which elements can be considered as a possible source of S^* with different levels of reliability. It is also natural to assume that the level of reliability decreases when we move closer to the boundaries of $Sp(S^*)$.

Therefore it seems reasonable and natural to construct the set $Sp(S^*)$ using a fuzzy set approach. The core of this fuzzy sets approach is as follows: A classical, but only partialy defined mapping $\Phi: \mathcal{X} \to \mathcal{Y}$ (or binary relation $\Phi \in Bin(\mathcal{X}, \mathcal{Y})$) is substituted by a completely defined but fuzzy mapping $\mathcal{F}\Phi: \mathcal{X} \to \mathcal{Y}$ (or fuzzy binary relation $\mathcal{F}\Phi \in \mathcal{F}Bin(\mathcal{X}, \mathcal{Y})$) with the kernel $Ker\mathcal{F}\Phi = \Phi_\mathcal{I}$. The choice of the fuzzy mapping $\mathcal{F}\Phi$, generally speaking, is not unique and is based on expert evaluations, which then are transformed to the appearence of the membership functions.

Fuzzy Sets Classification Algorithms 19

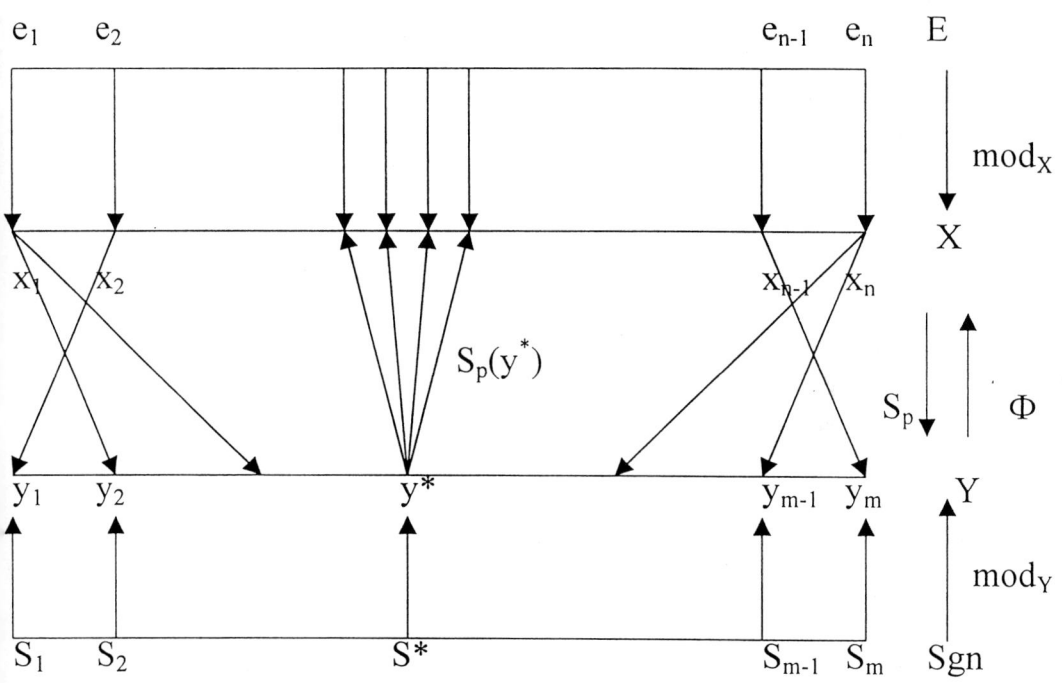

Figure 1.4: SPARS *decision* Chart of SPARS algorithm solutions..

Let us assume that we have choosen a fuzzy version $\mathcal{F}\Phi$ of a hypothetic mapping Φ. Therefore, for every point e from the considered region $\mathcal{E}$ and every signal s we have a measure over the set of signals Sgn of the fact that the signal s came from the point e. This measure is equal to the value of the membership function $\mu_{\mathcal{F}\Phi}$ on the pair $(mod_{\mathcal{X}} e, mod_{\mathcal{Y}} s)$. The existence of this measure allows us to construct a fuzzy analog $\mathcal{F}Sp$ of the mapping produced by SPARS (see 1.1).

For any $y^* \in \mathcal{Y}$, the value $\mathcal{F}Sp(y^*) = \mu_{\mathcal{F}\Phi}(x, y^*)$ is a fuzzy structure on $\mathcal{X}$ that presents a measure of the probability that the signal y^* comes from an earthquake with an epicentre in the point x.

This gives us the following decomposition of the space $\mathcal{X}$:

$$\mathcal{X} = \bigcup_\alpha \mathcal{X}_{y^*}(\alpha) \tag{1.33}$$

where $\mathcal{X}_{y^*} = \{x \in \mathcal{X} : \mu_{\mathcal{F}\Phi}(x, y^*) \geq \alpha\}$, $\mathcal{X}_{y^*}(\alpha) \subseteq \mathcal{X}_{y^*}(\beta)$, if $\beta < \alpha$. Establishing a threshold α^* (for example $\alpha^* = 0.5$) we can obtain a definite answer to the question concerning the appearance of the co-images of y^*.

Using SPARS to construct the set $Sp(S^*)$ assumes, in some sense, that the mapping "point" $\rightleftarrows$ "signal" is continuous. SPARS actually constructs $Sp(S^*)$ by analyzing how close $y^* = mod_y(S^*)$ is to the set $\mathcal{Y}_\mathcal{I}$ ([,]).

Furthermore, the existence of metrics in $\mathcal{X}$ and $\mathcal{Y}$ and the ability to form conclusions by analysing finite sets $\mathcal{X}_\mathcal{I} \subset \mathcal{X}, \mathcal{Y}_\mathcal{I} \subset \mathcal{Y}$ assumes some form of continuity of the mapping $\Phi: \mathcal{X} \to \mathcal{Y}$. If we know that the signal $y_j \in \mathcal{Y}_\mathcal{I}$ is registered from the point $x_i \in \mathcal{X}_\mathcal{I}$, then the closer a point x^* is to the point x_i in the metric space $(\mathcal{X}, d)$ and the closer a signal y^* is to the signal y_j in the metric space $(\mathcal{Y}, \rho)$, the more evidences we have that the signal y^* is registered from the point x^*. In formal terms this means that the mapping Φ should be a so-called "closed" function for which the following condition exists: if $x_n \to x$ in the metric space $(\mathcal{X}, d)$ and $\Phi(x_n) \to y$ in the metric space $(\mathcal{Y}, \rho)$ then $y = \Phi(x)$. The equivalent condition is: the graph $\Gamma_\Phi = \{x, \Phi(x)\}$ is a closed set in the direct product $(\mathcal{X} \times \mathcal{Y}, d \times \rho)$ of two metric spaces.

Following the goal of this chapter, let us formulate this statement using a fuzzy logic approach. First, we introduce a coordinate representation of the binary relation $\Phi_\mathcal{I}$ for the database $\mathcal{I}$. We consider a matrix $\|a_{ij}\|; i = 1, \cdots, n; j = 1, \cdots, m$, where

$$a_{ij} = 1, \text{ if } y_j \in \mathcal{F}(x_i)$$
$$a_{ij} = 0, \text{ if } y_j \notin \mathcal{F}(x_i)$$

Let us fix a pair (x_i, y_j). From the assumption of the agreement of mapping

Fuzzy Sets Classification Algorithms

Φ with the metrics d and ρ we obtain the fuzzy structure $\mu(x_i, y_j)$ on $\mathcal{X} \times \mathcal{Y}$. The function $\mu_{(x_i,y_j)}(x,y)$ is a "conditional measure" of the fact that a signal y may be registered from an earthquake with the epicentre at the point x under the condition that the signal y_j has been definitely registered from an earthquake with the epicenter at the point x_i. Possible versions of this measure are defined by choosing a form of the direct product of the distances d and ρ, and depends on the interpretation of the notion of fuzzy set products.

Here are some versions of the conditional measure :
If the direct product of the distances has the form

$$(\rho \times d)_1 = \max(\rho, d) \tag{1.34}$$

then the conditional measure is

$$\mu^p_{(x_i,y_j)}(x,y) = \frac{a_{ij}}{(1+d(x,x_i))(1+\rho(y,y_j))} \tag{1.35}$$

(μ^p-probabilistical version)
and

$$\mu^\ell_{(x_i,y_j)}(x,y) = a_{ij} \min\left(\frac{1}{(1+d(x,x_i))}; \frac{1}{(1+\rho(y,y_j))}\right). \tag{1.36}$$

(μ^ℓ-logistic version)

If the direct product has the form

$$(\rho \times d)_2 = \rho + d \tag{1.37}$$

Then the additive measure is

$$\mu^+_{(x_i,y_j)}(x,y) = \frac{a_{ij}}{(1+d(x,x_i))(1+\rho(y,y_j))} \tag{1.38}$$

It is natural that all the above measures are trivial if $a_{ij} = 0$, and have the pair (x_i, y_j) as a kernel if $a_{ij} = 1$.

The measures $\mu^p_{(x_i,y_j)}(x,y)$ and $\mu^\ell_{(x_i,y_j)}(x,y)$ can be illustrared by the following graph:

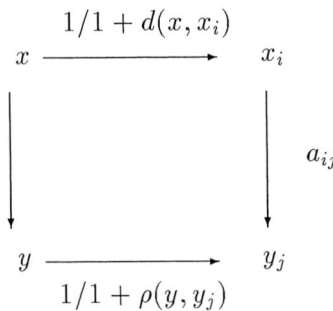

(1.39)

"Collecting" by disjunction (maximum) the measures $\mu_{(x_i,y_j)}$ for the whole database $\mathcal{I}$, $i = 1, \cdots, n$, $j = 1, \cdots, m$, we obtain the following functions that are natural to consider as measures of the relationship between the point $x \in \mathcal{E}$ on the Earth surface and the incoming signal $y \in Sgn$ defined using the database $\mathcal{I}$:

$$\mu_\mathcal{I}^p(x,y) = \bigvee_{i,j} \mu_{(x_i,y_j)}^p(x,y) = \max_{i,j} \mu_{(x_i,y_j)}^p(x,y), \quad (1.40)$$

$$\mu_\mathcal{I}^\ell(x,y) = \bigvee_{i,j} \mu_{(x_i,y_j)}^\ell(x,y) = \max_{i,j} \mu_{(x_i,y_j)}^\ell(x,y), \quad (1.41)$$

$$\mu_\mathcal{I}^+(x,y) = \bigvee_{i,j} \mu_{(x_i,y_j)}^+(x,y) = \max_{i,j} \mu_{(x_i,y_j)}^+(x,y), \quad (1.42)$$

The fuzzy set $(\mathcal{X} \times \mathcal{Y}, \mu_\mathcal{I})$ is the graph of completely defined fuzzy mapping $\mathcal{X} \to \mathcal{Y}$. It is a fuzzy extension $\mathcal{F}\Phi$ of the partially defined classical mapping $\Phi_\mathcal{I} \colon \mathcal{X}_\mathcal{I} \to \mathcal{Y}_\mathcal{I}$. Indeed the kernel $\mathcal{K}er\mu_\mathcal{I} = \mathcal{K}er\mathcal{F}\phi$ is exactly equal to the graph of the mapping $\Phi_\mathcal{I}$:

$$\mathcal{K}er\mu_\mathcal{I} = \mathcal{K}er\mathcal{F}\phi = \Gamma_{\Phi_\mathcal{I}} = \bigcup_{i=1}^{n}(x_i, \Phi(x_i)) \in \mathcal{X} \times \mathcal{Y}.$$

Another way to "collect" the functions $\mu_{(x_i,y_j)}$ is to consider them from the point of view of additive functions. For example consider the following measure defined by the initial database:

$$\mu_\mathcal{I}^\sigma(x,y) = \frac{1}{n \cdot m} \sum_{i,j} \mu_{(x_i,y_j)}(x,y).$$

This additive approach better reflects the existing distribution in the set $\{\mu_{(x_i,y_j)}(x,y)\}$, $i = 1, \cdots, n$, $j = 1, \cdots, m$. In general, the kernel $Ker\mu_{\mathcal{I}}^\sigma$ can equal the empty set. In other words, the measure $\mu_{\mathcal{I}}^\sigma$ is not necessarilly normal.

For the final stage we use a system analysis approach. Its core is the fact that the region $\mathcal{E}$ and the space Sgn are nopn randomly connected with $(\mathcal{X}, d)$ and $(\mathcal{Y}, \rho)$ respectively. Indeed, $(\mathcal{X}, d)$ is a model of $\mathcal{E}$ and $(\mathcal{Y}, \rho)$ is a model of Sgn.

Therefore, FSPARS a fuzzy version of SPARS $\mathcal{F}\Phi$ defines the free parameters of the above scheme. Different choices of the distance d in different levels of discretization of the region $\mathcal{E}$, give different relations $\mathcal{F}_{\mathcal{I}}$. We have the same situation with the Levenstein distance ρ, which depends on the weights (free parameters).

If $(\mathcal{X}_1, d_1), \cdots, (\mathcal{X}_N, d_N)$ are different metric models of the region $\mathcal{E}$ and $(\mathcal{Y}_1, \rho_1), \cdots, (\mathcal{Y}_M, \rho_M)$ are different metric models of the set of signals Sgn then, for every pair : (s,t), $s = 1, \cdots, N$; $t = 1, \cdots, M$ we obtain coordinate representation $\mathcal{I}_{st}$ of the information

$$\mathcal{I} = (\mathcal{X}_{\mathcal{I}}, \mathcal{Y}_{\mathcal{I}}, St_{\mathcal{I}} \in Bin(\mathcal{X}_{\mathcal{I}}, \mathcal{Y}_{\mathcal{I}})) \qquad (1.43)$$

about operations of the station St_i:

$$\mathcal{I}_{st} = (\mathcal{X}_{\mathcal{I}_{st}}, \mathcal{Y}_{\mathcal{I}_{st}}, (St_{\mathcal{I}})_{st}) \in Bin((\mathcal{X}_{\mathcal{I}})_{st}, (\mathcal{Y}_{\mathcal{I}})_{st}) \qquad (1.44)$$

Correspondingly, we obtain the fuzzy extension $(\mathcal{F}\Phi)_{st}$ for $(St_{\mathcal{I}})_{st}$ on the whole product $\mathcal{X}_s \times \mathcal{Y}_t$. The inverse fuzzy mapping is

$$\begin{aligned}(\mathcal{F}Spars)_{st} &: (\mathcal{F}SP)_{st} \in Bin(\mathcal{Y}_t, \mathcal{X}_s) \\ (\mathcal{F}Sp)_{st} &= (\mathcal{F}\Phi^{-1})_{st}\end{aligned} \qquad (1.45)$$

If S^* is a fixed signal in Sgn, and y_t^* is a model of S^* in $\mathcal{Y}_t \ni y_t^*$, then $(\mathcal{F}SP)_{st}(y_t^*)$ is a fuzzy structure in $\mathcal{E}$. (Reminder : by $\mathcal{F}SP$ we mean the mapping $\mathcal{F}SP: \mathcal{Y} \to \mathcal{X}$, as well as the mapping $\mathcal{Y} \xrightarrow{\mathcal{F}SP} \mathcal{X} \xrightarrow{mod_x^{-1}} \mathcal{E}$).

The most natural way to aggregate all fuzzy structures $(\mathcal{F}SP)_{st}(y_t^*)$ is by their fuzzy conjunction over $\mathcal{E}$. However, other types of such a disjunction "collection" can be considered. A vector approach to the "collection" gives the following vector order on $\mathcal{E}$:

$$e \longmapsto \{((\mathcal{F}SP)_{st}(y_t^*))(e)\}; \ s = 1, \cdots, N; \ t = 1, \cdots, M. \qquad (1.46)$$

Analyzing the subsets in $\mathcal{E}$ according to values of (1.46) we obtain solutions to the problem under consideration.

1.2.5 Fuzzy version of SPARS algorithm (FSPARS)

At first we introduce basic notions, being oriented towards seismological problems. By a region $\mathcal{X}$ we will understand both an area on the Earth surface and the body cone with foundation $\mathcal{X}$. We will use both euclidean (x, y, z) and (ω, φ, h) spheric coordinates, where h is the depth. Following seismological applications of this technique we symbolize by $\Delta_F = \Delta(omega, varphi)$ earthquake epicentral distance to seismic station F.

By $\mathcal{S}$ we denote the database of signals produced by a studied set of earthquakes. Below, we consider different "features" of signals $s \in \mathcal{S}$ which are seismograms, epicentral distances, depths, magnitudes etc.

Let us assume that to the moment T the station F has received the signals $s_1, \cdots, s_n$ with epicenters in the points $x_1, \cdots, x_n$ of the region $\mathcal{X}$. The mapping $\{x_i \mapsto s_i \mid_1^n\}$ is actually the result of the job produced by the station F. We denote $\mathcal{I} = \{x \mapsto s_i \mid_1^n\}$ and call $\mathcal{I}$ the main information kernal. It is convenient to consider $\mathcal{I}$ as a linear relation $F_\mathcal{I}$ between $\mathcal{X}$ and $\mathcal{S}$-projections of $F_\mathcal{I} : \mathcal{X}_\mathcal{I} = \{x_i \mid_1^n\}$, $\mathcal{S}_\mathcal{I} = \{s_i \mid_1^n\}$. Thus, $\mathcal{I} = (\mathcal{X}_\mathcal{I}, F_\mathcal{I}, \mathcal{S}_\mathcal{I})$. Representing the signal $s \in \mathcal{S}$ as a vector of its components $seism_F(s)$-seismogram recorded at the station F, $m_F(s)$-magnitude, $\Delta_F(s)$-epicentral distance, $h_F(s)$-depth, $t_F(s)$-time in the source, we obtain the appearence of $\mathcal{I}$ as a function of a given database.

$$(x, s) \mapsto (x, y, h_F, (s), seism_F(s), \Delta_F(s), m_F(s), t_F(s)) \quad (1.47)$$

We assume that the region $\mathcal{X}$ under consideration is divided into sub regions

$$\mathcal{X} = \bigcup_{i=1}^{K} \mathcal{X}_i, \ \mathcal{X}_i \bigcap \mathcal{X}_j = \emptyset. \quad (1.48)$$

The decomposition 1.48 leads to decomposition of the signals

$$\mathcal{S}_\mathcal{I} = \bigcup_{i=1}^{K} F_\mathcal{I}(\mathcal{X}_j) : F_\mathcal{I}(\mathcal{X}_j) = \{s \in \mathcal{S}_\mathcal{I} : \exists x_{i_j} \in \mathcal{X}_j \setminus x_{i_j} F s\} \quad (1.49)$$

If we have a signal s in a certain moment t_{n+1} the problem is to affiliate it to a subregion $\mathcal{X}_j$, $j = 1, \cdots, K$.

SPARS algorithm (see 1.1) does it using metric classification produced by Levenstein distance. Using nearest neighbours approach SPARS classifies the signal s to a subset $\mathcal{S}_\mathcal{I}(j) = F_\mathcal{I}(\mathcal{X}_j)$.

Following this classification its co-image is classifical to some $\mathcal{X}_j$.

Fuzzy Sets Classification Algorithms

This SPARS classification scheme can be described in the following terms. The map F is known on the subset $\{\mathcal{X}_i \mid_1^n\} \subset \mathcal{X}$, where $F: x_i \mapsto s_i$. The problem is to find a co-image $F^{-1}(s)$ for all $s \in \mathcal{S} \setminus \{s_1, \cdots, s_n\}$.

The metric solution produced by SPARS in these terms looks as follows. We introduce a metric d on $\mathcal{S}$ and choose $s_i^* \in \mathcal{S}$ as d-nearest element to s. The co-image $\mathcal{X}_i^*$ of s_i^* we then consider as co-image for s. As it is seen on the figure (scheme 1.39), the path from s to x_i^* consists in two parts (s, s_i^*) and (s_i^*, x_i^*). Let us count the length of (x_i, s_i) as 1, and the length (s_i, s) equal to $d(s_i, s)$. Thus, the path (s, s_i^*) will have the minimum length among all the possible paths (s, x_i^*).

Following this interpretation we want to assign a non-trivial weight to (s_i, x_i) and therefore, to obtain new methods of decision making concerning the co-image of s in terms of $\{x \mid_1^n\}$. Speaking more formaly, we want to assign the path (s, s_i) and (s_i, x_i) by the weights $\mu(s, s_i)$ and $\mu(s_i, x_i)$-correspondingly and to consider a problem of finding the co-image $F^1(s)$ in $\{x \mid_1^n\}$ as a problem of choice among the paths $(s, x_i) = (s, s_i) + (s_i, x_i)$. Such a choice is made towards an aggregated estimation $\psi(\mu(x_i, s_i); \mu(s_i, s))$.

The function ψ have the meaning "and" the statement: the path (s, x_i) is the path (s, s_i), "and" the path (s_i, x_i).

Thus, ψ corresponds to a fuzzy conjunction.

We construct $\mu(x_i, s_i)$ as follows. First of all we substitute the metrics d to a measure of similarity by putting a "lamp" in each of the points s_i and s. In this way a fuzzy technique is introduced into the construction. The examples of constructions of such "lamps" (which are free parameters of the algorithm) follows:

1. $\mu(s, s_i) = \dfrac{1}{1 + d(s, s_i)}$

2.
$$\mu(s, s_i) = \begin{cases} 1 - \dfrac{d(s, s_i)}{r}, & \text{if } d(s, s_i) \leq r \\ 0, & \text{if } d(s, s_i) \geq r \end{cases}$$

3. $\mu_{\ell r}(s, s_i) = e^{-\frac{d(s, s_i)}{r}}$

These three constructions are connected with the corresponding functions of potential type on $\Re^+$.

1) $f(x) = \dfrac{1}{1+x}$, 2) $f(x) = \begin{cases} 1 - \dfrac{x}{r}, & \text{if } x \leq r \\ 0, & \text{if } x \geq r \end{cases}$ 3) $f(x) = e^{-x/r}$

Let the light comes from s to s_i. If its intensity was equal to 1, it will be equal to $\mu(s, s_i)$ in this model. SPARS algorithm does not take into account a metrics on $\mathcal{X}$ (see 1.1). We also do not have it yet and that is why all the connections $s_i \mapsto x_i$ have the same weight equal to one. In other words there is no losses when the light goes from s_i to x_i. The co-image of s is the point x_i^* in which the light produced by s will be the most intensive. That means that:

$$x^* = \arg\max_i \psi(\mu(s, s_i), \mu(s_i, x_i)) \qquad (1.50)$$

Below we will use Zadeh conjunction $- \min(u, v)$ and probability conjunction $u \cdot v$. The function ψ will be a free parameter. As a metric on $\mathcal{X}$ we will consider an usual euclidean metrics. In this way, the direct product $\mathcal{X} \times \mathcal{S}$ will also be a metric space.

On the other hand, it is the space of elementary outcomes for our events earthquakes.

Herein is some mathematical information, which we will need for further constructions. Let $\mathcal{Z}$ be a subset of possible outcomes of some experiment and up to the moment we have fixed "n" results. This is that we consider as initial information $\mathcal{I} = \{z_i \mid_1^n\} \subset \mathcal{Z}$. The usual statistics the probability measure with $\mathcal{I}$.

$$P_\mathcal{I} : P_\mathcal{I} = \sum_i p_i \delta_{z_i},$$

where $p_i = \frac{\nu(z_i)}{n}$; δ_{z_i} ; $z_i \in \mathcal{Z}$. The measure $P_\mathcal{I}$ is concentrated only on $\mathcal{I}$. Therefore the probability forecast of future experiment development on the basis of its history looks as follows. In the future we can speak only about the points z_i, in which something has already occured. The level how prospective is a point z_i depends on the probability p_i.

Let us assume that a metric p is defined on $\mathcal{Z}$. We also assume that p "corresponds" to our experiment in the sense that if $z \bar{\in} \mathcal{I}$ is closer to $\mathcal{I}$, then more "prospectives" z has in the future of our experiment. A complicated point is to formalize that. Indeed z should be near to $\mathcal{I}$ in an integrated way. In other words this notion must depend on all points $z_i \in \mathcal{I}$. At the same time, apparently the standard measures $\min r(z, \mathcal{I})$ and $\max r(z, \mathcal{I})$ do not suite our situation.

Therefore, we may count on our potential functions approach based on the "lamp type" functions centered in the points z_i, $i = 1, \cdots, n$. Indeed, it allows us to intergrate the illumination and to obtain inthis way quantitative measure of the notion "nearer" using fuzzy disjunctions.

Fuzzy Sets Classification Algorithms

A transfer from the metrics r on $\mathcal{Z}$ to a potential $K(r)$ (measzure of similarity) leads to much more flexibility and clarity in the model. Two free parameters appear which are the potential itself (law of emination) and a law of its aggregation in a point (for example fuzzy disjunctions and conjunctions).

The intermediate conclusion can be formulated as follows. The existence of metrics on $\mathcal{Z}$ leads to a fuzzy distribution $mu_{\mathcal{I}}$ of outcomes (not probabilities !) on the basisof the information $\mathcal{I} = \{z_i \mid_1^n\}$.

Herein is the difference in conclusions between probabilistic and fuzzy approaches. Probabilistic approach "history-future" is based on the measure p_i on $\mathcal{Z}$.

$$P_{\mathcal{I}}(z) = \frac{D(z_i)}{w}, \text{ in particular sup } P_{\mathcal{I}} = \mathcal{I}$$

Fuzzy metrics r-conclusio "history-future" is the Sugeno measure [ ] on $\mathcal{X}$ (see 1.51), constructed using the potential "lamp function" $K(r)$

$$\mu_{\mathcal{I}}^r(z) = \text{fuzzy disjunction } (K(r)(z, z_i) \mid_1^n). \tag{1.51}$$

Construction of potential and way of its aggregation (see 1.51) are free parameters and $\sup \mu_{\mathcal{I}}^r \geq \mathcal{I}$.

To illustrate what has been said we come up with with the following examples.

Example 1 If the disjunction is max, then we consider the light from the nearest active elements.

Example 2. Let us consider $\mu(z) = \frac{1}{n}\sum_i K(z, z_i)$. Then the kernel of the fuzzy structure $\mu_{\mathcal{I}}$ is equal to zero. That means that the maximum lighted elements may, generally speaking, fiffer from the most active ones.

Numerical example 3. Let us assume that any point of the segment $[-1, 1]$ can be an outcome of one experiment. We also assume that to present the moment of time the following outcomes have been realized.

$$\{-1, -1/2, \cdots, -1/n, \cdots, 1/2, 1\}$$

Then the usual statistics gives the frequency distribution of probabilities:

$$p(-1) = p(1) = \cdots = p(1/n) = p(1/n) = 1/2n. \tag{1.52}$$

Therefore, the probability of all other outcomes will be equal to zero.

Now we assume that the distance on $\Re$ is defined in correspondance with our experiment. That means that the more nearer an outcome z to one which actually took place in past, the more possible z will be realized in the future. Therfore, it becomes clear that the outcome $z = 0$ has another possibility (not probability) to appear, rather than, for instance, $z = 3/4$. On the other hand (refspare) gives the probability zero in both cases.

To express it formaly we light the potential "lamp functions" in the points $-1, 1, \cdots, -1/n, 1/n$. Thus, any outcome z will be lighted. We define its possibility (agin, not probability) equal to the sum of its lightening from active points devided by $2n$.

Let $n=1$. Then we have two active points $\mathcal{I} = \pm 1$.

If $x \in [-1, 1]$, then $\mu_\mathcal{I} = \dfrac{1}{2}(\dfrac{1}{1+|x+1|} + \dfrac{1}{1+|x+1|})$ and the function $\mu_\mathcal{I}$ is even. Therefore, it is sufficient to analyse this function on the segment $[0, 1]$. In this case

$$\mu_\mathcal{I} = \frac{1}{2}(\frac{1}{2-x} + \frac{1}{2+x}), \quad \mu'_\mathcal{I} = \frac{4x}{(2-x)^2(2+x)^2} > 0$$

Thus $\mu_\mathcal{I}$ is an increasing function on $[0, 1]$ which increases from $\mu_\mathcal{I}(0) = 1/2$ upto $\mu_\mathcal{I}(1) = 2/3$. For example $\mu(3/4) > \mu_\mathcal{I}(0)$.

Let $n=3$. We have six active points $\mathcal{I} = \{\pm 1; \pm 1/2; \pm 1/3\}$. Calculating the possibility $\mu_\mathcal{I}$ for some points wed obtain

$$\mu_\mathcal{I}(-1) = \mu_\mathcal{I}(1) = \frac{1}{6}(\frac{1}{1+0} + \frac{1}{1+1/2} + \frac{1}{1+2/3}$$

$$+ \frac{1}{1+1/3} + \frac{1}{1+1/2} + \frac{1}{1+2}) = 0.57$$

$$\mu_\mathcal{I}(-1/2) = \mu_\mathcal{I}(1/2) = 0.66;$$

$$\mu_\mathcal{I}(-1/3) = \mu_\mathcal{I}(1/3) = \frac{4.03}{6} = 0.67 + \epsilon;$$

$$\mu_\mathcal{I}(0) = 0.64 \qquad (1.53)$$

For better understanding of the function $\mu_\mathcal{I}$ we also calculate the function in the points $x = 1/4$ and $x = 3/4$.

$$\mu_\mathcal{I}(-1/4) = \mu_\mathcal{I}(1/4) = 0.657; \quad \mu_\mathcal{I}(-3/4) = \mu_\mathcal{I}(3/4) = 0.6$$

To describe the behaviour of the function $\mu_\mathcal{I}$ in a close vicinity of possible maximums $x = \pm 1/3$, we also calculate the function in the points $x = 0.332$ and $x = 0.334$. We obtain that $\mu_\mathcal{I}(0.332) = \mu_\mathcal{I}(0.334) = \dfrac{4.021}{6} = 0.67$ and following (1.53).

$$\mu_\mathcal{I}(-1/3) = \mu_\mathcal{I}(1/3) > \mu_\mathcal{I}(0.332) = \mu_\mathcal{I}(0.334). \tag{1.54}$$

In particular $\mu_\mathcal{I}(3/4) < \mu_\mathcal{I}(0)$.

In the considered cases the maximum lighted points have been among the active ones (if $n = 1$, it is $z = \pm 1$, if $n = 3$, it is $\pm 1/3$.). This is not true in the general case. We shall operate now on the plane. As initial information $\mathcal{I}$ we consider the summits 1, 2, 3 of the correct triangle with variable radius "x".

Assuming that all the points of the plane are "possible" in the future, we calculate $\mu_\mathcal{I}(0)$, $\mu_\mathcal{I}(i)$, where 0 is the geometrical centre of the triangle. We obtain:

$$\mu_\mathcal{I}(1) = \mu_\mathcal{I}(2) = \mu_\mathcal{I}(3) = \frac{1}{3}\left(1 + \frac{2}{1 + \sqrt{3}x}\right) \text{ and } \mu(0) = x.$$

Solving the inequation

$$x > \frac{1}{3}\left(1 + \frac{2}{1 + \sqrt{3}x}\right)$$

$$3\sqrt{3}x^2 + (3 - \sqrt{3})x - 3 > 0 \text{ and}$$

$$\mathcal{D} = (3 - \sqrt{3})^2 + 43 \cdot 3\sqrt{3} > 0$$

we conclude that if x is sufficiently small, $x = 0$ is the most lighted point which is among the active ones.

Let us come back to our seismological model. In our case information $\mathcal{I}$ has three components $X_\mathcal{I} = \{x_i \mid_1^n\}$, points of the region $\mathcal{X}$ in which earthquakes are registred, $S_\mathcal{I} = \{s_i \mid_1^n\}$, signals from the earthquakes, registered by a station F and $F_\mathcal{I} \in Bin(X_\mathcal{I}, S_\mathcal{I})$-relations. By ρ-conclusion and $X_\mathcal{I}$ we induce an activity on S, by d-conclusion and $S_\mathcal{I}$ we induce activity on S and by $\rho \times d$-conclusion and $F_\mathcal{I}$-activity on the relations.

The activity on X reflects the seismic activity.

In these terms we call "activity of a point" its density. Thus the "activity of a region" is the average density of the points in the region.

The activity of a relation is the "level of trust" of this relation in the scheme

$$x_i \xrightarrow{\mu(x_i, s_i)} s_i \xrightarrow{\mu(s_i, s)} s.$$

Following the scheme, we can interpret the situation as follows. More "dense " a relation is, more trust we give to this relation. Thus, if $d(s, s_1) = d(s, s_2)$, (but the relation $x \mapsto s_1$ is "more dense" as it is seen on the figure ??) we give the priority to the way $s \to s_1 \to x_1$.

In the general case we make a choice using a criterium $\psi(\mu(s_i, x_i), \mu(s_i, s))$, where ψ is a conjunction. We will came back to this matter later.

Let us construct the weight $\mu_\mathcal{I}(x_i, s_i) = \mu(x_i, s_i)$.

The weight of the real relation (x_i, s_i) is calculate as follows.

First we choosed two parameters $r_2, r_3 \in (0, 1]$ and "lamp" potentials $\mathcal{K}_{r_2}$ on $\mathcal{X}$ and $\mathcal{K}_{r_3}$ on $\mathcal{S}$ induced by the function

$$f(x) = \begin{cases} 1 - \dfrac{x}{r}, & \text{if } x \leq r \\ 0, & \text{if } x > r. \end{cases}$$

Then we considered the product of this "lamp functions"

$$\mathcal{K}_{r_2, r_3}((x, s), (\bar{x}, \bar{s})) = \mathcal{K}_{r_2}(x, \bar{x})\mathcal{K}_{r_3}(s, \bar{s})$$

-probabilistic conjunction
  and

$$\mathcal{K}_{r_2, r_3}((x, s), (\bar{x}, \bar{s})) = \min(\mathcal{K}_{r_2}(x, \bar{x})\mathcal{K}_{r_3}(s, \bar{s}))$$

-Zadeh conjunction

Since the "potential-product" is defined on $\mathcal{X} \times \mathcal{S}$, the weight of the relations is established on the basis of the set of active points $\mathcal{F}_\mathcal{I} = \{(x_i, s_i)\,|_1^n\}$ as follows:

$$\mu_{r_2, r_3}(x, s) = \frac{1}{n} \sum_{j=1}^{n} \mathcal{K}_{r_2, r_3}((x, s), (x_j, s_j)) \qquad (1.55)$$

As a current conclusion we can state that (1.55) introduces another construction of the measure of possibility on $\mathcal{X} \times \mathcal{S}$.

Taking into account what information $\mathcal{I}$ means in our case we can rewrite the formula (1.55) as follows

$$\mu_{r_2,r_3}(x,s) = \text{possibility } (SOD(s) = x \mid \mathcal{I} = SOD(s_i) = x_i \mid_1^n) \quad (1.56)$$

where SOD ("Signal Origin Definition") is a conditional measure of the fact, that the signal s can arrive from the point x obtained on the basis of the information $\mathcal{I}$.
When then choose the function $\mathcal{F}^{-1}(s^*)$ for the coming signal s^* using the rule

$$x^* = \mathcal{F}^{-1}(s^*) = \arg\max \mu_{r_2,r_3}(x,s^*) \quad (1.57)$$

We notice that $x^* = x^*(r_2, r_3)$ is not necessary included into $\{x_i \mid_1^n\} = \mathcal{X}_{\mathcal{I}}$. Let us extend the initial information $\mathcal{F}_{\mathcal{I}} = \{(x_i, s_i) \mid_1^n\}$ by the incoming signal s_{n+1}. This can be done as follows: any relation (x_i, s_i) with s_{n+1} gains an additional weight $\mathcal{K}_{r_n}(s_{n+1}, s_i)$, where r_n is another free parameter from $[O, 1]$. Therefore the corresponding cones $\mathcal{K}_i(r_2, r_3)$ having the same basis will posess different hights. This leads to measure $\mu_{r_2 r_3 r_4}$, which corresponds well with the information $\mathcal{F}_{\mathcal{I}} + s_{n+1}$.

$$\mu_{r_2,r_3,r_4}(x,s) = \frac{1}{n}\sum_{j=1}^{n} \mathcal{K}_{r_2,r_3}((x,s),(x_j,s_j))\mathcal{K}_{r_n}(s_j,s_{n+1}). \quad (1.58)$$

In particular, the solution $SOD(s_{n+1}) = \mathcal{F}^{-1}(s_{n+1})$ will have an appearence $x^* = x^*(x_2, x_3, x_4); x^* = \arg\max \mu_{r_2 r_3 r_4}(x, s_{n+1})$
Another construction of the measure of possibility is the measure on graphic, which we symbolize by $\mu_{\mathcal{I}}^g(x,s)$. Its construction is quite apparent and follows from the fact that all is going on in the direct product $\mathcal{X} \times \mathcal{S}$.
Another construction is called relation measure and symbolized by $\mu_{\mathcal{I}}^c(x,s)$. This measure actually gives the generalization of the SPARS scheme with non-trivial relations $\mu(x_i, s_i)$. Thus

$$\begin{array}{ccc} \mu(x,x_i) & \mu(x_i,s_i) & \mu(s_i,s) \\ x\cdot \longrightarrow & x_i\cdot \longrightarrow & s_i\cdot \longrightarrow s\cdot \end{array}$$

$$\mu_{\mathcal{I}}^c(x,s) = \psi(\mu(x,x_i),\ \mu(x_i,s_i),\ \mu(s_i,s))$$

The relations $\mu(x, x_i) = \mu_{r_1}(x, x_i)$ and $\mu(s, s_i) = \mu_{r_n}(s_n, s_i)$ have been already described above. We define $\mu(x_i, s_i)$ using the following formula :

$$\mu(x_i, s_i) = \mu_{r_2, r_3}(x_i, s_i) = \frac{1}{n}\sum_{i=1}^{n} \mathcal{K}_{r_2 r_3}((x,s), (x_j, s_j))$$

Thus, classical SPARS corresponds to the situation $(0, 0, 0, r_n)$. Another version of SPARS, which we call SPARS 99 corresponds to $(r_1, 0, 0, r_4)$. The law ψ a composition $\psi_1 \circ \psi_2$ of disjunction ψ_1, that makes exterior agregation by $i = 1, \cdots, n$ and conjuntion ψ_2, which connect three relations $\mu(x, x_i)$, $\mu(x_i, s_i)$ and $\mu(s_i, s)$:

$$\mu_{\mathcal{I}}^c(x, s) = \psi_1(\psi_2(\mu(x, x_i), \ \mu(x_i, s_i), \ \mu(s_i, s) \mid_1^n)$$

Herein are some examples of $\mu_{\mathcal{I}}^c$.

1. $\psi_1 = \max$, $\psi_2 = \min$

$$\mu_{r_1, r_2, r_3, r_4}^c(x, s) = \max_{i=1,\cdots,n} (\min \mu(x, x_i), \ \mu(x_i, s_i), \ \mu(s_i, s))$$

2. $\psi_1 = \max$, ψ_2 is a product,

$$\mu_{r_1, r_2, r_3, r_4}^c(x, s) = \max_{i=1,\cdots,n} (\mu(x, x_i), \ \mu(x_i, s_i), \ \mu(s_i, s))$$

Thus, we have introduced two parametric families of fuzzy measures: using the graphic $\mu_{r_2, r_3, r_4}^g(x, s)$ and using relations $\mu_{r_1, r_2, r_3, r_4}^c(x, s)$. The second construction generalizes SPARS and, therefore, in the worse case, it gives a result equivalent to the classical SPARS.

To realize this model in practice we need "seismically grounded" metrics on signals $\mathcal{S}$, that somewhat reflects the region $\mathcal{X}$ in the set of signals $\mathcal{S}$. An example is cos-measure (correlation model). If the measure d on $\mathcal{S}$ corresponds to the measure ρ on $\mathcal{X}$ in the above defined sense, then the "light" from the incoming signal s_{n+1} produces the trace n-dimensional vector on $s_i \mid_1^n$:

$\overrightarrow{\mu}(s_{n+1}) = (\mu(s_{n+1}, s_i) \mid_1^n)$. We transfer this verctor down to $\mathcal{X}$ and try to find $x^* \in \mathcal{X}$, for which the light trace $\overrightarrow{\mu}(x^*)(\mu(x^*, x_i) \mid_1^n)$ is the "closest" to $\overrightarrow{\mu}(s_{n+1})$. In other words

$$\mu_{\cos}(x, s) = \cos(\overrightarrow{\mu}(s), \overrightarrow{\mu}(x)) = \frac{(\overrightarrow{\mu}(s), \overrightarrow{\mu}(x))}{|(\overrightarrow{\mu}(s)| \cdot |\overrightarrow{\mu}(x)|)} \quad (1.59)$$

Fuzzy Sets Classification Algorithms

and $x^* = \arg\max \mu_{\cos}(x, s)$.

Let us notice that the scalar product (1.59) is a free parameter of fuzzy SPARS as well as the model of light.

In conclusion, let us consider the following dynamical model. Let the events (earthquakes) (x_i, s_i) occured in the moment of time t_i, $t_i < t_{i+1}$, $i = 1, \cdots, n$. We assume that in the moment of time t_{n+1} the station has registered the signal s_{n+1}. We symbolize by $\mu(x, y)$ the light eminated by x into y following the choosen model on the corresponding space : $\mu(t, \bar{t})$ in time, $\mu(s, \bar{s})$ in signals, $\mu(x, \bar{x})$ in the region. We fix the pair (x, s). Then on the semi interval $[t_k, t_{k+1})$, $k \leq n$, the calculations are done by the formula:

$$\mu(x, s, t) = \frac{1}{n} \sum_{i=1}^{k} \mu(x_i, x)\mu(s_i, s)\mu(t, t_i) \quad (1.60)$$

It is not difficult to see that (1.60) describes the calculations on "k" cones $\mathcal{K}_{r_2, r_3}(x_i, s_i)$, introduced above. The summit $\mu(t, t_i)$ of the cones depend on t. Indeed, the farer an event (x_i, s_i) is from the moment "t", the smaller weight this event has. Thus, we have a function on $[t, t_{n+1})$ with the steps in the points $t_2, \cdots, t_n$, because in these moments of time the information extension $(x_i, s_i) \mid_1^k +(x_{k+1}, s_{k+1})$ takes place.

Let us now do the calculations in the moment of time t_{n+1}. Here we have information only about the signal s_{n+1}.

Since we cannot add the cone $\mathcal{K}_{n+1}$ in this case, we add $\mu(s_{n+1}, s)$ to $\mu(t_{n+1}, t_i)$ on the existing cones $\mathcal{K}_1, \cdots, \mathcal{K}_n$.

Therefore, for $t = t_{n+1}$ we obtain.

$$\bar{\mu}(x, s, t_{n+1}) = \frac{1}{n} \sum_{i=1}^{n} \mu(x_i, x)\mu(s_i, s)\mu(t_{n+1}, t)\mu(s_{n+1}, s). \quad (1.61)$$

As it follows from (1.61, $\bar{\mu}(x, s, t_{n+1}$ is a possibility of the fact that in the moment of time t_{n+1} the signal s will arrive from x. This conclusion is obtained from the information $(x_i, s_i) \mid_1^n + s_{n+1}$ in t_{n+1} i.e. s_{n+1} and s.

We are interested in $s = s_{n+1}$. For every point $x \in \mathcal{X}$ we have a function similar to a seismogram

$$\mu_x(t) = \begin{cases} \mu(x, s_{n+1}, t), & \text{for } t < t_{n+1} \\ \bar{\mu}(x, s_{n+1}, t_{n+1}), & \text{for } t = t_{n+1} \end{cases} \quad (1.62)$$

The function (1.62) opens new opportunities in the problem of $SOD(s_{n+1})$ calculations. These opportunities come from the theory of functions: step functions, type of approximation to moment of time t_{n+1}, etc.

Concluding, we can formulate that the basis of what has been said in this chapter is actually the following chain of implications.

The nearer the sources inside the Earth $\Longrightarrow^1$ more similar way to the station should make the signals produced by the sources $\Longrightarrow^2$ more similar corresponding seismograms should be.

This apparent simplification is based on the model of continuity, which can be not true in many special cases. In particular, except distance between locations other important features of the sources can differ: source mechanisms, magnitudes, etc, and the seismograms from very close events can differ significantly.

Thus, the follow up of what was described in this chapter is the problem to find the features $\Pi_1, \cdots, \Pi_k$ of seismograms, that regardless such difference will, however, reflect that the sources are nearby. The solution of this problem will allow to construct corresponding expert system, that will serve as a real fuzzy realization of Syntactic Pattern Recognition Approach (Fuzzy SPARS).

1.3 Clustering Algorithms

This section describes new constructions of clustering algorithms introduced in 2000-2001 by A. Gvishiani and S. Agayan (Schmidt Institute of Physics of the Earth in Moscow).

1.3.1 "Lighting" and clustering in finite metric spaces

Imagine that at every point of a finite metric space $\mathcal{X} = (x, d)$, there is a lamp emitting light. Let us see how a point is illuminated by the other points. Such an illumination can be considered as an important characteristic of a given point in the space $\mathcal{X}$: the greater is the outer lighting at a point, the more "active" it is in $\mathcal{X}$. On the other hand, a poorly illuminated point can be interpreted as "not active" and rather isolated in $\mathcal{X}$. Light has different characteristics, but we are mainly interested in so called density (activity, intensity) of the illumination. This is the basic notion from which the further constructions are made.

Clustering Algorithms

Light dispersion laws.

A law, δ_x, of light dispersion from a lamp at a point "x" is a free parameter. It is defined by a descending non-negative potential function φ over $[0, \infty]$: $\varphi(0) = 1$, $\varphi(t_1) \geq \varphi(t_2)$, for $t_1 < t_2$. A light dispersion law on $\mathcal{X}$ is defined by the formula:

$$\delta_x(y) = \delta_x^\varphi(y) = \varphi(d(x,y)) \quad \forall y \in \mathcal{X}. \tag{1.63}$$

Some important light dispersion laws are:

1. "Smooth descending illumination"

$$\begin{aligned}\varphi(t) &= \frac{1}{1+t} \\ \delta_x(y) &= \frac{1}{1+d(x,y)}\end{aligned} \tag{1.64}$$

In this case, the illumination smoothly falls with distance from $x \in \mathcal{X}$.

2. "Ball type illumination"

$$\varphi_x(t) = \begin{cases} 1 - t/r, & t < r \\ 0, & t \geq r \end{cases} \quad \delta_x(y) = \begin{cases} 1 - \frac{1+d(x,y)}{r} & ; d(x,y) < r \\ 0 & ; d(x,y) \geq r \end{cases} \tag{1.65}$$

Illumination in this case does not extend beyond the ball $\mathcal{B}(x,r) = \{y \in \mathcal{X} : d(x,y) \leq r\}$. Inside $\mathcal{B}(x,r)$, it propagates by the law of a circular cone.

3. "Gauss law Illumination"

$$\begin{aligned}\varphi(t) &= e^{-t/r} \\ \delta_x^\tau(y) &= e^{-\frac{d(x,y)}{r}}\end{aligned} \tag{1.66}$$

In this case illumination expands according to Gauss-law.

Outer illumination

For a point $x \in \mathcal{X}$ and an arbitrary subset $\mathcal{A} \subset \mathcal{X}$ we introduce the set

$$\mathcal{A}_x = \begin{cases} \mathcal{A} & \text{, if } x \notin \mathcal{A} \\ \mathcal{A} \setminus \{x\} & \text{, if } x \in \mathcal{A} \end{cases} \qquad (1.67)$$

and the natural number value functions

$$|\mathcal{A}|_x = |\mathcal{A}_x| = \begin{cases} |\mathcal{A}|, & \text{if } x \notin \mathcal{A} \\ |\mathcal{A}| - 1, & \text{if } x \in \mathcal{A} \end{cases}, \qquad (1.68)$$

where, as usual, $|\mathcal{A}|$ is the number of elements in the set $\mathcal{A}$.

Definition 1.3.1 :

By illumination of a point $x \in \mathcal{X}$ by the set $\mathcal{A} \subset \mathcal{X}$ we mean the sum of illuminations provided to x by all points $y \in \mathcal{A}$

$$\mathcal{O}_\mathcal{A}(x) = \sum_{y \in \mathcal{A}_x} \delta_y(x) \qquad (1.69)$$

Due to the equality $(\mathcal{A} \cap \mathcal{B})_x = \mathcal{A}_x \cap \mathcal{B}_x$, for every $x \in \mathcal{X}$ the illumination $\mathcal{A} \to \mathcal{O}_\mathcal{A}(x)$ is a measure on $\mathcal{X}$.
It is obvious from (1.69) that $\mathcal{O}_\emptyset(x) = 0$.
Furthermore,

$$\mathcal{O}_{\mathcal{A} \cup \mathcal{B}}(x) = \sum_{y \in (\mathcal{A} \cup \mathcal{B})_x} \delta_y(x) = \sum_{y \in \mathcal{A}_x} \delta_y(x) + \sum_{y \in \mathcal{B}_x} \delta_y(x) - \sum_{y \in \mathcal{A}_x \cap \mathcal{B}_x} \delta_y(x) =$$

$$\mathcal{O}_\mathcal{A}(x) + \mathcal{O}_\mathcal{B}(x) - \mathcal{O}_{\mathcal{A} \cap \mathcal{B}}(x). \qquad (1.70)$$

The function $|\mathcal{A}|_x$, defined by (1.68) is another measure on $\mathcal{X}$. To prove it, we can analyse sequentially all the cases which may occur:

a) $x \notin \mathcal{A}$, $x \notin \mathcal{B} \Rightarrow x \notin \mathcal{A} \cup \mathcal{B}$:

$$|\mathcal{A} \cup \mathcal{B}|_x = |\mathcal{A} \cup \mathcal{B}| = |\mathcal{A}| + |\mathcal{B}| - |\mathcal{A} \cap \mathcal{B}| =$$

$$|\mathcal{A}|_x + |\mathcal{B}|_x - |\mathcal{A} \cap \mathcal{B}|_x.$$

Clustering Algorithms

b) $x \in \mathcal{A}, x \notin \mathcal{B} \Rightarrow x \in \mathcal{A} \bigcup \mathcal{B}$:

$$|\mathcal{A} \bigcup \mathcal{B}|_x = |\mathcal{A} \bigcup \mathcal{B}| - 1 = |\mathcal{A}| - 1 + |\mathcal{B}| - |\mathcal{A} \bigcap \mathcal{B}| =$$

$$|\mathcal{A}|_x + |\mathcal{B}|_x - |\mathcal{A} \bigcap \mathcal{B}|_x .$$

c) The case $x \notin \mathcal{A}, x \in \mathcal{B}$, is analogous to b)

d) $x \in \mathcal{A}, x \in \mathcal{B} \Rightarrow x \in \mathcal{A} \bigcap \mathcal{B}$:

$$|\mathcal{A} \bigcup \mathcal{B}|_x = |\mathcal{A} \bigcup \mathcal{B}| - 1 = |\mathcal{A}| - 1 + |\mathcal{B}| - 1 - (|\mathcal{A} \bigcap \mathcal{B}| - 1) =$$

$$|\mathcal{A}|_x + |\mathcal{B}|_x - |\mathcal{A} \bigcap \mathcal{B}|_x .$$

Illumination intensity:

We are now ready to determine the main notion:

Definition 1.3.2:

The "illumination intensity" of a point $x \in \mathcal{X}$ created by a subset $\mathcal{A} \subset \mathcal{X}$ the intensity of the measure $\mathcal{O}_\mathcal{A}(x)$ with respect to the measure $|\mathcal{A}|_x$ on:

$$\mathcal{P}_\mathcal{A}(x) = \begin{cases} \frac{\mathcal{O}_\mathcal{A}(x)}{|\mathcal{A}|_x} & , \text{if } x \neq \mathcal{A} \\ 0 & , \text{if } x = \mathcal{A} \end{cases} \qquad (1.71)$$

Note:

In the case $\mathcal{A} = x$, we have $\mathcal{O}_\mathcal{X}(x) = \Sigma_\emptyset = 0$, $|x|_\mathcal{X} = 0$. Therefore, it is natural to assume that $\mathcal{P}_\mathcal{X}(x) = 0$ (the outer illumination intensity of "x" by "x" is equal to 0).

Being a measure, the illumination $\mathcal{O}_\mathcal{A}(x)$ is a monotonic function in the sense that, if $\mathcal{A} \subset \mathcal{B}$, then $\mathcal{O}_\mathcal{A}(x) \leq \mathcal{O}_\mathcal{B}(x)$ (Kirilov and Gvishiani [281]). This is also clear from our physical interpretation: the greater is a subset, the greater the number of lamps it contains. The illumination intensity $\mathcal{P}_\mathcal{A}(x)$ is a more sophisticated function of the pair $(x, \mathcal{A})$, and is not monotonous. Let us consider some examples, $\delta_x(y) = 1/(1 + d(x,y))$:

a) $\mathcal{X} = (-1, -1/2, 0, 1/2, 1)$; $\mathcal{A} = (-1, 0, 1)$; $x = 0$

$$\begin{aligned}\mathcal{P}_\mathcal{A}(0) &= \frac{1}{2}(\frac{1}{1+1}+\frac{1}{1+1}) = 1/2 \\ \mathcal{P}_\mathcal{X}(0) &= \frac{1}{4}(\frac{1}{1+1}+\frac{1}{1+1}+\frac{1}{1+1/2}+\frac{1}{1+1/2}) \\ &= 7/12 > 1/2\end{aligned} \Bigg\} \Rightarrow \mathcal{P}_\mathcal{A}(0) < \mathcal{P}_\mathcal{X}(0)$$

b) $\mathcal{X} = (-2, -1, 0, 1, 2)$; $\mathcal{A} = (-1, 0, 1)$; $x = 0$

$$\begin{aligned}\mathcal{P}_\mathcal{A}(0) &= 1/2 \\ \mathcal{P}_\mathcal{X}(0) &= \frac{1}{4}(\frac{1}{1+1}+\frac{1}{1+1}+\frac{1}{1+2}+\frac{1}{1+2}) \\ &= 5/12 < 1/2\end{aligned} \Bigg\} \Rightarrow \mathcal{P}_\mathcal{A}(0) > \mathcal{P}_\mathcal{X}(0)$$

c) Let $\mathcal{X}(z)$ be the set $(-1, -z, 0, z, 1)$ and again $\mathcal{A} = (-1, 0, 1)$, $x = 0$. We would like to know the relation between $\mathcal{P}_\mathcal{A}(0)$ and $\mathcal{P}_{\mathcal{X}(z)}(0)$. It is not complicated to show that $\mathcal{P}_{\mathcal{X}(z)}(0) = 1/4(1 + \frac{2}{1+z})$ and the relation

$$\mathcal{P}_\mathcal{A}(0) = 1/2 \gtrless \mathcal{P}_{\mathcal{X}(z)}(0) = 1/4 + 12(1+z)$$

is equivalent to the inequality $z \gtrless 1$.

In other words, if $\mid z \mid < 1$, then $\mathcal{P}_\mathcal{A}(0) = 1/2 < \mathcal{P}_{\mathcal{X}(z)}(0)$ and if $\mid z \mid > 1$, then $\mathcal{P}_\mathcal{A}(0) = 1/2 > \mathcal{P}_{\mathcal{X}(z)}(0)$.

d) To show this quality we consider a plane: $\mathcal{X}$-cross, $\mathcal{A} = (-1, 0, 1)$

$$\mathcal{P}_{\mathcal{X}}(0) = 1/4(\frac{1}{1+1} + \frac{1}{1+1} + \frac{1}{1+1} + \frac{1}{1+1}) = 1/2 = \mathcal{P}_{\mathcal{A}}(0).$$

The above examples show the capability of the introduced intensity to "catch" clustering and to quantitatively interpret the integral concentration of subset $\mathcal{A}$ around the point "x". The higher the intensity $\mathcal{P}_{\mathcal{A}}(x)$ at the point x the more evidence we have that "x" may be "a center" for the subset $\mathcal{A}$ (even if it is not one in a geometrical sense, especially if $x \notin \mathcal{A}$). The construction $\mathcal{P}_{\mathcal{A}}(x)$ depends both on "x" and "$\mathcal{A}$".
The following fact is fundamental.

Theorem 1.3.1. The mapping $\mathcal{A} \to \mathcal{P}_{\mathcal{A}}(x)$ is linear in the following sense :

$$\begin{aligned} \exists \alpha &= \alpha(\mathcal{A}, \mathcal{B}, \mathcal{X}) \\ \beta &= \beta(\mathcal{A}, \mathcal{B}, \mathcal{X}) \\ \gamma &= \gamma(\mathcal{A}, \mathcal{B}, \mathcal{X}) : \end{aligned}$$

$$\forall x \in \mathcal{X}, \exists \mathcal{A}, \mathcal{B} \subset \mathcal{X} \Rightarrow \mathcal{P}_{\mathcal{A} \bigcup \mathcal{B}}(x) = \alpha(\mathcal{A}, \mathcal{B}, \mathcal{X})\mathcal{P}_{\mathcal{A}}(x) + \beta(\mathcal{A}, \mathcal{B}, \mathcal{X})\mathcal{P}_{\mathcal{B}}(x)$$
$$-\gamma(\mathcal{A}, \mathcal{B}, \mathcal{X})\mathcal{P}_{\mathcal{A} \bigcup \mathcal{B}}(x)$$

Proof:

$$\mathcal{P}_{\mathcal{A} \cup \mathcal{B}}(x) = \frac{\mathcal{O}_{\mathcal{A} \cup \mathcal{B}}(x)}{\mid \mathcal{A} \cup \mathcal{B} \mid_x} = \frac{\mathcal{O}_{\mathcal{A}}(x) + \mathcal{O}_{\mathcal{B}}(x) - \mathcal{O}_{\mathcal{A} \cap \mathcal{B}}(x)}{\mid \mathcal{A} \cup \mathcal{B} \mid_x} =$$

$$\frac{\mid \mathcal{A} \mid_x}{\mid \mathcal{A} \cup \mathcal{B} \mid_x}\mathcal{P}_{\mathcal{A}}(x) + \frac{\mid \mathcal{B} \mid_x}{\mid \mathcal{A} \cup \mathcal{B} \mid_x}\mathcal{P}_{\mathcal{B}}(x) - \frac{\mid \mathcal{A} \cap \mathcal{B} \mid_x}{\mid \mathcal{A} \cup \mathcal{B} \mid_x}\mathcal{P}_{\mathcal{A} \cap \mathcal{B}}(x), \quad (1.72)$$

Therefore,

$$\begin{aligned} \alpha(\mathcal{A}, \mathcal{B}, \mathcal{X}) &= \frac{\mid \mathcal{A} \mid_x}{\mid \mathcal{A} \cup \mathcal{B} \mid_x} \\ \beta(\mathcal{A}, \mathcal{B}, \mathcal{X}) &= \frac{\mid \mathcal{B} \mid_x}{\mid \mathcal{A} \cup \mathcal{B} \mid_x} \\ \gamma(\mathcal{A}, \mathcal{B}, \mathcal{X}) &= \frac{\mid \mathcal{A} \cap \mathcal{B} \mid_x}{\mid \mathcal{A} \cup \mathcal{B} \mid_x} \end{aligned}$$

In particular, if $\mathcal{A} \cap \mathcal{B} = \emptyset$, then

$$\mathcal{P}_{\mathcal{A}\cup\mathcal{B}}(x) = \frac{\mid\mathcal{A}\mid_x}{\mid\mathcal{A}\mid_x + \mid\mathcal{B}\mid_x}\mathcal{P}_{\mathcal{A}}(x) + \frac{\mid\mathcal{B}\mid_x}{\mid\mathcal{A}\mid_x + \mid\mathcal{B}\mid_x}\mathcal{P}_{\mathcal{B}}(x) \qquad (1.73)$$

If we also assume that $x \in \mathcal{A}$, then

$$\mathcal{P}_{\mathcal{A}\cup\mathcal{B}}(x) = \frac{\mid\mathcal{A}\mid - 1}{\mid\mathcal{A}\mid + \mid\mathcal{B}\mid - 1}\mathcal{P}_{\mathcal{A}}(x) + \frac{\mid\mathcal{B}\mid}{\mid\mathcal{A}\mid + \mid\mathcal{B}\mid - 1}\mathcal{P}_{\mathcal{B}}(x) \qquad (1.74)$$

Formula (1.74) espresses the intensity of a point in relation to the whole space $\mathcal{X}$ and to its subspace. If $x \in \mathcal{A} \subset \mathcal{C} \subset \mathcal{X}$, then $\mathcal{C} = \mathcal{A} \cup (\mathcal{C} \setminus \mathcal{A})$ and

$$\mathcal{P}_{\mathcal{C}}(x) = \frac{\mid\mathcal{A}\mid - 1}{\mid\mathcal{C}\mid - 1}\mathcal{P}_{\mathcal{A}}(x) + \frac{\mid\mathcal{C}\mid - \mid\mathcal{A}\mid}{\mid\mathcal{C}\mid - 1}\mathcal{P}_{\mathcal{C}\setminus\mathcal{A}}(x) = \mu\mathcal{P}_{\mathcal{A}}(x) + \nu\mathcal{P}_{\mathcal{C}\setminus\mathcal{A}}(x). \qquad (1.75)$$

It follows from (1.75) that the value of $\mathcal{P}_{\mathcal{C}}(x)$ lies on the segment with the ends $\mathcal{P}_{\mathcal{A}}(x)$, $\mathcal{P}_{\mathcal{C}\setminus\mathcal{A}}(x)$, which is why "non-Archimediarity" takes place. The equality (1.73) shows the "non-Archimedial" behavior of our model (Gvishiani, Agajan and Troussov) and opens an important connection with the principle of this algorithmic theory with the theory of non-Archimedial metric spaces.

At the same time it gives the necessary and sufficient condition for inequality $\mathcal{P}_{\mathcal{C}} \gtrless \mathcal{P}_{\mathcal{A}}(x)$:

Statement 1.3.1:

Let $x \in \mathcal{A} \subset \mathcal{C} \subset \mathcal{X}$, then

$$\cap \mathcal{P}_{\mathcal{C}}(x) \lessgtr \mathcal{P}_{\mathcal{A}}(x) \iff \mathcal{P}_{\mathcal{C}\setminus\mathcal{A}}(x) \gtrless \mathcal{P}_{\mathcal{A}}(x). \qquad (1.76)$$

Let us prove that $\mathcal{P}_{\mathcal{C}}(x) > \mathcal{P}_{\mathcal{A}}(x)$ if and only if $\mathcal{P}_{\mathcal{C}\setminus\mathcal{A}} < \mathcal{P}_{\mathcal{A}}(x)$ (The opposite inequality can be shown in a similar way). Thus,

$$\frac{\mid\mathcal{A}\mid - 1}{\mid\mathcal{C}\mid - 1}\mathcal{P}_{\mathcal{A}}(x) + \frac{\mid\mathcal{C}\mid - \mid\mathcal{A}\mid}{\mid\mathcal{C}\mid - 1}\mathcal{P}_{\mathcal{C}\setminus\mathcal{A}}(x) > \mathcal{P}_{\mathcal{A}}(x),$$

then

$$(\mid\mathcal{A}\mid - \mid\mathcal{C}\mid) > (\mid\mathcal{A}\mid - \mid\mathcal{C}\mid)\mathcal{P}_{\mathcal{C}\setminus\mathcal{A}}(x)$$

and

$$\mathcal{P}_{\mathcal{A}}(x) > \mathcal{P}_{\mathcal{C}\setminus\mathcal{A}}(x).$$

Thus, the increasing $\mathcal{P}_{\mathcal{C}}(x) > \mathcal{P}_{\mathcal{A}}(x)$ is related to the existence in $\mathcal{C}$ of points closely located to "x", but not belonging to $\mathcal{A}$.

Example:

$\mathcal{A} = (-1, 0, 1); \mathcal{C} = (-1, -1/2, 0, 1),$
then
$$\mathcal{P}_\mathcal{A}(0) = 1/2 < \mathcal{P}_\mathcal{C}(0) = 5/9 \text{ and } \mathcal{P}_{\mathcal{C}-\mathcal{A}}(0) = \mathcal{P}_{-1/2}(0) = 2/3 > \mathcal{P}_\mathcal{A}(0).$$

Furthermore, if $\mathcal{A} \subset \mathcal{C} \subset \mathcal{X}$ the difference $\mathcal{P}_\mathcal{C}(x) - \mathcal{P}_\mathcal{A}(x)$ can be written as:

$$\begin{aligned}\mathcal{P}_\mathcal{C}(x) - \mathcal{P}_\mathcal{A}(x) &= \tfrac{|\mathcal{A}|-1}{|\mathcal{C}|-1}\mathcal{P}_\mathcal{A}(x) + \tfrac{|\mathcal{C}|-|\mathcal{A}|}{|\mathcal{C}|-1}\mathcal{P}_{\mathcal{C}\setminus\mathcal{A}}(x) - \mathcal{P}_\mathcal{A}(x) \\ &= \tfrac{|\mathcal{C}|-|\mathcal{A}|}{|\mathcal{C}|-1}(\mathcal{P}_{\mathcal{C}\setminus\mathcal{A}}(x) - \mathcal{P}_\mathcal{A}(x)).\end{aligned} \quad (1.77)$$

We will call this difference the "step". Formula (1.77) shows that step depends naturally on the location of the point "x", the location and configuration of the set $\mathcal{C} - \mathcal{A}$, and the power of the normalized addition $|\mathcal{C}| - |\mathcal{A}|$. The function $\mathcal{P}_\mathcal{C}(x)$ is linear (1.74), but we may consider more general models where $\mathcal{P}_\mathcal{C}(x)$ behaves arbitrarily.

The case shown below graphically is of a particular interest
Now imagine that the set $\mathcal{A}$ consists of the points $x \in \mathcal{A}$ such that their intensity with respect to $\mathcal{A}$ is not lower than with respect to any space containing $\mathcal{A}$. It follows that $\mathcal{A}$ in any $\mathcal{C}$ containing $\mathcal{A}$ should be an accumulating set (or cluster), because the inequality $\mathcal{P}_\mathcal{A}(x) \geq \mathcal{P}_\mathcal{C}(x)$, $x \in \mathcal{A}$ means (as it was presented above) that there are no points in $\mathcal{C} \setminus \mathcal{A}$ close to "x". Therefore, a cluster is a subset with higher density, resistant to outer (surrounding) "actions".
Above condition enables us to introduce the formal definition of a cluster:

Definition 1.3.3.

We call $\mathcal{A} \subset \mathcal{X}$ a cluster in $\mathcal{X}$ if

$$\mathcal{P}_\mathcal{A}(x) \geq \mathcal{P}_{\mathcal{A}\cup\mathcal{B}}(x) \quad \forall x \in \mathcal{A} \text{ and } \mathcal{B} \text{ such as } \mathcal{B} \cap \mathcal{A} = \emptyset. \quad (1.78)$$

If $\mathcal{B} = y$, then (1.78), due to statement 1.3.1, is equivalent to the inequality:

$$\mathcal{P}_\mathcal{A}(x) \geq \mathcal{P}_{\{y\}}(x) = \delta_y(x). \tag{1.79}$$

The inequality (1.79) is valid for any $\mathcal{A} \subset \mathcal{X}$, $x \in \mathcal{A}$ and $y \in \mathcal{X} \setminus \mathcal{A}$ which satisfies the condition (1.78). This is important because, to recognize a cluster, it is much more convenient to use (1.79) than (1.78).

Let $\mathcal{B} = (y_1 + y_2 + \cdots + y_k)$. Then the linearity (1.74) gives the equality

$$\mathcal{P}_\mathcal{B}(x) = \mathcal{P}_{y_1+(y_2+\cdots+y_k)}(x) = \tfrac{1}{k}\mathcal{P}_{y_1}(x) + \tfrac{k-1}{k}\mathcal{P}_{y_2+\cdots+y_k}(x) = \cdots$$
$$= \tfrac{1}{k}[\mathcal{P}_{y_1}(x) + \cdots + \mathcal{P}_{y_k}(x)],$$

and therefore

$$\mathcal{P}_{\mathcal{A}+\mathcal{B}}(x) = \frac{|\mathcal{A}|-1}{|\mathcal{A}|+k-1}\mathcal{P}_\mathcal{A}(x) + \frac{k}{|\mathcal{A}|+k-1}\mathcal{P}_\mathcal{B}(x) = \frac{|\mathcal{A}|-1}{|\mathcal{A}|+k-1}\mathcal{P}_\mathcal{A}(x)+$$

$$\frac{1}{|\mathcal{A}|+k-1}[\mathcal{P}_{y_1}(x) + \cdots + \mathcal{P}_{y_k}(x)] \leq \frac{|\mathcal{A}|-1}{|\mathcal{A}|+k-1}\mathcal{P}_\mathcal{A}(x)+$$

$$\frac{1}{|\mathcal{A}|+k-1}[\mathcal{P}_\mathcal{A}(x) + \cdots + \mathcal{P}_\mathcal{A}(x)] = \mathcal{P}_\mathcal{A}(x). \tag{1.80}$$

Using (1.78) we come to the following conclusion: a cluster is a set, any point of which has greater inner illumination rather than the light that can come from any outer source. In term of distances (1.78) it means that the average distance to $x \in \mathcal{A}$ is not less than the illumination of x produced by any point y which does not belong to $\mathcal{A}$. In other words

$$\forall x \in \mathcal{A}, \forall y \in \mathcal{X} \setminus \mathcal{A}; \quad \frac{1}{|\mathcal{A}|-1} \sum_{z \in \mathcal{A}_x} \delta_z(x) \geq \delta_y(x). \tag{1.81}$$

Remarks:

1. By definition, a cluster includes at least two points. The space $\mathcal{X}$ itself is a cluster, if it is not trivial ($|\mathcal{X}| \geq 2$), as well as all the pairs $\{x,y\} \subset \mathcal{X}$, where the metrics $d(x,y)$ are minimum.

2. The example below demonstrates that in certain cases a point can be considered as a cluster: a point x is a cluster in $\mathcal{X}$, if its complement $(\mathcal{X} \setminus x)$ is a cluster in $\mathcal{X}$. In this case x is necessarily isolated in $\mathcal{X}$ and so it can be considered as a cluster. Note that in this case the condition $|\mathcal{X}| \geq 3$ is always true.

Clustering Algorithms

3. It is clear that the "quality" of a point x as a cluster is higher than the quality of the complementary set $\mathcal{X} \setminus x$. The point x is more "monolithic" than $\mathcal{X} \setminus x$. One possible formalisation of this impression is given by the following definition.

Definition 1.3.4.

Let $\mathcal{A} \subset \mathcal{X}$ be a subset. We call the function:

$$\mathcal{P}(\mathcal{A}) = \begin{cases} 1, & \text{if } \mid \mathcal{A} \mid = 1 \\ \min_{x \in \mathcal{A}} \mathcal{P}_{\mathcal{A}}(x), & \text{if } \mid \mathcal{A} \mid > 1 \end{cases} \qquad (1.82)$$

the quality or level of illumination.
In general, the inclusion $\mathcal{A} \subset \mathcal{B}$ is followed by the inequality
$\mathcal{P}(\mathcal{A}) \geq \mathcal{P}(\mathcal{B})$, i.e. a smaller set is more "monolithic" than a bigger one.

Examples:

Let $\mathcal{X} = (-2, -1, -3/4, -1/2, -1/4, 0, 1/4, 1/2, 3/4, 1, 2)$.

a) $\mathcal{A} = (-1, 1)$; $\delta_x(y) = \dfrac{1}{1 + d(x,y)}$;

therefore $\mathcal{P}(\mathcal{A}) = 1/2$

b) $\mathcal{B} = (-2, -1, 1, 2)$.

In this case

$$\mathcal{P}(\mathcal{B}) = \min\{\mathcal{P}_{\mathcal{B}}(-2) = \mathcal{P}_{\mathcal{B}}(2) \; ; \; \mathcal{P}_{\mathcal{B}}(-1) = \mathcal{P}_{\mathcal{B}}(1)\} = \mathcal{P}_{\mathcal{B}}(-2),$$

because

$$\mathcal{P}_{\mathcal{B}}(-2) = 1/3(\frac{1}{1+1} + \frac{1}{1+3} + \frac{1}{1+4}) = 0.95/3 < 1/3 < \mathcal{P}(\mathcal{A})$$
$$\mathcal{P}_{\mathcal{B}}(-1) = 1/3(\frac{1}{1+1} + \frac{1}{1+2} + \frac{1}{1+3}) > \mathcal{P}_{\mathcal{B}}(-2)$$

c)
$$\mathcal{B} = (-1,\ -3/4,\ -1/2,\ -1/4,\ 0,\ 1/4,\ 1/2,\ 3/4,\ 1).$$

In this case

$$\begin{aligned}\mathcal{P}(\mathcal{B}) &= \min\{\mathcal{P}_\mathcal{B}(-1) = \mathcal{P}_\mathcal{B}(1)\ ;\ \mathcal{P}_\mathcal{B}(0)\ ;\ \mathcal{P}_\mathcal{B}(-3/4)\} = \mathcal{P}_\mathcal{B}(3/4),\\ &\quad \mathcal{P}_\mathcal{B}(-1/2) = \mathcal{P}_\mathcal{B}(1/2),\ \mathcal{P}_\mathcal{B}(-1/4) = \mathcal{P}_\mathcal{B}(1/4))\\ \mathcal{P}_\mathcal{B}(-1) &= 1/8(4/5 + 4/6 + 4/7 + 4/8 + 4/9 + 4/10 + 4/11 + 4/12)\\ &= 0.51 > \mathcal{P}(\mathcal{A}).\end{aligned}$$

The above situation is impossible if $\mathcal{A}$ and $\mathcal{B}$ are clusters in $\mathcal{X}$. In this case the inclusion $\mathcal{A} \subset \mathcal{B}$ is necessarily followed by the inequality $\mathcal{P}(\mathcal{A}) \geq \mathcal{P}(\mathcal{B})$. In fact, $\mathcal{P}_\mathcal{A}(x) \geq \mathcal{P}_\mathcal{B}(x)\ \forall x \in \mathcal{A}$, and

$$\mathcal{P}(\mathcal{A}) = \min_{x \in \mathcal{A}} \mathcal{P}_\mathcal{A}(x) \geq \min_{x \in \mathcal{A}} \mathcal{P}_\mathcal{B}(x) \geq \min_{x \in \mathcal{B}} \mathcal{P}_\mathcal{B}(x) = \mathcal{P}(\mathcal{B}).$$

Conclusion:

$\mathcal{P}(\mathcal{A})$ is a measure of the compaction and irreductibility of a cluster. The higher is the value $\mathcal{P}(\mathcal{A})$, the more "monolithic", compact and irreducible is the cluster $\mathcal{A}$.

1.3.2 Algorithm "RODIN":

The RODIN Algorithm searches for a cluster of given quality in the initial space $\mathcal{X}$, lopping off all the excess. This explains its name. The theoretical background of the algorithm is based on the quality of the density function. Further on we present its block-algorithm.

To define a cluster we used the dependence of construction $\mathcal{P}_\mathcal{A}(x)$ on $\mathcal{A}$; with "x" fixed. Now assume "$\mathcal{A}$" is fixed and "x" is variable. We consider the density function $x \to \mathcal{P}_\mathcal{A}(x) \in \mathcal{A}$:

Statement 6:

If the "lamp" is a potential-induced $\varphi = 1/1 + t$, the following inequality is true for any $x, y \in \mathcal{A}$

Clustering Algorithms

$$|\mathcal{P}_A(x) - \mathcal{P}_A(y)| \le \frac{|\mathcal{A}|-2}{|\mathcal{A}|-1} d(x,y) < d(x,y) \qquad (1.83)$$

Proof:

We have

$$\mathcal{P}_A(x) = \frac{1}{|\mathcal{A}|_x} \sum_{z \in \mathcal{A}_x} \delta_z(x) = \frac{1}{|\mathcal{A}|-1} \sum_{z \in \mathcal{A}_x, z \ne y} \delta_z(x) + \frac{1}{|\mathcal{A}|-1} \delta_y(x)$$

$$\mathcal{P}_A(y) = \frac{1}{|\mathcal{A}|_y} \sum_{z \in \mathcal{A}_y} \delta_z(y) = \frac{1}{|\mathcal{A}|-1} \sum_{z \in \mathcal{A}_y, z \ne x} \delta_z(y) + \frac{1}{|\mathcal{A}|-1} \delta_x(y)$$

But $\delta_y(x) = \delta_x(y)$, so

$$|\mathcal{P}_A(x) - \mathcal{P}_A(y)| \le \frac{1}{|\mathcal{A}|-1} \sum_{z \in \mathcal{A}, z \ne x, y} |\delta_z(x) - \delta_z(y)|$$

$$= \frac{1}{|\mathcal{A}|-1} \sum_{z \in \mathcal{A}, z \ne x, y} \left| \frac{1}{1+d(x,z)} - \frac{1}{1+d(x,z)} \right|$$

$$= \frac{1}{|\mathcal{A}|-1} \sum_{z \in \mathcal{A}, z \ne x, y} \left| \frac{d(x,z) - d(y,z)}{(1+d(x,z))(1+d(y,z))} \right|$$

$$\le \frac{|\mathcal{A}|-2}{|\mathcal{A}|-1} d(x,y) \text{ (triangle inequality } : |d(x,z) - d(y,z)| \le d(x,y))$$

Note

Clearly, one can obtain corresponding variants of such an estimate for the other "lamps" (linear and Gauss) defined in the beginning of this section. From the stated inequality it follows that the density is "continuous" close to the weak and strong neighbours. Consequently, while eliminating weak points from the initial space, one can seek other weak points to eliminate in the close neighbourhood of the current point. However this should be done with caution. We show why with one more statement describing density change in elimination.

If there is a set $\mathcal{A}$ and the point "x" is eliminated from it, how will the density of any $y \in \mathcal{A} - x$ in $\mathcal{A} - x$ change?

From the linearity $\mathcal{P}_A(y)$ it follows that

$$\mathcal{P}_{\mathcal{A}}(y) = \frac{|\mathcal{A}|-2}{|\mathcal{A}|-1}\mathcal{P}_{\mathcal{A}-x}(y) + \frac{1}{|\mathcal{A}|-1}\delta_x(y)$$

Now we express $\mathcal{P}_{\mathcal{A}-x}(y)$:

$$\frac{|\mathcal{A}|-2}{|\mathcal{A}|-1}\mathcal{P}_{\mathcal{A}-x}(y) = \mathcal{P}_{\mathcal{A}}(y) - \frac{1}{|\mathcal{A}|-1}\delta_x(y) \Rightarrow$$

$$\mathcal{P}_{\mathcal{A}-x}(y) = \frac{|\mathcal{A}|-1}{|\mathcal{A}|-2}\mathcal{P}_{\mathcal{A}}(y) - \frac{1}{|\mathcal{A}|-2}\delta_x(y)$$

Then the conditions $\mathcal{P}_{\mathcal{A}}(y) \lessgtr \mathcal{P}_{\mathcal{A}-x}(y)$ take the form

$$\mathcal{P}_{\mathcal{A}}(y) \lessgtr \frac{|\mathcal{A}|-1}{|\mathcal{A}|-2}\mathcal{P}_{\mathcal{A}}(y) - \frac{1}{|\mathcal{A}|-2}\delta_x(y) \Rightarrow$$

$$\frac{1}{|\mathcal{A}|-2}\delta_x(y) \lessgtr \left(\frac{|\mathcal{A}|-1}{|\mathcal{A}|-2} - 1\right)\mathcal{P}_{\mathcal{A}}(y) = \frac{1}{|\mathcal{A}|-2}\mathcal{P}_{\mathcal{A}}(y)$$

or

$$\mathcal{P}_{\mathcal{A}}(y) \lessgtr \mathcal{P}_{\mathcal{A}-x}(y) \iff \delta_x(y) \lessgtr \mathcal{P}_{\mathcal{A}}(y) \tag{1.84}$$

Let us formulate the statement thus obtained

Statement 7:

If the light from the point "x" to the point "y" is stronger (weaker) than the density of "y" in $\mathcal{A}$, by eliminating "x", the density of "y" in $\mathcal{A} - x$ will decrease (increase); the set $\mathcal{A} - x$ is divided into two parts : $\mathcal{A} - x = \mathcal{A}_x^- \cup \mathcal{A}_x^+$:

$$\mathcal{A}_x^- = \left\{ \begin{array}{l} y \in \mathcal{A} - x \colon \delta_x(y) > \mathcal{P}_{\mathcal{A}}(y) \quad \text{—all the points in } \mathcal{A} \text{ whose densities} \\ \phantom{y \in \mathcal{A} - x \colon \delta_x(y) > \mathcal{P}_{\mathcal{A}}(y)} \text{decrease by eliminating ``} x\text{''} \\ \text{or} \quad \mathcal{P}_{\mathcal{A}-x}(y) < \mathcal{P}_{\mathcal{A}}(y). \end{array} \right\}$$

$$\mathcal{A}_x^+ = \left\{ \begin{array}{l} y \in \mathcal{A} - x \colon \delta_x(y) > \mathcal{P}_{\mathcal{A}}(y) \quad \text{—all the points in } \mathcal{A} \text{ whose densities} \\ \phantom{y \in \mathcal{A} - x \colon \delta_x(y) > \mathcal{P}_{\mathcal{A}}(y)} \text{do not decrease by eliminating ``} x\text{''} \\ \text{or} \quad \mathcal{P}_{\mathcal{A}-x}(y) \geq \mathcal{P}_{\mathcal{A}}(y). \end{array} \right\}$$

The dynamics of density change becomes clear in deciding on the elimination strategy. The delicacy consists in the fact that an initially weak point may become stronger every time, resulting in a cluster of rather high quality under the defined elimination strategy.

We will now describe the "Rodin" algorithm : the quality level "α" of the eliminated cluster is defined at the beginning. Denote by $\mathcal{K}_n$ the current version of the cluster.

1. Block $\boxed{\text{Initialisation}}$: set the quality level "α" and the starting cluster $\mathcal{K}_0 = \mathcal{X}$, pass to 2

2. Block $\boxed{\text{Quality}}$: Compute $\mathcal{P}(\mathcal{K}_n)$, further compare with α

    $\boxed{\text{Logic of the block}}$:

    $$\text{if } \begin{array}{l} \mathcal{P}(\mathcal{K}_n) \geq \alpha, \text{ then go to 4)} \\ \mathcal{P}(\mathcal{K}_n) < \alpha, \text{ then go to 3)} \end{array}$$

3. Block $\boxed{\text{P-generation}}$: generation based on quality-choosing x_{n+1}:

    $$x_{n+1} = \arg\min \mathcal{P}_{\mathcal{K}_n}(x)$$

    The point x_{n+1} is the weakest in $\mathcal{K}_n$. From a new version $\mathcal{K}_{n+1}$:

    $$\mathcal{K}_{n+1} = \mathcal{K}_n - x_{n+1}$$

    and test its quality (go to 2).

4. Block $\boxed{\text{Cluster}}$: Check $\mathcal{K}_n$ using criterion (1.77) : $\mathcal{K}_n = \mathcal{X} - (x_1 + x_2 + \cdots + x_n)$. For any $y \in \mathcal{K}_n$ in this block the outer lighting is calculated by

    $$\mathcal{S}_n(y) = \min_{i=1}^{n}(\mathcal{P}_{\mathcal{K}_n}(y) - \delta_{x_i}(y))$$

    at the same time from the subset $\mathcal{A}_n$ of points in $\mathcal{K}_n$ that have not satisfied the test ;

    $$\mathcal{A}_n = \{y \in \mathcal{K}_n : \mathcal{S}_n(y) < 0\}$$

Logic of the block :

if $\begin{array}{l}\mathcal{A}_n = \emptyset, \\ \mathcal{A}_n \neq \emptyset,\end{array}$ $\begin{array}{l}\text{then } \mathcal{K} = \mathcal{K}_n. \text{ This is the end, because} \\ \text{the cluster of required quality is obtained.} \\ \text{then go to 5)}\end{array}$

5. Block Cl-generation : choosing x_{n+1} :

$$x_{n+1} = \arg\{\min \mathcal{S}_n(y) \colon y \in \mathcal{A}_n\}$$

is the weakest point in $\mathcal{K}_n$ (from the cluster-test point of view) .

From a new version $\mathcal{K}_{n+1} = \mathcal{K}_n - x_{n+1}$ test its quality (go to 2).

"Rodin" : **free parameters and potentialities:**

The algorithm stops after coming across a cluster of given quality "α". Another cluster inside the first one is possible and, as was shown above, its quality should be higher than α. We may come back to the same cluster or obtain a new one. To avoid this we can move further : having found the cluster $\mathcal{K}$, we eliminate it from the space $\mathcal{X}$, i.e. we pass from $\mathcal{X}$ to $\mathcal{X} - \mathcal{K}$ and use "Rodin" again.

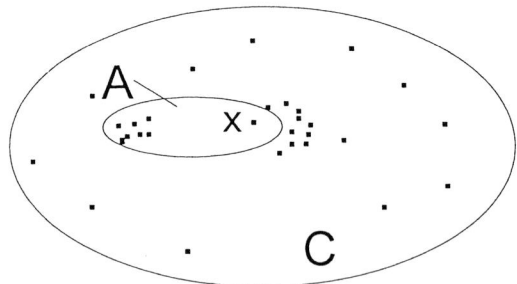

Figure 1.5: RODIN *clustering*.

The Rodin algorithm results in a quasi-clusterization of $\mathcal{X}$: dividing $\mathcal{X} = \bigcup_{i=1}^{n} \mathcal{K}_i$, where $\mathcal{K}_i$ is a cluster in $\mathcal{X} - \bigcup_{j=1}^{i-1} \mathcal{K}_j \ \forall \ i = 1, \cdots, n$ (assume that $\mathcal{K}_0 = \emptyset$). We can consider a more revolutionary variant of "Rodin" : instead

Clustering Algorithms

of consequent one-point elimination we cut subsets that are closely packed together. This algorithm is based on qualities and density characteristics $\mathcal{P}_A(x)$ presented below.

We know that $\mathcal{P}_\mathcal{X}(y) \geq \mathcal{P}_\mathcal{X}(x) - d(x,y)$ (1.83). If the set $\mathcal{A}$ contains a ball $d_\mathcal{X}(x,r)$ such that the impact of every point $y \in \mathcal{X} - \mathcal{A}$ on "x" does not exceed $1/1+r$ (again, we consider $\delta_x(y) = 1/1 + d(x,y)$), then one can find a threshold of radius "r", such that, according to statement 3, $\mathcal{P}_{\mathcal{X}-\mathcal{A}}(x) \leq \mathcal{P}_\mathcal{X}(x)$, but $\mathcal{P}_{\mathcal{X}-\mathcal{A}}(x) \leq 1/1+r$, consequently it will suffice to satisfy the inequality

$$1/1 + r \leq \mathcal{P}_\mathcal{X}(x) \Rightarrow 1 + r \geq 1/\mathcal{P}_\mathcal{X}(x) \Rightarrow r \geq 1/\mathcal{P}_\mathcal{X}(x) - 1$$

It is clear that the greater the density $\mathcal{P}_\mathcal{X}(x)$, the smaller the ball required for the set $\mathcal{A}$ to light "x" not less than $(\mathcal{P}_\mathcal{A}(x) \geq \mathcal{P}_\mathcal{X}(x))$. In this case the density $\mathcal{P}_\mathcal{X}(y)$ of each point "y" from the ball $d_\mathcal{X}(x, 1/\mathcal{P}_\mathcal{X}(x) - 1)$ will not be less than $\mathcal{P}_\mathcal{X}(x) - 1/\mathcal{P}_\mathcal{X}(x) + 1$. The same concerning $\mathcal{A}$:

$$\mathcal{P}_\mathcal{A}(y) \geq \mathcal{P}_\mathcal{X}(x) - 1/\mathcal{P}_\mathcal{X}(x) + 1$$

But the relation between $\mathcal{P}_\mathcal{A}(y)$ and $\mathcal{P}_\mathcal{X}(y)$ is unknown ; we can only estimate $\mathcal{P}_\mathcal{A}(y)$ through $\mathcal{P}_\mathcal{X}(x)$.

Conclusion:

1. The function $\alpha - 1/\alpha + 1$ shows, all the estimates are interested for $\mathcal{P}_\mathcal{X} > (\sqrt{5} - 1)/2$.

2. for $\alpha \in (0, 1)$ the union

$$\mathcal{X}_\alpha = \bigcup_{x \in \mathcal{X} : \mathcal{P}_\mathcal{X}(x) \geq \alpha} d_\mathcal{X}(x, 1/\mathcal{P}_\mathcal{X}(x) - 1)$$

has a density $\mathcal{P}(\mathcal{X}_\alpha \geq \alpha - 1/\alpha + 1$. That is why, according to the preceding formula $\mathcal{P}_{\mathcal{X}_\alpha}(x) \geq \mathcal{P}_\mathcal{X}(x) \geq \alpha$, that is why for any $y \in d_\mathcal{X}(x, 1/\mathcal{P}_\mathcal{X}(x) - 1)$, due to $\alpha \to \alpha - 1/\alpha + 1$ is a monotonous function

$$\mathcal{P}_{\mathcal{X}_\alpha}(y) \geq \mathcal{P}_{\mathcal{X}_\alpha}(x) - 1/\mathcal{P}_{\mathcal{X}_\alpha}(x) + 1 \geq \alpha - 1/\alpha + 1$$

3. If α is big enough $\mathcal{X}_\alpha$ is a rather tight subspace in $\mathcal{X}$ and may be a cluster-candidate.

The intervening review: by varying the quality level «α», elimination rythm (character), strategies (i.e. choosing x_{n+1} in blocks $\boxed{\text{P-generation}}$ and $\boxed{\text{Cl-generation}}$) so as by «cutting», "Rodin" can find quite enough clusters in $\mathcal{X}$, if they exist in $\mathcal{X}$. "Rodin" as if diagnoses the space $\mathcal{X}$ searching cluster-thickenings in it.

Further variations of "Rodin" are possible by varying the rest of the constructions and notions in it. We analyse them consequently:
$\boxed{\text{Potential}}$: the clusters depend on the light travel law, i.e. on the potential «φ».

Example:

Let $\mathcal{X}(z) = (0, 1/4, 1/2, 3/4, 1, z > 1)$. It is clear that for $z > 5/4$ the subset $\mathcal{A} = (0, 1/4, 1/2, 3/4, 1)$ should be considered as a cluster in $\mathcal{X}(z)$. Let us see what happens for potential $\varphi(t) = 1/1 + t$: $\mathcal{P}(\mathcal{A}) = \mathcal{P}_\mathcal{A}(0) = \mathcal{P}_\mathcal{A}(1) = 1/4(1/1 + 1/4 + 1/1 + 2/4 + 1/1 + 3/4 + 1/1 + 4/4) = 0.63$.
The light from point «z» is the greatest in 1 is equal to $1/(1+(z-1)) = 1/z$. Then $\mathcal{A}$ is a cluster in $\mathcal{X}(z)$ for a potential $\varphi(t) = 1/1+t$, only if $1/z < 0.63$, i.e. $z > 1.59$. This does not completely correspond to our concept: $\mathcal{A}$ should be a cluster in $\mathcal{X}(z)$ for $z > 1.25$. It will go in place if we choose the finite linear potential «φ_r» for $r < z - 1$.
$\boxed{\text{Cluster}}$: one can also isolate $\mathcal{A}$ in $\mathcal{X}(z)$ by modifying the cluster definition:

a) **β-cluster:** by adding to condition (1.77) a parameter $\beta > 0$ we obtain an extension of clusters shown as a β-cluster:

   Definition 8: Let $\beta > 0$, $\mathcal{A}$ is a subset in $\mathcal{X}$. We call $\mathcal{A}$ a β-cluster in $\mathcal{X}$ if

$$\mathcal{P}_\mathcal{A}(x) \geq \beta \delta_y(x) \quad \forall x \in \mathcal{A},\ y \in \mathcal{X} - \mathcal{A} \qquad (1.85)$$

   Thus, for $\beta = 1$ we obtain the common cluster definition; for $\beta < 1$ we obtain its decay and for $\beta > 1$ we obtain its growth. Certainly, there exists a "β- Rodin".

   Turn back to the example : $\mathcal{A}$ will be a β-cluster in $\mathcal{X}(z)$ with $z = 1.25$ for $0.65 > 1.25\beta$, i.e. $\beta < 0.5$. Note exactly that β should be treated with caution, not to weaken the cluster too much. The choice of «β» depends on the task at hand.

b) **local cluster** : for this potential the presentation is more clear if we assume that $\varphi(t) = 1/1+t$

Let $d_{\mathcal{X}}(x)(r)$ be a ball in $\mathcal{X}$ with center at "x" and radius "r". Since for all $y \in d_{\mathcal{X}}(x)(r)$ $(y \notin d_{\mathcal{X}}(x)(r))$, $d(x,y) \leq r (d(x,y) > r)$:

$$\mathcal{P}_{d_{\mathcal{X}}(x)(r)} \geq \frac{1}{1+r} \text{ and } \mathcal{P}_{\mathcal{X} - d_{\mathcal{X}}(x)(r)}(x) < \frac{1}{1+r}$$

From statement 3 we may conclude that function $\mathcal{P}_{\mathcal{X}}(x)(r)$:

$$\mathcal{P}_{\mathcal{X}}(x)(r) \stackrel{\text{def}}{=} \mathcal{P}_{d_{\mathcal{X}}(x)(r)}(x)$$

(which we call the local r-density of the point "x" in the space $\mathcal{X}$), decreases when $r \in (0, \text{diam}\mathcal{X})$ increases. In particular

$$\mathcal{P}_{d_{\mathcal{X}}(x)(r)}(x) \geq \mathcal{P}_{\mathcal{X}}(x) \qquad \forall x \in \mathcal{X} \qquad (1.86)$$

Remember that we defined a cluster as a subset $\mathcal{A}$ in $\mathcal{X}$, each point "x" of which is not interior in $\mathcal{A}$ than in any $\mathcal{B}$ that contains $\mathcal{A}$: $\mathcal{P}_{\mathcal{A}}(x) \geq \mathcal{P}_{\mathcal{B}}(x) \; \forall x \in \mathcal{A}$.

We will show that this quality is locally transmitted to any point of a cluster: let $\mathcal{A}$ be a cluster in $\mathcal{X}$, $r \in (0, \text{diam}\mathcal{X}]$, $x \in \mathcal{A}$, $\mathcal{A} \subset \mathcal{B}$ and $\mathcal{P}_{\mathcal{A}}(x)(r)$ $(\mathcal{P}_{\mathcal{B}}(x)(r))$ be local r-densities of the point "x" in $\mathcal{A}(\mathcal{B})$, then the following statement is true:

Statement 9

(Hereditary principle):

$$\mathcal{P}_{\mathcal{A}}(x)(r) \geq \mathcal{P}_{\mathcal{B}}(x)(r) \qquad \forall x \in \mathcal{A}$$

Proof:

Assume the contrary. In this case the contrary inequality $\mathcal{P}_{\mathcal{A}}(x)(r) < \mathcal{P}_{\mathcal{B}}(x)(r)$ due to (1.77), means that there exists a "y" $\in d_{\mathcal{B}}(x)(r) - d_{\mathcal{A}}(x)(r)$, that illuminates the point "x" better than the ball $d_{\mathcal{A}}(x)(r)$: $\delta_y(x) > \mathcal{P}_{\mathcal{A}}(x)(r)$. But assuming in (1.86) that $\mathcal{X} = \mathcal{A}$ we obtain

$$\mathcal{P}_{\mathcal{A}}(x)(r) \geq \mathcal{P}_{\mathcal{A}}(x) \Rightarrow \delta_y(x) > \mathcal{P}_{\mathcal{A}}(x)$$

so $\mathcal{A}$ can't be a cluster. This contradiction proves the statement.
The hereditary principle gives (together with β-cluster) another generalization of a cluster.

Definition 10:

$\mathcal{A}$ is a local cluster in $\mathcal{X}$ if

$$\mathcal{P}_\mathcal{A}(x)(r) \geq \mathcal{P}_\mathcal{B}(x)(r) \quad \forall x \in \mathcal{A} \text{ and } \forall \mathcal{B} \colon \mathcal{A} \subset \mathcal{B}.$$

Note that we have not proved the cluster-character of $d_\mathcal{A}(x)(r)$ in $d_\mathcal{X}(x)(r)$. It remains open but, nevertheless, gives one more local generalization of a cluster.

Definition 11:

Let us call $\mathcal{A}$ strong, a local r-cluster in $\mathcal{X}$ if $d_\mathcal{A}(x)(r)$ is a cluster in $d_\mathcal{X}(x)(r)$ for any $x \in \mathcal{A}$.

Due to the given definitions we should change all the blocks of "Rodin" to corresponding local analogues. In this case we need one more parameter "r". The local"Rodin" has more flexibility (depending on "r") than the generic one.

For example, one may find a ring "$\mathcal{K}$", while the common "Rodin" gives its inner part as well. Turning back to our example $\mathcal{X}(z)$, we find $\mathcal{A}$ for $1/4 < r < z - 1$.

In conclusion, it may be said that common global clusters, built up on r-section φ_r:

$$\varphi_r(t) = \begin{cases} \varphi(t), & t \leq r \\ 0, & t > r \end{cases}$$

are closely connected with local (strong local) r-clusters that are built up on potential "φ". They differ in that: the computing global density denominator of $\mathcal{P}_\mathcal{A}(x)$ is constant and equal to $\mid \mathcal{A} \mid_\mathcal{X} = \mid \mathcal{A} \mid -1$, while the denominator of the local density (equal to $\mid d_\mathcal{A}(x)(r) \mid -1$) depends on "$x$". The strong local case is more rigid; the inequality $\mathcal{P}_d(r) \geq \alpha$ must satisfy any ball $d(r)$ of radius "r" in $\mathcal{A}$; but in φ_r-law only the integral middle estimate of densities $\mathcal{P}_{d(r)}(x)$ should be more than "α".

 Density : Besides the local method of density generalization, there is another method based on the initial construction $\mathcal{P}_\mathcal{A}(x)$.

Remember that there are two expressions for $\mathcal{P}_\mathcal{A}(x)$: the first one as a density measure :

$$\mathcal{P}_\mathcal{A}(x) = \frac{\mathcal{O}_\mathcal{A}(x)}{\mid \mathcal{A} \mid_\mathcal{X}}$$

Clustering Algorithms

and the second as the average of spot sources $\delta_y(x)$:

$$\mathcal{P}_\mathcal{A}(x) = \frac{1}{|\mathcal{A}|-1} \sum_{y \in \mathcal{A}_\mathcal{X}} \delta_y(x).$$

We can generalize either of them:

a) Let us set two measures on $\mathcal{X}$: the first, $\mu(x)$ is the mass of point "x", the second $\nu(x)$ is the charge force at point "x". Then for an arbitrary subset $\mathcal{A} \subset \mathcal{X}$, the μ-version $\mathcal{O}_\mathcal{A}^\mu(x)$ and the ν-version of illumination of point "x" and its mass, as a result $\mu - \nu$-version $\mathcal{P}_\mathcal{A}^{\mu,\nu}(x)$:

$$\left. \begin{array}{l} \mathcal{O}_\mathcal{A}^\mu(x) = \int_{\mathcal{A}_\mathcal{X}} \delta_y(x) d\mu(y) = \sum_{y \in \mathcal{A}_\mathcal{X}} \delta_y(x) \mu(y) \\ \nu(\mathcal{A}_\mathcal{X}) = \int_{\mathcal{A}_\mathcal{X}} d\nu(y) \quad \sum_{y \in \mathcal{A}_\mathcal{X}} \nu(y) \end{array} \right\} \Rightarrow \mathcal{P}_\mathcal{A}^{\mu,\nu}(x) = \frac{\mathcal{O}_\mathcal{A}^\mu(x)}{\nu(\mathcal{A}_\mathcal{X})}$$

b) We may obtain another generalization by replacing the averaging operator $1/|\mathcal{A}_\mathcal{X}| \sum_{\mathcal{A}_\mathcal{X}} (\delta_y(x)|y \in \mathcal{A}_\mathcal{X})$ with any generalized Kolmogorov average or by a fuzzy disjunction Ψ. The same procedure gives a Ψ-version of the spot density:

$$\mathcal{P}_\mathcal{A}^\Psi = \Psi(\delta_y(x)|y \in \mathcal{A}_\mathcal{X})$$

Example: $\Psi = \max$. $\mathcal{P}_\mathcal{A}^{\max}(x) = \max(\delta_y(x)|y \in \mathcal{A}_\mathcal{X})$ is the brightest light, i.e. the light from the nearest element in $\mathcal{A}$ to "x".

c) The previous ideology can be applied to generalize the density of the set $\mathcal{P}(\mathcal{A})$. Remember that the conjunction $\mathcal{P}(\mathcal{A}) = \min_{x \in \mathcal{A}} \mathcal{P}_\mathcal{A}(x)$ participated in its definition. By replacing min by an arbitrary fuzzy conjunction or generalized average j, we obtain a corresponding j-generalization of density $\mathcal{P}^j(\mathcal{A})$:

$$\mathcal{P}^j(\mathcal{A}) = j(\mathcal{P}_\mathcal{A}(x) : x \in \mathcal{A})$$

Example: If j is common average $\sum$, then $\mathcal{P}^\sum(\mathcal{A})$ is the average density in $\mathcal{A}$:

$$\mathcal{P}^\sum(\mathcal{A}) = \frac{1}{|\mathcal{A}|} \sum_{x \in \mathcal{A}} \mathcal{P}_\mathcal{A}(x)$$

By joining b) and c) we obtain the $j - \Psi$-generalization of density

$$\mathcal{P}^{j,\Psi}(\mathcal{A}) = j(\mathcal{P}_\mathcal{A}^\Psi(x)|x \in \mathcal{A}) = j(\Psi(\delta_y(x)|y \neq x|x \in \mathcal{A})$$

The goal: "Rodin" searches for clusters of a defined density level. Other limitations connected with different notions of quality may be imposed on the cluster; in this case a corresponding block should be added to the algorithm. We pass now to the connectivity. The connectivity approach is based on the local density: we know that $\mathcal{P}_A(x)(r)$ decreases as "r" increases. The nature of this decrease tells a good deal about "holes" (or "gaps") in $\mathcal{A}$ and depends on the distribution of points of space $\mathcal{A}$ with radius of concentric circles $\rho(x)(r) = \{y | d(x,y) = r\}$.

Let us fix a subset $\mathcal{A} \subset \mathcal{X}$ and a point "x" in $\mathcal{A}$. WE denote the numerical set of already ordered values $\{d(x,y) | y \in \mathcal{A}\}$ by $\Gamma(\mathcal{X}, \mathcal{A})$:

$$\Gamma(\mathcal{X}, \mathcal{A}) = \{r_1 < r_2 < \cdots < r_{m=m(x)}\}$$

In other words, $\Gamma(\mathcal{X}, \mathcal{A})$ are the radii of all the balls in $\mathcal{A}$ with centers in "x". The value of the function $\mathcal{P}_A(x)(r)$ is calculated as follows: if $n_k = |\mathcal{C}_A(x, r_k)|$, $k = 1, \cdots, m(x)$, then

$$\mathcal{P}_A(x)(r_1) = \frac{1}{1 + r_1}$$

$$\mathcal{P}_A(x)(r_2) = \mathcal{P}_{\mathcal{C}(x,r_1) + \mathcal{C}(x,r_2)}(x) = \frac{n_1}{n_1 + n_2} \mathcal{P}_{\mathcal{C}(x,r_1)}(x)$$

$$+ \frac{n_2}{n_1 + n_2} \mathcal{P}_{\mathcal{C}(x,r_2)}(x) = \frac{n_1}{n_1 + n_2} \frac{1}{1 + r_1} + \frac{n_2}{n_1 + n_2} \frac{1}{1 + r_2}$$

$$\mathcal{P}_A(x)(r_3) = \mathcal{P}_{d(x,r_1) + \mathcal{C}(x,r_3)}(x)$$

$$= \frac{n_1 + n_2}{n_1 + n_2 + n_3} \mathcal{P}_A(x)(r_2) + \frac{n_3}{n_1 + n_2 + n_3} \mathcal{P}_{\mathcal{C}(x,r_3)}(x)$$

$$= \frac{n_1}{n_1 + n_2 + n_3} \frac{1}{1 + r_1} + \frac{n_2}{n_1 + n_2 + n_3} \frac{1}{1 + r_2} + \frac{n_3}{n_1 + n_2 + n_3} \frac{1}{1 + r_3}$$

Finally: for $k = 1, \cdots, m(x)$

$$\mathcal{P}_A(x)(r_k) = \frac{1}{n_1 + \cdots + n_k}\left(n_1 \frac{1}{1 + r_1} + \cdots + n_k \frac{1}{1 + r_k}\right).$$

In particular

Clustering Algorithms

$$\mathcal{P}_\mathcal{A}(x) = \mathcal{P}_\mathcal{A}(x)(r_{m(x)}) = \frac{1}{|\mathcal{A}|-1}(\frac{n_1}{1+r_1} + \frac{n_2}{1+r_2} + \cdots + \frac{n_k}{1+r_k})$$

The formula for step is:

$$\mathcal{P}_\mathcal{A}(x)(r_{k+1}) - \mathcal{P}_\mathcal{A}(x)(r_k) = \frac{1}{n_1+\cdots+n_{k+1}}(\frac{n_1}{1+r_1} + \cdots + \frac{n_{k+1}}{1+r_{k+1}})$$

$$- \frac{1}{n_1+\cdots+n_k}(\frac{n_1}{1+r_1} + \cdots + \frac{n_k}{1+r_k})$$

$$= (\frac{1}{n_1+\cdots+n_{k+1}} - \frac{1}{n_1+\cdots+n_k})(\frac{n_1}{1+r_1} + \cdots + \frac{n_k}{1+r_k})$$

$$+ \frac{n_{k+1}}{n_1+\cdots+n_{k+1}}\frac{1}{1+r_{k+1}} = \frac{n_{k+1}}{n_1+\cdots+n_{k+1}}\frac{1}{1+r_{k+1}}$$

$$- \frac{n_{k+1}}{(n_1+\cdots+n_{k+1})(n_1+\cdots+n_k)}(\frac{n_1}{1+r_1} + \cdots + \frac{n_k}{1+r_k})$$

$$= \frac{nk+1}{n_1+\cdots+n_{k+1}}\left(\frac{1}{1+r_{k+1}} - \frac{(\frac{n_1}{1+r_1}+\cdots+\frac{n_k}{1+r_k})}{n_1+\cdots+n_k}\right)$$

Conclusion:

The local density $\mathcal{P}_\mathcal{A}(x)(r)$ is locally a constant function. It is dereasing on $[r_1, r_m(x)]$ and it is continuous at right. Its steps take place in the points $\Gamma(x,\mathcal{A}) - r_1$ and depend on the differences $r_{k+1} - r_k$ and n_{k+1}; the greater are these parameters, the greater is the step down.

Choosing the threshold $\epsilon \in (0,1)$, we call $\mathcal{A}$ ϵ-connected, if steps of local density $\mathcal{P}_\mathcal{A}(x)(r)$ donot exceed ϵ for all $x \in \mathcal{A}$.

The formula for step presented above may be the basis for the block $\boxed{\text{Connectivity}}$ in corresponding modifications of the "Rodin" algorithm.

1.3.3 Fuzzy clustering

Introduction

Before analysing the transformation "fuzzy to non-fuzzy" in the "Rodin" algorithm, we discuss the reasons for such a transformation or the character of the apparent "fuzziness". If the space $\mathcal{X}$ is a set of alternatives to be chosen or a set of possible results of an experiment, any point $x \in \mathcal{X}$ has its particular importance (weight) $\psi(x)$ for the person taking the decision (PTD). In the first case the PTD estimates the possibility of the alternative "x" exceeding $\psi(x)$ quantitatively ; In the second case the PTD estimates the possibility that "x" is the result of a given experiment.

The above assumes that PTD can modify $\varphi(x)$, the starting estimate of $\psi(x)$ and decrease it so that: $\varphi(x) \leq \psi(x)$. This allows us to trade off between the choice between "x" and its weight. Such a choice must be a fuzzy cluster that is compact modification, compatible with the starting PTD estimate. Fuzziness may also appear in the second starting component-metrics "d" on the space $\mathcal{X}$ in the following cases :

1. If the starting fuzziness of "d" is large. For example, in the case of Levenstein-distance on seismograms or if the dependance is stochastic, "the experimental result $\rightarrow$ is possible exit $\in \mathcal{X}$" : any result (test) corresponds here to a fuzzy structure on $\mathcal{X}$, that is why the distance between two results is necessarily a fuzzy structure on a numerical set $\Gamma_{\mathcal{X}} = \{d(x,y) \mid x, y \in \mathcal{X}\}$;

2. If it means the ignorance of PTD concerning starting non-fuzzy distance $d(x^*, y^*)$ between fixed $x^*, y^* \in \mathcal{X}$. In this case (the same above) instead of $d(x^*, y^*)$ PTD introduces a fuzzy structure that describes his estimate of $d(x^*, y^*)$ on $\Gamma_{\mathcal{X}}$.

All above makes possible to consider that fuzziness in our context may be, at least, of 3 types :

- We are not sure about the points of the space $\mathcal{X}$ (classic metrics, fuzzy space);

- We are not sure about the distance $d(x,y)$ between the points (fuzzy metrics, classic space);

- We are not sure about both of them (fuzzy metrics, fuzzy space).

Clustering Algorithms

Let us consider the first case: the model of light difusion $\delta_y(x)$, connected with metrics $d(x,y)$ fixed in space $\mathcal{X}$: we choose a fuzzy structure $\psi(x)$ describing the importance (weight) of the elements of the starting space. The fuzziness "ψ" must be introduced into all notions and constructions necessary to define any common cluster. We shall do it using the S-principle (seriousness):

S-principle: We follow a S-principle if, while defining a notion or building a construction in a point "x", we consider the weight $\psi(x)$.

For the fuzzy cluster version considered, we will follow the S-principle and show its expression at every stage. Using the S-principle we expose fuzzy variants of all the stages of cluster-definition. As a result we have a version of a fuzzy cluster and a fuzzy variant of the "Rodin" algorithm that is connected with it. We represent it graphically, where $\xrightarrow{s}$ means a transformation according to the S-principle:

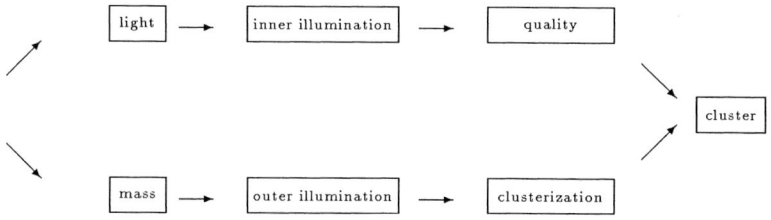

Light. Following the S-principle, we consider the unit charge at the point "x" equal to $\psi(x)$. In other words, the light force sent from "y" to "x" in the set $(\mathcal{X}, \psi)$ is equal to

$$\delta_y^\psi = \delta_y(x)\psi(y) \qquad (1.87)$$

Illumination: $\mathcal{O}_\psi(x)$ of the point "x" by the fuzzy set ψ is a sum:

$$\mathcal{O}_\psi(x) = \sum_{y \in \mathcal{X}} \delta_y^\psi(x) = \sum_{y \in \mathcal{X}} \delta_y(x)\psi(y) \qquad (1.88)$$

Mass : the mass is naturally the overall power of charges $\mu(\psi)$:

$$\mu(\psi) = \sum_{y \in \mathcal{X}} \psi(y) \qquad (1.89)$$

for the simplest expression of the S-principle the charge at the point "x" coincides with its mass and is equal to $\psi(x)$ (see note 3 at the end).

Inner illumination: we are interested in the density:

$$\mathcal{P}_\psi(x) = \frac{\mathcal{O}_\psi(x)}{\mu(\psi)} = \frac{\sum_y \delta_y(x)\psi(y)}{\sum_y \psi(y)} \qquad (1.90)$$

Remember that a subset is generally considered a cluster if the density of inner illumination at its every point within the subset is higher than the outer illumination, coming from any point outside the subset. To complete the definition of a fuzzy cluster we should describe outer fuzzy illumination.
<u>Outer illumination</u>: the subset $\mathcal{B}$ in $\mathcal{X}$ is generally considered outer to the subset $\mathcal{A}$ in $\mathcal{X}$ if $\mathcal{A} \cap \mathcal{B} = \emptyset$ and $\mathcal{A} \cup \mathcal{B} \subset \mathcal{X}$. Speaking in the language of functions this conjunction is equivalent to the inequality $\mathcal{X}_A + \mathcal{X}_B \leq \mathcal{X}_\mathcal{X} = l_\mathcal{X}$ or $\mathcal{X}_B \leq l_\mathcal{X} - \mathcal{X}_A$. The latter formula allows us to define additional fuzzy structures.

Definition 1. The fuzzy structure μ on $\mathcal{X}$ is outer to the fuzzy structure φ on $\mathcal{X}$ in the fuzzy set $(\mathcal{X}, \psi)$ if $\varphi + \mu \leq \psi$. The difference $\psi - \varphi$ is called the "addition to φ in ψ".
If we fix a point "x" in $\mathcal{X}$, $\varphi(x)$ is the absolute "inner" measure of belonging of "x" to the subset $(\mathcal{X}, \varphi)$, and $\psi(x) - \varphi(x)$ is the absolute "outer" measure of non-belonging of "x" to the subset $(\mathcal{X}, \varphi)$ in $(\mathcal{X}, \psi)$.

Example 2. If $\mathcal{A} \subset \mathcal{X}$ is the usual situation, the outer structure ψ coincides with $l_\mathcal{X}$ and the inner φ with $\mathcal{X}_A$. Then the measure of belonging of "x" to the structure
$$\mathcal{X}_A = l \Leftrightarrow \mathcal{X}_A(x) = l \Leftrightarrow x \in \mathcal{A}$$
and the measure of non-belonging of "x" to the structure $\mathcal{X}_A$ is
$$l_\mathcal{X} = l \Leftrightarrow \mathcal{X}_A(x) = 0 \Leftrightarrow x \in \mathcal{A}$$
The density of outer illumination (that in the one-point case coincides with the outer illumination $\delta_y(x)$ due to the equality between the unit of the charge and the mass in every point of the starting space) participates in the usual outer criteria. It follows that in the fuzzy case we should also compare the inner density $\mathcal{P}_\varphi(x)$ with the density of the singular outer illumination. The latter construction should

1. depend on the pair of points "x" and "y" so as on the fuzzy structures "φ" and "ψ" ; and

2. be equal to zero for $x \in \mathcal{A}$ (outer illumination outer point = 0) and for $y \in \mathcal{A}$ (outer light from inner point = 0) projected on the usual case (speaking the language of functions for $\mathcal{X}_A(x) = 0$ and $\mathcal{X}_A(y) = 0$).

Thus if we denote it by $\mathcal{P}_y^{\varphi,\psi}(x)$, then

$$\mathcal{P}_y^{\varphi,\psi}(x) = 0, \quad \begin{array}{l} \text{if } \varphi(x) = 0 \\ \text{if } \varphi(y) = \psi(y) \end{array}$$

Non formally, $\varphi(x) = 0$ means that the point supposed as inner becomes outer ; and $\varphi(y) = \psi(y)$ means that point supposed as outer becomes inner. We can construct $\mathcal{P}_y^{\varphi,\psi}(x)$ on the S-principle as follows :

a

the point "y" is considered as outer $\implies$ its absolute charge is equal to $\psi(y)$.

b

the charge $\psi(y)$ should be considered according the measure of non-belonging "y" to φ $\implies$ the charge (light) that really comes from "y" to "x" is equal to $\psi(y)(\psi(y) - \varphi(y))$.

c

the point "x" is considered as inner $\implies$ its measure of perception is is equal to $\varphi(x)$

d

the light is transformed according to the law $\delta_y(x)$ $\implies$ the quantity of light (illumination) perceiving in the point "x" is equal to $\mathcal{O}_y^{\varphi,\psi}(x) = \varphi(x)\delta_y(x)(\psi(y) - \varphi(y))$

e

the charge mass in the point "y" is equal to $\psi(y)$ $\implies$ the illumination density required in the point "x" is
$\mathcal{P}_y^{\varphi,\psi}(x) = \varphi(x)\delta_y(x)(\psi(y) - \varphi(y))$

Example 3. Let $\varphi = l_\mathcal{A}$, $\psi = l_\mathcal{X}$ (with the same conditions as example 2). Then

$$\mathcal{P}_y^{\varphi,\psi}(x) = \mathcal{X}_\mathcal{A}(x)\delta_y(x)(l_\mathcal{X}(y) - l_\mathcal{A}(y)) = \begin{cases} \delta_y(x), & \text{if } x \in \mathcal{A} \wedge y \notin \mathcal{A} \\ 0, & \text{if } x \notin \mathcal{A} \vee y \in \mathcal{A} \end{cases}$$

Thus the density of the outer fuzzy illumination is defined and everything is ready to definite the fuzzy cluster

Definition 4. A fuzzy subset $(\mathcal{X}, \varphi)$ is a cluster in $(\mathcal{X}, \psi)$, if

$$\mathcal{P}_\varphi(x) \geq \mathcal{P}_y^{\varphi,\psi}(x) \qquad \forall\ x,\ y\ \in \mathcal{X}. \tag{1.91}$$

Let us denote by $\mathcal{P}_y^{\varphi,\psi}(x)$ the maximum of the outer fuzzy illumination of the point "x":

$$\mathcal{P}_y^{\varphi,\psi}(x) = \max_{y\in\mathcal{X}} \delta_y(x)(\psi(y)-\varphi(y)) \qquad (1.92)$$

This gives the inequality (1.91) the following form of (1.93), in which the S-principle is clearly visible:

$$\mathcal{P}_\varphi(x) \geq \varphi(x)\mathcal{P}^{\varphi\psi}(x) \qquad \forall\, x \in \mathcal{X} \qquad (1.93)$$

We consider the point "x" in the frame of the fuzzy structure "φ".

Fuzzy Rodin: The condition (1.91) is constructive, so it can serve as a basis for the block "Cluster" in the fuzzy version of the "Rodin" algorithm. We usually eliminate a cluster from the given set by omitting its weak points. In function language, we trivialize the corresponding characteristic function at these points. In the fuzzy version a cluster can be eliminated from the fuzzy structure ψ given on $\mathcal{X}$ by trivializing the structure at the weak points. This can lead to more delicate clusters that cannot be found by such a strategy.

Conclusion: it is desirable to make the strategy as smooth as possible. We can find such a strategy, but we must start with the second component of the algorithm - the quality of a fuzzy set.

Quality. It is easier to start with the fuzzy analogous $\mathcal{P}(\mathcal{A})$. In the usual case we have a chain of equivalences:

$$\mathcal{P}(\mathcal{A}) \geq \alpha \iff \mathcal{P}_\mathcal{A}(x) \geq \alpha \quad \forall\, x \in \mathcal{A} \iff \mathcal{P}_\mathcal{A}(x) \geq \mathcal{X}_\mathcal{A}(x)\alpha \quad \forall x \in \mathcal{X}.$$

The latter shows how the S-principle looks and suggests how it should be used in the general case.

Definition 5 We denote the relation $\mathcal{P}_\varphi(x)/\varphi(x)$ as a local quality of fuzzy structure "φ" in the point "x" and the global quality $\mathcal{P}_\varphi$ as their minimum

$$\mathcal{P}_\varphi \stackrel{det}{=} \min_{x\in\mathcal{X}} \frac{\mathcal{P}_\varphi(x)}{\varphi(x)} \qquad (1.94)$$

Consequently, in the fuzzy case the block "Quality" of the "Rodin" algorithm for the given "α" should examine the inequality

$$\mathcal{P}_\varphi(x) \geq \alpha\varphi(x) \quad \forall\, x \in \mathcal{X}. \qquad (1.95)$$

We will start with quality-based elimination.

Quality-based elimination. Let us denote by $i_x(y)$ a fuzzy structure on $\mathcal{X}$, bound with "x":

Clustering Algorithms

$$i_x(y) = \begin{cases} 1, & \text{if } y = x \\ 0, & \text{if } y \neq \end{cases}$$

then $(\varphi - \lambda i_x)(y)$ is a result of eliminating from "φ" in "x" with the step "λ".

By assigning some desired quality level "α" we can correct the weak points by elimination. Our task is to find in the structure "φ" (i.e. with the inequality $\mathcal{P}_\varphi(x) < \alpha\varphi(x)$) the "$\lambda$"s that make α-weak point "x" α-normal in a qualitative sense (i.e. with the inequality $\mathcal{P}_{\varphi-\lambda i_x}(x) \geq \alpha(\varphi - \lambda i_x)(x)$) in the structure $\varphi - \lambda i_x$ after elimination $\varphi \to \varphi - \lambda i_x$. One way to do this, considering the equality $(\varphi - \lambda i_x)(x) = \varphi(x) - \lambda$, is to put $\lambda = \varphi(x)$ i.e trivialization of the structure "φ" in the point "x". This is done by the usual "Rodin". But there is another more adaptable way that takes advantage of fuzziness. We consider it here.

We need evident expression of the quality of point "x" in the structure $\varphi - \lambda i_x$ through the starting structure "φ", that is done by using the following relations:

$$\begin{aligned}
(\varphi - \lambda i_x)(x) &= \varphi(x) - \lambda \\
\mathcal{O}_{(\varphi-\lambda i_x)}(x) &= \sum_{y \in \mathcal{X}} (\varphi - \lambda i_x)(y)\delta_y(x) = \mathcal{O}_\varphi(x) - \lambda \sum_{y \in \mathcal{X}} i_x(y)\delta_y(x) \\
&\qquad\qquad\qquad\qquad\qquad\qquad\quad = \mathcal{O}_\varphi(x) - \lambda \\
\mu(\varphi - \lambda i_x) &= \sum_{y \in \mathcal{X}} (\varphi - \lambda i_x)(y) = \mu(\varphi) - \lambda \\
\frac{\mathcal{P}_{(\varphi-\lambda i_x)}(x)}{(\varphi - \lambda i_x)(x)} &= \frac{\mathcal{O}_{(\varphi-\lambda i_x)}(x)}{(\varphi - \lambda i_x)(x)} = \frac{\mathcal{O}_\varphi(x) - \lambda}{\mu(\varphi - \lambda)(\varphi(x) - \lambda)}
\end{aligned}$$
(1.96)

Let $\mathcal{A} = \varphi(x)$, $\mathcal{B} = \mathcal{O}_\varphi(x)$, $\mathcal{C} = \mu(\varphi)$ and note that $0 \leq \mathcal{A} < \mathcal{B} < \mathcal{C}$. The dependance (1.96) of quality "x" on elimination "λ" from "φ" is a function $(\mathcal{B} - \lambda)/(\mathcal{A} - \lambda)(\mathcal{C} - \lambda)$ on the segment $[0, \mathcal{A}]$. Let us find its derivative and show that under our condition on $\mathcal{A}, \mathcal{B}, \mathcal{C}$ it increases on the segment $[0, \mathcal{A}]$ from $\mathcal{B}/\mathcal{BC}$ to ∞:

$$\left(\frac{\mathcal{B} - \lambda}{(\mathcal{A} - \lambda)(\mathcal{C} - \lambda)}\right)' = \frac{\lambda^2 + 2\lambda(\mathcal{A} + \mathcal{C} - \mathcal{B}) + \mathcal{AB}}{(\mathcal{A} - \lambda)^2(\mathcal{C} - \lambda)^2} \geq 0, \text{ because } \mathcal{C} - \mathcal{B} + \mathcal{A} \geq 0$$

Proposition 6: If "α"-is the necessary quality level in the algorithm, one can always find $\lambda(\alpha) \in [0, \varphi(x)]$ so that

$$\frac{\mathcal{P}_{\varphi-\lambda(\alpha)i_x}(x)}{(\varphi-\lambda(\alpha)i_x)(x)} = \alpha$$

This is the minimum (the most smooth) elimination "φ" with the necessary quality level in "x":

$$\lambda(\alpha) = \lambda(\alpha,\ x,\ \varphi) \qquad (1.97)$$

<u>Elimination in cluster.</u> It remains to theoretically justify the last block of fuzzy "Rodin", *i.e.* to find the elimination strategy that eliminates cluster-condition (1.93) in the point "x": $\mathcal{P}_\varphi(x) \geq \varphi(x)\mathcal{P}^{\varphi,\psi}(x)$.

Here everything is analogous to the previous case except that, instead of "α" we take the maximum outer light $\mathcal{P}^{\varphi,\psi}(x)$ and set

$$\lambda = \lambda(\mathcal{P}^{\varphi,\psi}(x)) \qquad (1.98)$$

The block-scheme of fuzzy "Rodin" is analogous to the usual one, which is why we only consider the following blocks.

1. $\boxed{\text{The beginning}}$: set up the starting data, such as: the set $\mathcal{X}$, the light illumination law $\delta_y(x)$, the fuzzy structure ψ, and the quality level α. Denoting by ψ_n the current fuzzy version of the cluster, we set $\psi_0 = \varphi$ and pass to 2

2. $\boxed{\text{Quality}}$: here we calculate the quality ψ_n:

    if $\mathcal{P}_{\psi_n}(x) \geq \alpha\psi_n(x) \quad \forall x \in \mathcal{X}$, then we pass to 4

    Logic of the block:

    if $\mathcal{P}_{\psi_n}(x^*) < \alpha\psi_n(x^*)$ for some $x^* \in \mathcal{X}$, then we pass to 3

3. $\boxed{\mathcal{P}\text{-generation}}$: choose, in a quality sense, the weakest point

    $x_{n+1} \in \mathcal{X}$ about the structure ψ_n:

    $$x_{n+1} = \arg\min_\mathcal{X} \frac{\mathcal{P}_{\psi_n}(x)}{\psi_n(x)}$$

    and the elimination level λ_{n-1}:

    $$\lambda_{n+1} = \lambda(\alpha,\ x_{n+1},\ \psi_n)$$

    Generate ψ_{n+1}

    $$\psi_{n+1} = \psi_n - \lambda_{n+1}i_{x_{n+1}}$$

    and pass to 2

Clustering Algorithms

4. Cluster : implement the outer criterium test (1.91) : let "x" be a variable point in $\mathcal{X}$, where x is considered to be ordered

   if $\mathcal{P}_{\psi_n}(x) \geq \mathcal{P}_y^{\psi_n,\psi}(x)$ $\forall y \in \mathcal{X}$, then pass to the next point

   Logic of the block: in $\mathcal{X}$, if "x"-is the last point, then ψ_n-is the desired cluster

$$\text{if } \mathcal{P}_{\psi_n}(x) < \sup_y \mathcal{P}_y^{\psi_n,\psi}(x) = \varphi(x)\mathcal{P}^{\psi_n,\psi}(x), \text{ then pass to 5}$$

5. Cl-generation : choose $x_{n+1} \in \mathcal{X}$ the weakest point in a cluster sense regarding the structure ψ_n:

$$x_{n+1} = \arg\min_{\mathcal{X}} \frac{\mathcal{P}_{\psi_n}(x)}{\psi_n(x)\mathcal{P}^{\psi_n,\psi}(x)}$$

and the elimination level λ_{n+1} :

$$\lambda_{n+1} = \lambda(\mathcal{P}^{\psi_n,\psi}(x_{n+1}))$$

Generate ψ_{n+1} :

$$\psi_{n+1} = \psi_n - \lambda_{n+1} i_{x_{n+1}}$$

and pass to 2

Notes

1. The cluster obtained in step 5 is a fuzzy analog of the usual global 1-cluster. Other analogs include the β-cluster, the local cluster. Note that β is naturally considered as a variable that depends on "x" : $\beta = \beta(x)$. Thus fuzziness appears when the usual set $\mathcal{X}$ is clustered with the variable "β".

2. Using the S-principle in one of the stages (light, mass, inner illumination, outer illumination, quality) is of fundamental importance, because only the S-principle gives the necessary monotony to proposition 6; if it is abandoned elimination may be done only by trivialisation.

3. The dependance between mass and charge in the point may be more complicated than that given. In particular, mass and charge may be independent of one another. The charge was found above in the construction $\mathcal{P}_{\mathcal{A}}^{\nu,\mu}(x)$ through generalization of the usual "Rodin" algorithm.

1.4 Linear and Circle Structures Recognition Algorithms

1.4.1 Differential operations in GIS environment

Herein, the Geographical Information System (GIS) [57, 58] is used to develop an automated procedure for circular and linear feature recognition in different types of geophysical data sets. The procedure is based on the differential operation approach (derivating) in the GIS environment, and consists of two steps: pre-processing of data sets to produce images suitable for further analysis, and processing the images to extract linear and circular features. In chapter 2, we will apply the technique developed here to bathymetry data collected during a multipurpose geophysical cruise in the Wharton Basin of the equatorial Indian Ocean.

Analyse of dynamical Earth processes are constrained by statistical parameters and distribution functions determined from geophysical data sets. Features of interest are often be observed in various forms of imagery, so automatic extraction of these features would considerably increase the volume of data processed and the reliability of these parameters. Here we focus on features such as fractures and faults, most of which are approximately linear. Another possible focus on circular structures, which occur on various scales within images, but which can be masked by a large amount of noise from other natural features as well as from sampling artifacts.
The usual procedure for recognizing simple geometric features in an image is to perform edge detection (gradient, Laplacian, Canny, etc) followed by thresholding and then linking edge pixels of the processed image. However,

images of geophysical data sets essentially contain features on all scales, so achieving a structured edge description map is highly unlikely.

The initial objects are defined in the GIS environment ATTILA [57, 58], and additional code covers both pre-processing (acquisition of data sets as images) and processing (recognition of linear features in images). Herein we describe methods and algorithms for recognition of linear and circular features in images of topographic altimetry as well as other types of geophysical and geological data.

Linear and circular features in an image can be recognized as **edges**, *i.e.* discontinuities, or boundaries between regions with relatively distinct values. This idea follows [174, 205], where I. Gelfand, A. Gvishiani *et al.* proposed to establish lineaments in the zones of clear distinction of geomorphological parameters for morphostructural zoning. Considering such image as a set of spatially organized pixels, edges can be defined as strongly contrasting pixels in an image.

Three types of discontinuities occur in an image: gradient, step and bar. A gradient discontinuity occurs where the gradient of the pixel values changes across some line. This type of discontinuity can be classed as ramp or roof, convex or concave edges, by noting the sign of the gradient component perpendicular to the edge on either side of the edge. Ramp edges have the same signs in the gradient components on either side of the edge, while roof edges have opposite signs.

A step discontinuity occurs where the pixel values themselves change rapidly across some line.

A bar discontinuity occurs where the pixel values increase rapidly then decrease again (or *vice versa*) across some line.

Many edge detection techniques can be broken into two distinct phases:

- Finding pixels in the image where edges are likely to occur by looking for discontinuities. Candidate points for edges in the image are usually referred to as *edge points*, or *edge pixels*.

- Linking these edge pixels in some way to produce descriptions of edges as lines, curves,*etc.*

Since edges correspond to strong gradient values, they can be highlighted by calculating the first and second derivatives of the image.

The usual approach is to simply define edges as step discontinuities in the image function.

The position of the edge can then be estimated using the local maximum of the first derivative or with the zero crossing of the second derivative. This idea was first suggested by Marr (1982) [338] and later developed by Marr and Hildreth (1980), Canny (1983, 1986) [339, 86, 87] and others (Deriche, 1987, Deriche and Faugeras, 1990 [120, 121]).

In image processing, edge detection is usually implemented by *convolving* the image function with some form of linear filter that approximates the first or the second derivative operator. An odd symmetric filter approximates the first derivative operator, and peaks in the convolution output will correspond to step edges in the image. An even symmetric filter approximates the second derivative operator. Zero crossings in the convolution output correspond to step edges, and maximums will correspond to bar edges.

In two-dimensional image processing terms, representing the convolution of two functions $f(x,y)$ and $h(x,y)$ by the symbol $\otimes$, the continuous convolution integral $g(x,y)$ may be expressed as

$$g(x,y) = f(x,y) \otimes h(x,y) = \int_{-\infty}^{+\infty} \int_{-\infty}^{+\infty} f(\tau_u, \tau_v) h(x - \tau_u, y - \tau_v) d\tau_u d\tau_v, \tag{1.99}$$

where $f(x,y)$ is the image function and $h(x,y)$ is the response of a particular filter. The function $h(0 - \tau_u, 0 - \tau_v)$ is simply the function $f(\tau_u, \tau_v)$ rotated by 180 degrees about the origin. The function $h(x - \tau_u, y - \tau_v)$ is the function translated to move the origin of the function g to the point (x,y) in (M, N) plane. The functions are then pointwise multiplied and the product function is integrated over two dimensions. Convolution for discrete images is analogous to that for continuous images.

$$g(x,y) = f(x,y) \otimes h(x,y) = \sum_{\tau_u} \sum_{\tau_v} f(\tau_u, \tau_v) h(x - \tau_u, y - \tau_v), \tag{1.100}$$

In the image processing context the kernel h is a matrix of pixels much smaller than the image. The coordinates x and y are substituted with the numbers i and j of the corresponding pixel row and column. The convolution is performed by sliding the kernel over the image, generally starting at the top left corner, and moving it through all the positions where the kernel fits entirely within the boundaries of the image. Each kernel position corresponds to a single output pixel, the value of which is calculated by multiplying the image pixel value by the kernel pixel value for each of the pixels in the kernel, and then summing up all the resulting values. Thus, if the input image $I(i,j)$ has M rows and columns, $m \ll M, n \gg N$, then the output image $O(i,j)$ will have $M - m + 1$ rows and $N - n + 1$ columns

$$O(i,j) = I(i,j) \otimes K(m,n) = \sum_{k=1}^{m}\sum_{l=1}^{n} I(i+k-1,\ j+l-1)K(k,l), \quad (1.101)$$

where $i = 1, M - m + 1$ and $j = 1, N - n + 1$. Many convolution implementations produce an output image the same size as the input image, by relaxing the constraint that the kernel can only be moved to positions where it fits entirely within the input image.

Various discrete difference convolution kernels are used for edge detection, such as Prewitt (1970) [420], Robert and Sobel (1990) [475], operators.

Edges are often defined to be pixels in an image where values vary rapidly. To estimate value change and characterize the nature of the change, it is often useful to calculate the partial derivatives $g_x = \dfrac{\partial f(x,y)}{\partial x}$ and $g_y = \dfrac{\partial f(x,y)}{\partial y}$ of the image function $f(x,y)$.

Zero crossing detector. Several edge detection algorithms use to locate zero crossings of second derivative of an image to find the desired edge pixels (Marr, 1982 [338]). Such points often occur at edges and in other places. A zero crossing detector is more a feature detector than a specific edge detector. This approach is often used for image post-processing.

Gradient edge detector. Edges are often defined to be pixels in an image with locally maximal gradient magnitude. The gradient of the image function $f(x,y)$, denoted by $\nabla f(x,y)$, is given by the vector

$$\nabla f(x,y) = [g_x,\ g_y] = \left[\frac{\partial f(x,y)}{\partial x},\ \frac{\partial f(x,y)}{\partial y}\right]. \quad (1.102)$$

The absolute magnitude $g(x,y)$ of this gradient is given by

$$|g(x,y)| = \sqrt{g_x^2 + g_y^2} = \sqrt{\left(\frac{\partial f(x,y)}{\partial x}\right)^2 + \left(\frac{\partial f(x,y)}{\partial y}\right)^2}, \quad (1.103)$$

and its direction $\theta(x,y)$ is estimated by

$$\theta(x,y) = \tan^{-1}(g_y/g_x) = \tan^{-1}\left(\frac{\partial f(x,y)}{\partial y} \Big/ \frac{\partial f(x,y)}{\partial x}\right). \quad (1.104)$$

The gradient edge detector involves convolving the image with two discrete difference kernels: in the x direction, g_x, and in the y direction g_y.

To identify edge pixels, the gradient magnitude image must then be thresholded, assuming that all pixels having a local gradient magnitude above the threshold represent an edge.

Laplacian edge detector. The Laplacian is a two-dimensional isotropic measure of the second derivative of an image, *i.e.* it is linear and rotationally symmetric. The Laplacian $\nabla^2 f(x,y)$ of the image function $f(x,y)$ is defined as:

$$\nabla^2 f(x,y) = \frac{\partial^2 f(x,y)}{\partial x^2} + \frac{\partial^2 f(x,y)}{\partial y^2}, \qquad (1.105)$$

The Laplacian can be approximated using a single convolution with an appropriate discrete difference kernel. Edge pixels can therefore be defined by the zero crossings of the Laplacian. To reduce high frequency noise in the second derivative of an image, the image is often Gaussian smoothing before applying the Laplacian. The two-dimensional Gaussian $G(x,y)$ with the standard deviation σ is defined by

$$G(x,y) = \frac{1}{2\pi\sigma^2} e^{-\frac{x^2+y^2}{2\sigma^2}}. \qquad (1.106)$$

The operation of edge detection after smoothing of the original image is given by the following formula $\nabla^2[G(x,y) \otimes f(x,y)]$. In other words the edge pixels can be detected by zero crossings of the image Laplacian smoothed by the Gaussian. This operation can be reduced to convolving the original image function $f(x,y)$ with the laplacian of the Gaussian $(L \otimes G)$ operator $\nabla^2[G(x,y)]$ (Marr and Hildreth, 1980 [339])

$$\nabla^2\left[G(x,y) \otimes f(x,y)\right] = \nabla^2\left[G(x,y)\right] \otimes f(x,y). \qquad (1.107)$$

Canny edge detector. This procedure targets the maximum partial derivative of the image function $f(x,y)$ in the direction orthogonal to the edge direction, and smooths the signal along the edge direction (Canny, 1983, 1986, [86, 87]). Thus, the Canny edge detector looks for the maximum of the function

$$\frac{\partial^2}{\partial n^2}\left[G(x,y) \otimes f(x,y)\right], \qquad (1.108)$$

where $n = \frac{\nabla[G(x,y) \otimes f(x,y)]}{|\nabla[G(x,y) \otimes f(x,y)]|}$. Usually, algorithmic implementations of the Canny edge detector approximate this procedure in multiple stages

Fuzzy Sets Classification Algorithms

(Canny, 1986, Deriche, 1987 [87, 120]). At first, a two-dimensional Gaussian *smoothing* is applied to the image. Then, a two-dimensional *differentiation* is performed on the smoothed image to produce the gradient magnitude image. In fact, the first derivative $g(x,y)$ of the original image function $f(x,y)$ smoothed with the Gaussian $G(x,y)$

$$g(x,y) = \bigtriangledown [G(x,y) \otimes f(x,y)] \qquad (1.109)$$

is equivalent to the original image function convolved with the first derivative of the Gaussian $\bigtriangledown [G(x,y)]$

$$g(x,y) = \bigtriangledown [G(x,y)] \otimes f(x,y) . \qquad (1.110)$$

Therefore, it is possible to combine smoothing and differentiation stages. A local maximum occurs at peaks in the gradient magnitude function. *Non-maximum suppression* is then performed by tracking along the edges and setting to zero all non-maximum pixels perpendicular to the edge direction. The tracking involves *thresholding*, which uses a method called "hysteresis controlled by two gradient magnitude thresholds : $T1 > T2$. For a line segment, pixels that have values greater than the upper threshold, $T1$ are immediately accepted as valid edge points, and pixels that have values between the two thresholds are accepted if and only if they are connected to valid edge points. Hysteresis ensures that noisy edges are not split into multiple edge fragments. The recommended ratio of upper to lower threshold is usually in the range of two or three to one, based on the predicted signal-to-noise ratio.

Linking edge pixels. Edge detectors yield edge pixels in an image. the next step is to connect the edge pixels together, replacing many points on edges with a few edges themselves. In pratice this problem is difficult, because small edge segments may be missing, small edge segments may appear to be present due to noise where there is no real edge, *etc*.

Most edge detectors yield information about local gradient magnitude $g(x,y)$ at an edge pixel and, more importantly, the edge direction $\theta(x,y)$ in the pixel locality. Edge linking methods usually start at some arbitrary edge pixel and consider pixels in a local neighbourhood for similarty of edge directions. Edge pixels in a neighbourhood having similar gradient directions are likely to lay on the same edge. If the pixels satisfy the direction similarity constraints then they are added to the current edge set. A gradient magnitude threshold must be set to identify which edges are strong enough to track.

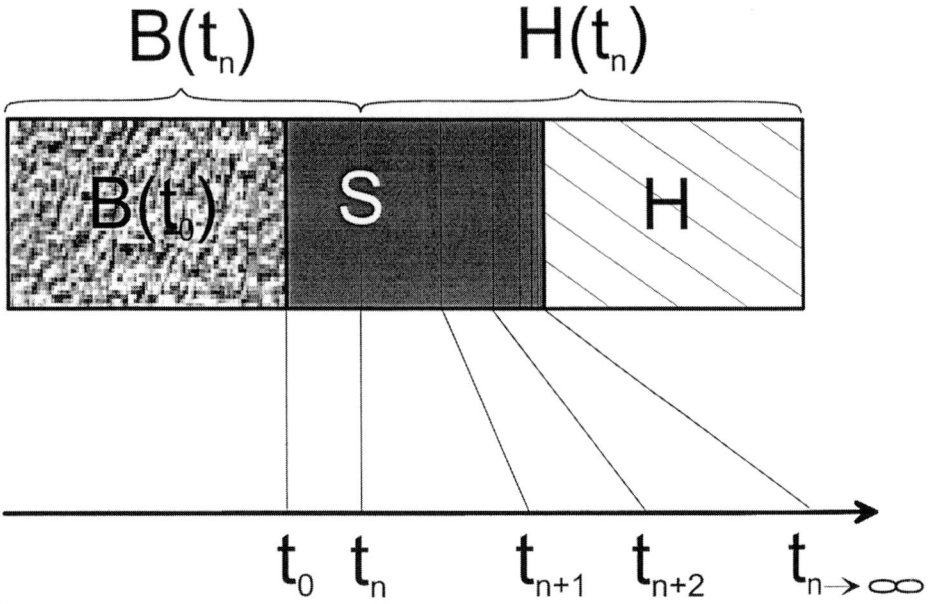

Dynamic pattern recognition problem - t—time scale, $\mathcal{W} = \mathcal{W}_0 = \mathcal{B}(t_0) \bigsqcup [\mathcal{S} \cup \mathcal{H}]$ - initial decomposition (learning material), $\mathcal{W} = \mathcal{B}(t_n) \bigsqcup \mathcal{H}(t_n)$ - current decomposition, $\mathcal{W} = \lim_{t_n \to \infty}[\mathcal{B}(t_n) \bigsqcup \mathcal{H}(t_n)] = [\mathcal{B}(t_0) \cup \mathcal{S}] \bigsqcup \mathcal{H} = \mathcal{B} \bigsqcup \mathcal{H}$ - final classification (prediction) (after Dubois and Gvishiani, 1998).

Chapter 2

Applications to Physics of the Earth, Seismology and Engineering Seismology

2.1 Syntactic Classification of Engineering Seismology Data

2.1.1 Strong ground motion earthquake data base

One of the principal tasks of Russian Academy of Sciences (RAS) Center of Geophysical Data Studies and Telematics Applications (CGDS) of the Institute of Physics of the Earth is the development of a world-wide Strong Ground Motion Database (SMDB[1]). As is usual in seismology, by strong ground motions we mean accelerations in the near field of strong earthquakes. At the 1994 General Assembly of European-Mediterrenean Seismo-

---

[1]This database has been launched in 1991 in the framework of CEA-EMSC-CGDS project devoted to artificial intelligence applications to strong motion data analysis. J. Bonnin (CIPGS, France), B. Mohammadioun (CEA, France), A. Gvishiani and M. Zhizhin (CGDS, Russia) started the project together working in Strasbourg. Since 1995, an important contribution to the project has been provided by A. Mikoyan and A. Burtsev (CGDS).

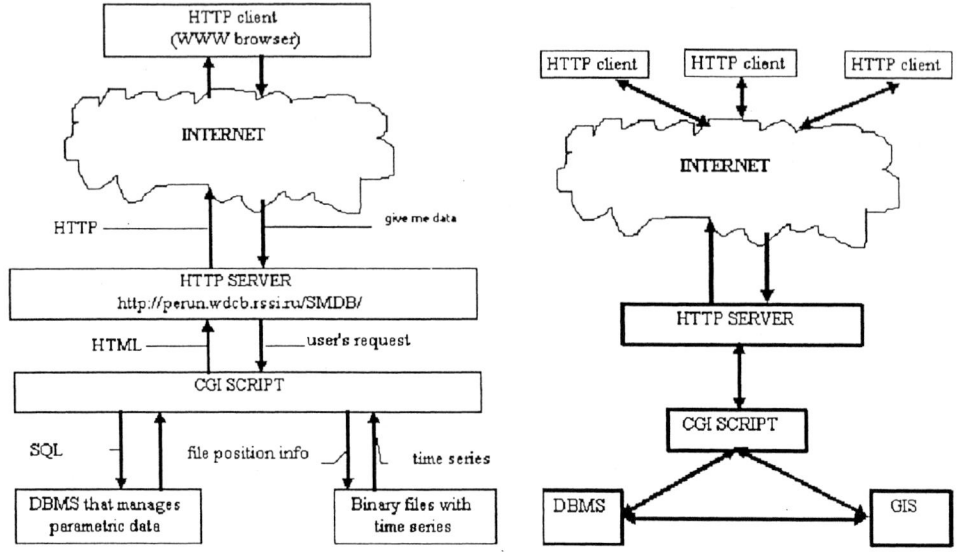

Figure 2.1: *Actual and Future SMDB data flow sketches* The left sketch represents the SMDB data flow. On the right is a sketch of future information flow.

logical Centre (EMSC) in Rome, this data-base project was defined as one of the major activities of the CGDS in its capacity as a key nodal member of the EMSC.

The structure and the contents of the SMDB in 1994 are described in detail in [548]. In 1995, the database was significantly extended and made accessible on-line. In 1996, the CGDS devoted considerable effort to study, implement and develop new methodologies for the creation, maintenance and access of the database. The database technologies now follow a World-Wide Web client server approach (see figure 2.1), which ensures openness, upward compatibility with future improvements, and computer platform independence, all of which are essential for the users.

The SMDB includes two types of data-parametric information: database information (waveforms, events, stations and recording instruments) and

the digital time series themselves [548]. The solution falls into three major parts (see figure 2.1).
First, since the data transmission is conducted through an HTTP (hyper-text transmission protocol) server, the client side uses a WWW browser (HTTP client), such as Netscape or Microsoft Internet Explorer. Such a browser must support HTML (hypertext mark-up language) tables.
The data transmission is done by HTTP server, of which there are a broad variety, distributed free of charge over the internet. The popular Apache server is used, which supports the required CGI (Common Gateway Interface) programming standard.
At the second stage the parametric data are managed by the (DBMS)-POSTGRES95 data management software which is SQL (Structured Querry Language) compatible.
This software is available from the authors together with all libraries needed to interface with different programming languages (such as C, $C++$, PERL, TCL/TK). The structure of the parametric data follows [548], in which data are stored in several relations which can be joined together by using unique keys in SQL statements.
The third major part consists of several CGI program (written in C, $C++$ and C shell) that interact between the HTTP server, the database management software and binary files containing digital waveform data. These program query the database, return results in HTML format to be transmitted by the HTTP server, dynamically create waveforms plots for user preview, and perform basic user authentification. If this scheme needs the above requierements, it is sensitive to the software used. Should another DBMS be used, only a small part of the CGI program need be changed.
The Internet adress for the SMDB is http://perun.wdcb.ru/SMDB
Users can query the database by filling out the dynamically created HTML forms in their standard WWW browser and submitting them to the HTTP server. The latter passes these forms on to the CGI program that creates the appropriate SQL statement, and this statement is forwarded to the DBMS. The results of the query are passed in the opposite direction.
Users from all over the Internet can employ different selection criteria, retrieve values for different fields, and preview drawings of the waveforms with no restrictions. However, one can only access the time serie by logging with one's password.
More CGI scripts will be developed to perform Fourier transform and sonogram calculation on request [221].
The next steps planned are to use the platform independent Java programming language and to use Geographical Information Systems GIS (see figure

2.1).

The Java technology will improve the user interface, provide better opportunities for data processing, and facilitate handling of dynamic objects such as 2-3D simulations. GISs will allow users to deal with spatial data, dynamically creating vector and pixel maps to help analyse strong ground motion information.

The contents of the SMDB on july 2001 are represented in table 2.1. It is available on Internet at http://socrates.wdcb.ru/SMDB

Region	Records	Ev.	Stat.	Date	Magn.	Dist. km
Former USSR	604	54	35	17/05/76-29/04/91	3.1-7.2	1.83-126.92
Europe	1324	99	70	04/11/73-23/05/94	3.3-7.2	0-205
Asia	3138	162	146	14/02/56-19/10/91	2.3-7.9	0-910
USA	16548	418	864	11/03/33-17/01/94	1.7-7.7	0-1006
North Amer.	354	12	29	30/01/73-25/12/85	4.8-8.1	6-465
Centr. Amer.	99	9	4	18/11/67-31/03/73	4.4-6.2	3-30
South Amer.	444	12	36	31/01/51-09/04/85	5.3-7.9	12-373
Africa	18	1	2	29/10/89-29/10/89		
Australia, Pac.	648	19	49	05/10/80-08/09/91	2.7-7.7	1-270
Total	23168	786	1235	11/03/33-23/05/94	1.7-8.1	0-1006

Table 2.1: M-range of magnitudes ; R-range of epicentral distances

2.1.2 Classification of strong motion records according to geotectonic regions

Strong motion record source identification can be used for fast event location as part of an early alarm system. It can be also used to select appropriate waveforms within the SMDB for application in engineering seismology. For these applications we need to distinguish the classes of strong motion records in the SMDB database (described in section 2.1,) a problem to which the syntactic pattern recognition classification described in section 1.1 is applicable and relevant.

If strong motion classes correspond well to geotectonic regions, then we have an efficient instrument to select strong ground motions data sets for seismic engineering projects. For example, if an important engineering construction should be studied from the point of view of its seismic vulnerability, the strong motion records have to be taken from the class which corresponds to the geotectonic region where the construction is located.

Artificial intelligence techniques can be used to facilitate the above identification. One of these technologies is SPARS (see 1.1.3), which is an efficient tool for many different recognition tasks.

Similar studies for regions in Italy, California and Japan, were described in [219] and [67] with a simpler parametrisation used for the strong motion records (smoothed energy envelope of the waveforms and zero-crossing rate in the time window). Here for the first time we use data from China and we parametrize strong-motion records with time-frequency diagrams (see figure 1.2).

As a knowledge base A number of well-defined records (vertical components) were used as a knowledge base (Table 2.2).

This knowledge base consisted of 19 records from California earthquakes (California class), and 13 records from China (Tangshan and Haicheng earthquakes and aftershocks, China I class). The test set consisted of 9 records of aftershocks of a Luquan (China) earthquake (China II class).

Two experiments were conducted on these data. In the first experiment the quality of discrimination for the California and China I sets was tested. Each record in these two sets was discriminated into with the rest used as a knowledge base (*"leave-one-out"* error rate estimator [226]). In the second experiment, the test set (China II) was used to assess the quality of SPARS classification with the first two sets as a knowledge base for the classes. The graph below (figure 2.2) shows the results of the classification.

The results are shown in figure 2.2, giving a robust classification of 75-90 % for the structural differences of the strong-motion records from the two different regions.

The major reason for the structural difference between waveforms from various geographical regions may be the different fine structure of the earthquake sources and the difference in the stress distribution on broad asperities. The latter may be reflected in the regression of the corner frequency versus the magnitude. Our comparison of the classification quality with such a regression has shown similar tendencies: regions with different co-frequency/magnitude regressions are well discriminated by SPARS [67].

The methods used in SPARS constitute the unusual step at modeling structures as a series of spring-mass filters in the spectral analysis of seismic records, as compared to the more routinely used response spectrum analysis. From the seismic engineering point of view this study is significant for

Record ID	MO	Dist. (km)	Event depth	Class	Record ID	MO	Dist. (km)	Event depth	Class
Coy.28	5.9	12	9	Calif.	IB20A096	5.5	23	0	China I
Coy.31	5.9	10	9	Calif.	IB24A108	5.7	7	14	China I
CF001	6.5	29	7	Calif.	IB28A120	4.5	9.5	0	China I
CF004	6.5	39	7	Calif.	IB29A123	5.0	11	12	China I
CF007	6.5	35	7	Calif.	IB37A146	3.5	13	16	China I
CF01	6.6	70	12	Calif.	IB39A152	5.0	15	15	China I
CF04	6.6	28	12	Calif.	2B20A091	5.4	23	12	China I
CF07	6.6	28	12	Calif.	2B25A113	5.3	19	11	China I
H073	6.0	2	16	Calif.	2B29A137	3.6	14	15	China I
I001	7.1	7	16	Calif.	2B31A151	3.9	16	9	China I
P103	5.5	55	10	Calif.	2B40A207	3.9	20	6	China I
P106	5.5	54	10	Calif.	GUQ13UD	4.8	?	4.1	China II
P109	5.5	61	10	Calif.	GUQ20UD	3.5	?	9.6	China II
P112	5.5	72	10	Calif.	MAJ13UD	4.8	?	4.1	China II
S151	6.5	43	8	Calif.	MAJ18UD	3.2	?	9.4	China II
S169	6.5	36	8	Calif.	MAJ20UD	3.5	?	9.6	China II
S202	6.5	34	8	Calif.	SYL20UD	3.5	?	9.6	China II
S316	6.5	35	8	Calif.	ZHL13UD	4.8	?	4.1	China II
S328	6.5	31	8	Calif.	ZHL18UD	3.2	?	9.4	China II
IB02A003	7.8	154	11	Ch.I	ZHL20UD	3.5	?	9.6	China II
IB14A078	4.8	11	4.6	Ch.I					

Table 2.2: Records classified by SPARS

two reasons. First of all, it may contribute to solving classical problems of seismic engineering which are difficult to solve by the standard analysis of the strong motion record parameters (peak-values, peak integral parameters, Fourier and response spectra analysis). Among those problems are the recognition of geographical regions, the classification of macroseismic intensities, and the investigation of source complexity. Moreover, the parametrisation used here could be employed to predict the energy distribution in the time-frequency domain for different event classes. It is also important that the parametrisation is reversible, which allows us to synthesize seismological records, although this is more a subject for theoretical seismology than for

Strong Motion Records

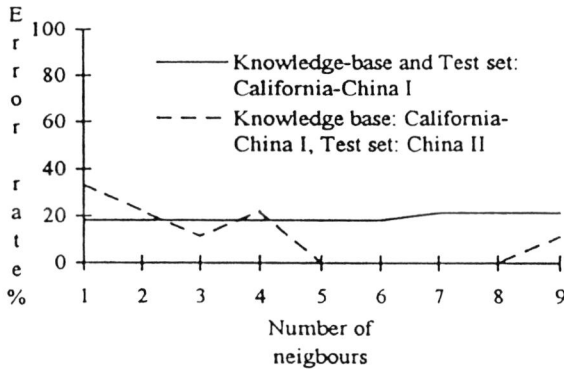

Figure 2.2: *SPARS classification* Results of applying SPARS to strong-motion records.

practical.

ISF Iterated function system resulting from affine transformations in $\Re^2$ according to the Barnsley's ISF code for the Black Speenwort fern (after Barnsley, 1988).

2.2 Seismological Data Classification

2.2.1 Seismic observation networks and databases

The Seismic Stations in CIS Countries (SSCC) Internet Accessible Database is a telematics product that presents on the Internet the seismological stations of the United System of Seismic Observations within the CIS. The data base describes the location and technical specifications of the instruments.

This product is an example of a modern Internet application for handling seismological databases. Similar telemetics techniques can be used to develop databases of in many other kinds of geophysical and environmental activities.

The SSCC is the joint project of the European-Mediterranean Seismological Centre (EMSC) and the Centre of Geophysical Data Studies and Telematics Application in Moscow, which is one of the key-nodal members of the EMSC. Approximately 100 years ago the first seismological observatories began making timed recordings of ground motions. Since that time the number of stations has dramatically increased into the thousands, leading to the need to systematize them and to unify their observations.

In 1908 the international seismological centers started publishing station lists. The catalogue of the Bureau Central de l'Association de Seismologie (BCIS), which included around 200 stations, was the first. Such lists must be frequently reviewed due to the increasing number of seismological stations. The staff of the International Seismological Summary at Kew (ISS) under the International Association of Seismology regularly publishes the updated station lists that operate around the world and transmit data to ISS. The stations are arranged in alphabetic order and their description contains the country of their location and their geographical coordinates (latitude, longitude, elevation). For example, the ISS list [277], published in 1961, described 755 stations. Such data are extremely important for finding the parameters of the earthquake's epicenter and are used for different kinds of seismological research.

The beginning of the 60s marked dramatic changes in seismology: the analysis of seismological data, as well as the registration of seismological events, started to be based on computers. At the same time the Worldwide Standard Seismograph Network WWSSN was set with 120 stations with standard

technical facilities launched in 60 countries around the world. Both the Canadian Network and the USSR's Unified System of Seismic Observation (USso) joined the WWSSN some time later, expanding the number of observations and raising considerably the accuracy of epicenter determinations.
At the same time the three-letter code names were introduced for the seismological stations by the U.S. Cost and Geodetic Survey to simplify the treatment of seismological data by computer programs. These codes were then introduced in the Preliminary Determination of Epicenters and the Earthquake Data Reports, published by the National Earthquake Information Center (NEIC) of the U.S. Geological Survey and in the Seismological Bulletin of the International Seismological Centre (ISC) in the United Kingdom. Very soon the three-letter codes were used world-wide.
The lists of all world seismological stations for which NEIC in cooperation with ISC has assigned an international code, are published on a regular basis by the NEIC, and the ISC [419, 260]. Such lists are continually expanded with information about the new stations and are regularly corrected by introduction of more precise station coordinates, network maintenance informations, etc. Numerous stations are already closed, but the information about them is still necessary for seismological research. The increasing number of stations led to the introduction of four-symbol codes. The information about seismological stations is now available on Internet [378].
Up to the 90s, the territory of Former Soviet Union was a "white spot" in this system as far as real locations and descriptions of seismic stations are concerned. The Internet-accessible online database presented in this book is the first attempt to close this gap.
The USSR's Unified System of Seismic Observations (USSO) was set up by the USSR Academy of Sciences in 1965, summarized:

- the base stations with medium- and short- period instruments. These stations were intended to record earthquakes with surface-wave magnitudes of $M \geq 4.0$ and to acquire data on global seismicity.

- the regional stations with short period instruments responsible for recording weak local and near earthquakes not recorded by the base stations.

Between the creation of the USSO and the moment thato the USSR stopped existing the development of the Network had been based on the construction of new stations, mostly regional ones, in zones of considerable seismic activity. In 1965 the USSO included 71 stations, by 1990 their numbers had grown to 468 with 123 base and 345 regional stations.

The list of Usso network base stations is updated annually and is published by the Geophysical Survey of the Russian Academy of Sciences (Ras) in Obninsk. The list includes the names, codes, station coordinates, the list of equipment with their parameters and instrument amplitude and phase characteristics [484].

The lists of regional stations were published in the seismological bulletins, of the different Ussr research centers, responsible for these stations [290, 398]. For numerous reasons a single catalogue of all Ussr seismic stations with their coordinates and detailed equipment description did not exist. In 1963-1964 the department of the Seismic Service of the Institute of Physics of the Earth issued two books "The parameters and frequency characteristics of Ussr seismological stations equipment". But, these publications were never updated or repeated.

The first significant publication of Ussr station lists (Kondorskaya and Fedorova, 1996, [285]) appeared after the Ussr's collapse. It contains descriptions of the Usso seismological stations as of January 1, 1990. The considerable advantage of this publication is that coordinates are given to within 0.001 degree and the elevations to within 1 meter for the first time.

The seismological stations in the catalogue are grouped in the following regions: the Carpathians, Crimea, Caucasus, Kopetdag, Central Asia and Kazakhstan, Altai and Sayan, Baylkal, Amur, Sakhalin, Kuril-Okhotsk, Kamchatka and Komandorsky Islands, North-East, Yakutia, Artic and Antartic, Baltic Shield, European part of Ussr, Ural. The stations are arranged in alphabetic order for each of these regions. The information about each station contains the station name, the Ussr code (if it exists), the station type (base or regional), the launch year, the coordinates (latitude, longitude, elevation), details about the equipment such as its type, recorded components, maximum magnification, maximum period and the information about the institutional affiliation of the station.

In early 1998, the Centre of Geophysical Data Studies and Telematics Applications Ras, World Data Center B for Solid Earth Physics, and Emsc launched the data base on Seismic Stations in Cis Countries (Sscc), accessible on Internet. The project was completed in the end of 1999.

As a first step, the data from [285] were transformed into computer form. Then these computer data were verified and expanded by introducing the international station codes (if available). The station codes were corrected to the list of the International Seismological Center [260].

The elaboration of the Sscc on-line data base, using the verified and collected station information, extended the already existing Strong Motion Data Base (Smdb, (see section 2.1.1)) [358, 359].

Seismology 81

The SSCC data base was made accessible on Internet using Java technology at the adress: **http://socrates.wdcb.ru/SSCC/**.
The data base is located on a Windows NT platform under the standard NT web server (Internet Information Server). Microsoft jet is used as the database Management System (DBMS). The data base is maintained using MSAccess 95/97 on the local host.
SSCC includes a dynamic map showing the location of all USSO seismological stations. Selecting part of the map by zoom window, the user finds the enlarged picture of the selected territory with its stations enumerated in a table which also contains all details about the station location and a link to table for each station containing equipment information and the timing of its activity.
Internet users can also find the needed information about the stations by filling out a query form and defining search criteria: the name of the station, the region and the country of location, the coordinates, the type of instruments, the date of the station's creation, etc. To find the English spelling of the station's, region's or country's name the user can use the table of correspondences. Both the query form and Java scripts pass the user query to a server script, which makes a request to the database and creates an HTML page to be sent to client browser. In this way the user can receive a station list. The stations can be seen on the map, and one station can be selected to get its detailed description.
Initially based on the publication [285], the database was considerably enlarged by the information about presently closed seismological stations and updated by introducing additional stations outside the Usso network. These additional stations provide seismic data for the investigation of regional seismicity in areas where there are few or no USSO stations and for investigation of seismicity in and near cities where strong and destructive earthquakes have occured. Incorporating these stations into the database increased the total number of stations from 465 to 531. The parameters for these additional stations were collected from different sources and were put in correspondence with the list of the International Seismological Center [260]. However, some details about their equipment are still unavailable.

2.2.2 Syntactic classification of seismograms

Modern seismological networks such as NEIC, IRIS, GEOSCOPE, ORFEUS produce huge amounts of high-quality digital data which are more and more difficult to process using traditional analytical methods. This is an ideal field for artificial intelligence methods to incorporate modern geophysical knowledge into the robust procedures of raw data processing and to acquire knowledge from the automated structural analysis of seismograms.

Recent developments in digital signal processing, wavelet transform theory and multiresolution analysis, along with breakthroughs in machine learning theory, make possible an expert system suitable for this problem. As a prototype of such an expert system, we use the Syntactic Pattern Recognition Scheme (SPARS see 1.1) based on the wavelet transform for time-frequency decomposition and resynthesis of the time series and syntactic methods (string distances) for the description of the waveform structure. The hypothesis to test is that the complexity of the earthquake source and of the geotectonic structure are reflected in the structure of the seismological waveforms.

The results presented in this section can be the basis for real (or nearly real) time system for early earthquake alarm and/or tsunami warning.

As described in section 1.1, the *Syntactic Pattern Recognition Scheme* (SPARS) is based on cluster analysis of the dissimilarity matrix between seismic waveforms. The dissimilarity matrix is constructed by non-linear alignment of the wavelet transforms of the traces using dynamic programming techniques. The knowledge source in SPARS is a collection of well identified seismograms, which can be considered as a set of examples for the automatic phase picker and nearest neighbour classifier.

We construct a pattern of a seismic record

$$x_j = x(t_0 + (j-1)\Delta t), \quad 1 \leq j \leq n_x, \ x(t) \in \Re \quad (2.1)$$

as an energy diagram in the time-frequency domain by assigning instantaneous spectra of the wavelet coefficients to each record fragment.

$$F_i(\delta T) = \{x(t) : t \in [t_i, t_{i+1}]\}, \ i = 1, \cdots, N_x, \ t_i = t_0 + (i-1)\delta T, \quad (2.2)$$

with fixed duration $\delta T : \Delta t \leq \delta T \leq T$. Here Δt is a sampling interval, and $T = n_x \Delta t$ is the record length.

Wavelet transform (Morlet *et al.*, 1982; Daubechies, 1990 [369, 115]) is defined as the projection of signal $x(t)$ onto a family of analyzing functions:

$$(Wx)(a,b) = \langle x, \ h_{a,b}\rangle = \frac{1}{\sqrt{|a|}} \int_{-\infty}^{+\infty} x(t)\bar{h}\left(\frac{t-b}{a}\right) dt \quad (a \neq 0) \quad (2.3)$$

where the family of functions $h_{a,b}(t) = \dfrac{1}{\sqrt{|a|}} h\left(\dfrac{t-b}{a}\right)$ is generated from a

basic modulated Gaussian wavelet $h(t) = e^{t^2/\Delta t_0^2} e^{2\pi f_0 t}$ by time translations by b (to select which part of the signal to analyze) and dilations or contractions using a scale parameter a (in order to focus on a given range of oscillations). Note, that the wavelet transform, unlike the short-time Fourier transform, treats frequency in a logarithmic way. If the basic wavelet $h(t)$ is focused in time around $t_0 = 0$ and in frequency around f_0 with the RMS duration Δt_0 and bandwidth Δf_0, then the wavelet $h_{a,b}(t)$ is focused in time around b with duration $\Delta t = a \cdot \Delta t_0$ and in frequency around f_0/a with bandwidth $\Delta f = \Delta f_0/a$. This corresponds to a constant relative bandwidth as opposed to fixed bandwidth filtering. We used the squared modulus of (2.3) to provide an *energy density distribution* (EDD) $Ex(a,b) = |(Wx)(a,b)|^2$.
In practice the time resolution of EDD is limited by the sampling interval ΔT and the finite record length, thus the frequency resolution is limited to, say, M octaves. We compress the EDD by changing the time step from the sampling interval ΔT to the experimentally defined translation step (fragment length) δt. At the same time we restrict possible values of the dilation step to $2^{1/K}$ for K "voices" per octave. This leads to a discrete decomposition (pattern)

$$(Ex)_{\min}(\delta T) = \left| \left\langle x, 2^{-m/2K} h\left(\frac{t - n\delta T}{2^{m/K}}\right) \right\rangle \right|^2, \qquad (2.4)$$
$$m, n \in \mathcal{Z},\ 0 \le n < T/\delta T,\ 0 \le m < KM$$

where $\langle \cdot, \cdot \rangle$ stands for L_2 convolution. For a fixed n the values $(Ex)_{\cdot n}(\delta T)$ give an instantaneous spectrum at the fragment F_n denoted by E_{F_n}. Additional packing of information in the pattern results from the clustering without learning of the parametrized fragments. The finite alphabet of the cluster labels (letters) allows us to represent a pattern as a word (sequence of letters).
Suppose that the records $x(t)$ and $y(t)$ are represented by the patterns $\mathcal{F}_x = E_{F_1}, \cdots, E_{F_{N_x}}$, and $\mathcal{G}_y = E_{G_1}, \cdots, E_{G_{N_y}}$; we call $\mathcal{F}_x$ a source pattern and $\mathcal{G}_y$ a target pattern.
We introduce a syntactic dissimilarity measure between the seismic records using a non-linear alignment of their EDD's. The purpose of the non-linear alignment is to find a montonic transform of the time scales of the two records, such that they show simultaneous and similar onsets of the waveform structural phenomena, such as P- and S-arrivals. The principles of the procedure were proposed by Levenstein (1965) [305]. The initial data for the alignment of the patterns $\mathcal{F}_x$ and $\mathcal{G}_y$ are the matrix of local spectral dissimilarities of their fragments

$$d(\mathcal{F}_i, \mathcal{G}_j) = \left\| E_{F_i} - E_{G_j} \right\|^2, \quad i = 1, \cdots, N_x, \; j = 1, \cdots, N_y. \quad (2.5)$$

The result is a pair of functions $k(\cdot), m(\cdot) : N \to N$, satisfying the following

Local constraints	Global constraints
$i = k(l), \; k(l+1) - k(l) = 0$ or 1,	$k(l) - m(l) \leq \delta$ (a deviation from the main diagonal),
$j = m(l), \; m(l+1) - m(l) = 0$ or 1,	$k(l) + m(l) = l_0 \geq 0$ (a shift of the first fragments),
$l = 1, \cdots, L, \; L \leq N_x + N_y$,	$k(L) + m(L) = l_1 \leq N_x + N_y$ (a shift of the last fragments),

which minimizes the accumulated dissimilarity between the patterns

$$\mathcal{D}_L(x, y) = \min_{k(\cdot), m(\cdot)} \left[\sum_l d^\star(\mathcal{F}_{k(l)}, \mathcal{G}_{m(l)}) \right]. \quad (2.6)$$

The sum is over the path $k(\cdot), m(\cdot)$ including possible insertions and deletions of the fragments, and $d^\star$ stands for the weights of the editing operations. Let NULL be the "empty" fragment (gap):

- if $k(l) = k(l+1)$, then we insert the NULL fragment into the target pattern, this called *insertion* of $\mathcal{G}_{m(l)}$;

- if $m(l) = m(l+1)$, then we insert the NULL fragment into the source pattern; this is called *deletion* of $\mathcal{F}_{k(l)}$;

- otherwise $k(l+1) - k(l) = m(l+1) - m(l) = 1$; this is called *substitution* of the fragments $\mathcal{F}_{k(l)}$ and $\mathcal{G}_{m(l)}$.

The result of non-linear alignment may be represented by extended patterns $\mathcal{F}_x^\star = \mathcal{F}_1^\star, \cdots, \mathcal{F}_{N_x'}^\star$ and $\mathcal{G}_y^\star = \mathcal{G}_1^\star, \cdots, \mathcal{G}_{N_y'}^\star$ now including gaps which are necessary to match the phases arriving with different time delays. For example, the path:

$$\begin{array}{lllllll} \mathcal{F}_x^\star = & \mathcal{F}_1, & \mathcal{F}_2, & NULL & \mathcal{F}_3, & \mathcal{F}_4 & \mathcal{F}_5 \\ & subst & subst & ins & subst & del & subst \\ \mathcal{G}_y^\star = & \mathcal{G}_1, & \mathcal{G}_2, & \mathcal{F}_3, & \mathcal{F}_4 & NULL & \mathcal{G}_6, \end{array} \quad (2.7)$$

We search for the optimal path $k(\cdot), m(\cdot)$ using a dynamic programming technique.

The definition of the weights of the pattern editing operations is essential for the alignment procedure (cf [309], where these weights were not used at all). Possible general formulae are :

insertion / deletion: $d^\star(\mathcal{F}_i, NULL) = d^\star(NULL, \mathcal{F}_i) = q_1 + q_2 \|E_{F_i}\|^2$,

substitution: $d^\star(\mathcal{F}_i, \mathcal{G}_j) = d(\mathcal{F}_i, \mathcal{G}_j) = \left\|E_{F_i} - E_{G_j}\right\|^2$.

where q_1, q_2 are constants and $i = k(l)$, $j = m(l)$. In figure ? we illustrate the non-linear alignment of two synthetic seismograms with the constants $q_1 = 0$, $q_2 = 1$.

We consider this dissimilarity measure between the patterns of seismic records (Equation 2.6) to be a generalisation of the well known linear crosscorrelation similarity $\mathcal{D}_C(x,y) \max_\tau | \rho_{xy}(\tau) |$ for

$$\rho_{xy}(\tau) = \frac{\sum_{i=1}^{N-\tau} \langle E_{F_1}, E_{G_{i+\tau}} \rangle}{\frac{N-\tau}{N} \sum_{i=1}^{N} \|E_{F_1}\|^2 \sum_{i=1}^{N} \|E_{G_1}\|^2}, \qquad (2.8)$$

where $N = \min(N_x, N_y)$, and $\langle \cdot, \cdot \rangle$ stands for the inner product of the parameter vectors of two fragments [267]. An important difference between the two methods is that in the non-linear case local changes are possible in the time scales of the patterns.

Classification. We define a syntactic dissimilarity $\mathcal{D}_K(\mathcal{F}, \mathcal{C})$ between a pattern $\mathcal{F}$ and a class of patterns $\mathcal{C}$ for a given K by the formula:

$$\mathcal{D}_K(\mathcal{F}, \mathcal{C}) = \min_{\mathcal{D}_i \in \mathcal{C}} \frac{1}{K} \left[\sum_{i=1}^k \mathcal{D}_L(\mathcal{F}, \mathcal{D}_i) \right]. \qquad (2.9)$$

Let the two classes of patterns $\mathcal{C}_1$ and $\mathcal{C}_2$ be given. For a pattern $\mathcal{F}$ from an unknown class we use the K-nearest neighbors decision rule

$$\mathcal{F} \in \begin{matrix} \mathcal{C}_1 \\ \mathcal{C}_2 \end{matrix}, \text{ if } \mathcal{D}_K(\mathcal{F}, \mathcal{C}_1) \begin{matrix} \leq \\ > \end{matrix} \mathcal{D}_K(\mathcal{F}, \mathcal{C}_1). \qquad (2.10)$$

The result is a projection of the test set of records at the nominal scale according to the *a priori* known classification in the Knowledge Base [226]. This "jack knife" estimation is used to refine the Knowledge Base and to test the classification stability.

Real Data Applications

Keeping in sight the possible application of this research, we tested the preliminary version of the SPARS algorithms on selected strong motion and regional seismic data. The results of applying these algorithms to strong

Dynamic Classification Approach

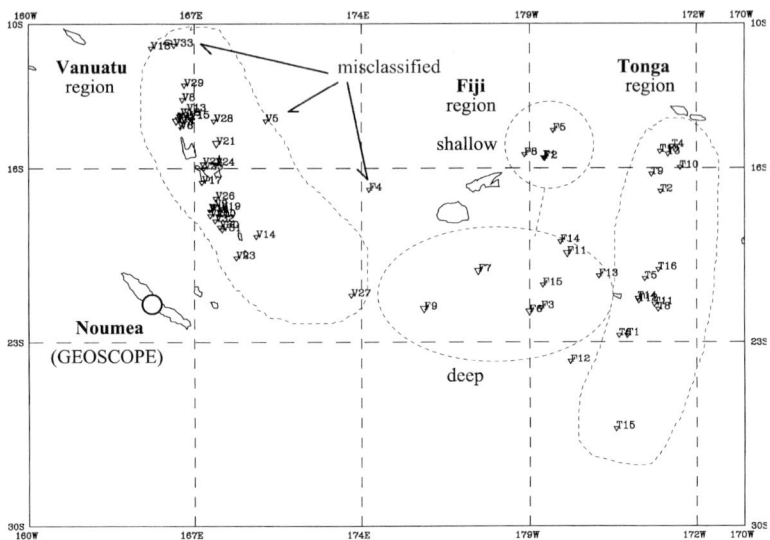

Figure 2.3: *SPARS classification* Epicenters of the Vanuatu, Fiji and Tonga events.

motion data are reported in [66, 220, 547]. Tests using these algorithms to identify earthquake regional structure look even more efficient than the application to strong motions. We may be approaching the most interesting stage of the SPARS based applications.

To test the capabilities of the method we analysed the vertical component of broad-band records from the GEOSCOPE / ORSTOM seismic station at Noumea (New Caledonia). The so called LGPL component, recorded at a 1 Hz sampling rate was selected for a group of 3 seismic regions: the Vanuatu - Loyalty region, the Fiji region, and the Tonga region, each of them including shallow, intermediate, and deep earthquakes. The Knowledge Base consists of collection of 61 events, 31 in Vanuatu, 15 in Fiji and 15 in Tonga. The records from these three data sets have comparable epicentral distances and magnitude ranges. Epicentral distances for the Vanuatu region vary from 447 to 1235 km, for the Fiji region from 1027 to 1462 km, and for the Tonga region from 1497 to 2310 km. The event magnitudes m_b range from 4.3 to 6.4. In this first approach, using only one component, we distinguished the regional nature of each record (Figure 2.3).

To do this, each seismogram (10 min duration) was divided into set of non-overlapping fragments 16 seconds long. Each record was filtered by a set of 10 Gaussian narrow band filters logarithmically centered between 10 to 50 sec (4 "voices" per octave) and "prewhitened" to have a dynamic range limited to 60 dB in the following manner:

$$(Rx)_{mn}(\delta T) = 20\log_{10}\frac{(Ex)_{mn}(\delta T)}{E_{\max}}, \text{ where } E_{\max} = \max_{m,n}(Ex)_{mn}(\delta T)$$

$$(\hat{R}x)_{mn} = \begin{cases} 60 + (Rx)_{mn}(\delta T), & \text{if } (Rx)_{mn}(\delta T) \geq -60 dB \\ 0, & \text{if } (Rx)_{mn}(\delta T) < -60 dB \end{cases}$$

Then we estimated the matrix of pairwise syntactic distances $\mathcal{D}_L(x,y)$ (Equation 2.6) for the parameter $\hat{R}_{F_1}$ with the weights of the editing operations $q_1 = 0$, $q_2 = 1$. Figure 2.3 gives an example of seismogram ordering according to syntactic distance.

Using the one nearest neighbour decision rule we obtained only 3 errors among the 61 analysed events; meaning a 95 % correct classification rate. The 3 "misinterpreted" events are in fact at the boundaries of the considered seismic regions. Two events are situated at the so called back arc of the Vanuatu subduction zone, and one event is at the northern part of the Vanuatu arc, close to the Solomon seismic region.

It is interesting to notice that within each seismic region the deep events are well associated between themselves, which can be seen on the average linkage hierarchical clustering dendrogram. For shallow events, which form the bulk of the data used (42 events), 33 events are well associated by SPARS as well as by epicentral location (pairwise distances less than 150 km), and 22 among them are perfectly associated, meaning that their epicenters are less than 50 km from each other.

2.3 Recognition of Magnetic Anomalies along the Mid-Atlantic Ridge

We present here an application of syntactic recognition for magnetic data. This application has been obtained in 1997 by M. Zhizhin, J. Dubois, A. Gvishiani and L. Hongre [552].

An efficient way to understand the geodynamical mechanism of oceanic lithosphere accretion is by detailed observations of variations of the accretion rate in space and time [304]. Since Vine and Matthews (1963) [528], we know that the oceanic crust formed at accreting plate boundaries is magnetized in the direction of the Earth's magnetic field at the time the crust was formed. Observation and identification of these magnetic field anomalies is sometimes hiden by high noise levels(mostly due to spreading asymmetry and axis discontinuities). Loncarevic and Parker (1971) [311] examined the problem of magnetic noise in an area of slow spreading and rough topography, and proposed a qualitative technique to extract the signals from the noise.

We developed this idea here using a syntactic pattern recognition procedure (see 1.1) [547] which consists of identifying the main features on a magnetic profile near the crest of the accretion boundary by comparison with the theoretical constant spreading rate profile. It is then possible to show temporal variations in spreading rate, lateral variations between profiles, and eventually, asymmetry of a profile with respect to the accretion center. By its systematic and automatic applicability, the method can strongly improve the study of temporal and spatial variations of the spreading rate.

2.3.1 Method

Recognition of magnetic anomalies is usually done in two steps [401]. First, theoretical profiles are created using a chronological scale of magnetic inversions [84]. These models are dependent on the spreading rate, bathymetry, localization and orientation of the rift: since the location of the anomalies is precisely known and so the models are dated. In the second stage, the measured profiles are copared with the models to find the position of the anomalies and to date the crust in accordance with these anomalies (figure 2.4) More generally, this path joins synchronous parts of the modeled and measured profiles. We assume that the complete path between the manually obtained points is continuous and should be non-linear because of temporal variations of the spreading rate.

Our goal is to automatically construct alignment paths similar that shown

Magnetic Anomalies

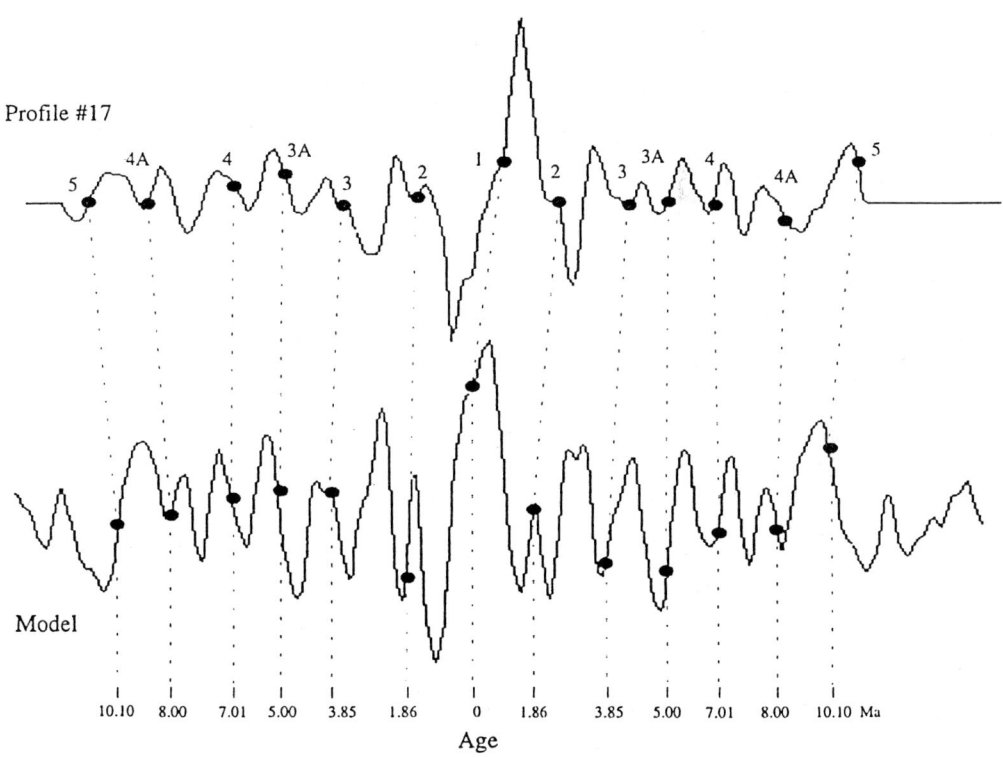

Figure 2.4: *Manual identification* It is done on profile 17 (measured at the SARA area) using a magnetic model. The model is calculated for a spreading rate of 12.5 mm/yr and a magnetized layer thickness of 400 m (Sloan and Patriat, 1993). The position of the anomalies, indicated by black points, are chosen at the boundary between certain reversed and normal blocks so that their age is precisely known.

in figure 2.5, which will allow continuous dating of the oceanic crust. We use the following notations

- $s(m)$, $-M \leq m \leq M$, is a theoretical anomaly profile where the index m, considered as the distance from the center of the ridge, is related to the geological time scale t by the simple relation $m(t) = v \times t$ for a spreading rate v of the crust, assumed to be constant in the model;

- $p(n)$, $-N \leq n \leq N$, is the measured magnetic anomaly profile where n, the observation index, equals 0 in the center of the ridge and is generally a non-linear function $n(t)$ of the geological time scale t.

Using this notation, the desired alignment path is the graph of the parametric function $P = (n(t), m(t))$ on the grid relating the part $p(n(t))$ of the observed profile with its synchronous theoretical counterpart $s(m(t))$. $m(t)$ and $n(t)$ are *monotonic* (non-decreasing), *continuous* and *non-linear* because the distance $m(t)$ in the model is proportional to the time scale t. In practice we assume that crustal spreading is continuous (without contraction), and that the instantaneous spreading rate $v'(t) = (d/dt)v(t)$ varies for different geological periods. As opposed to linear correlation ($n = m + C$), which permits only diagonal segments in the path on (figure 2.5), non-linear alignment also allows vertical and horizontal segments, corresponding respectively to relative contractions and dilatations of $p(n)$ with respect to $s(m)$.

If linear correlation is used, the best alignment is obtained along the path $n = m + C$ which minimizes the difference

$$\bar{D}(C) = \sum_m [p(m+C) - s(m)]^2 . \tag{2.11}$$

The fact that the instantaneous spreading velocity $v'(t)$ is in a limited neighborhood of the average velocity v selected for the model, specifies the local shape of the path. In figure 2.6 we draw examples of the minimal sets of 'elementary' segments necessary to construct the path for different constraints on the instantaneous velocity variation. The use of such segments avoids excessive and unrealistic contractions or dilatations of profiles.

Below we consider the first non-trivial case: $1/2v \leq v'(t) \leq 2v$ (figure 2.6).

If we denote the segments in this case by ↓, 0 and ↑, reflecting decreasing, constant, and increasing accretion, then the whole path, being a composition of the segments, may be written as a string of these symbols, e.g. $P = 00 \uparrow\uparrow 0 \downarrow\downarrow\downarrow 00$. This gives us a qualitative picture of the evolution of oceanic crust accretion.

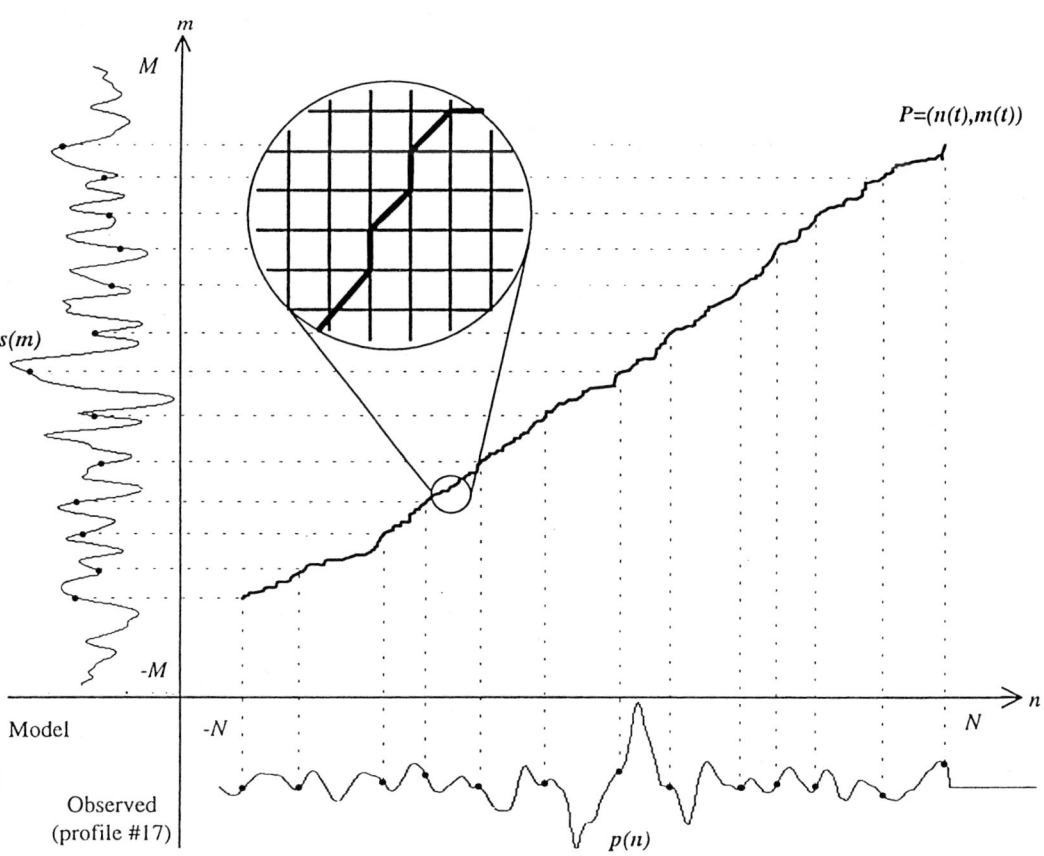

Figure 2.5: *Alignment-path joining points of both profiles.* This path is continuous and usually non-linear because of the local spreading rate variations. The path also joins points resulting from the manual identification done on figure (intersection of dashed lines). The inset shows a possible shape of the alignment path on the integer grid.

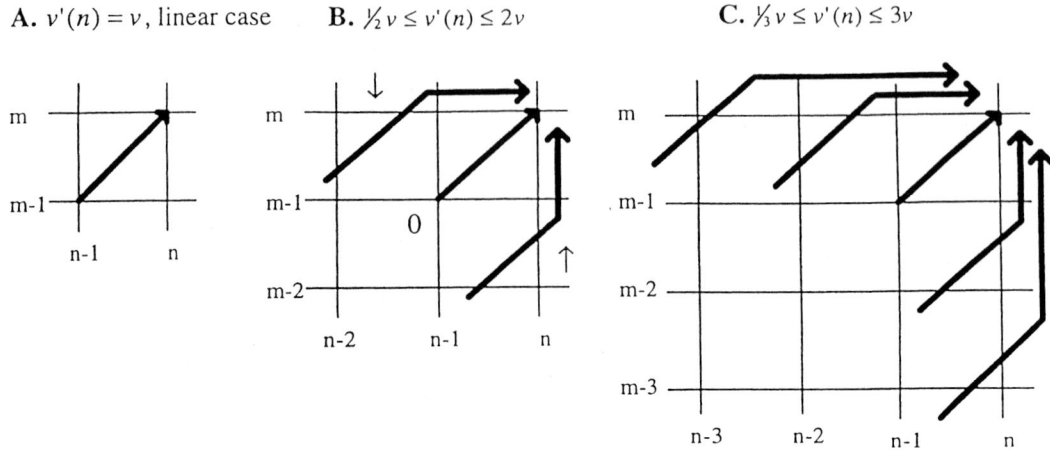

Figure 2.6: *Minimal sets of elementary segments.* The segments are used to construct the alignment path for different constraints on the instantaneous velocity variation.

Here, in the non-linear case, we generalize formula 2.11 defining 'local' differences for the segments ↓, 0 and, ↑ (figure 2.6):

$$\begin{aligned}
d_\downarrow(n,m) &= \tfrac{1}{2}[(p(n)-s(m))^2 + (p(n)-s(m-1))^2] \\
&\quad \text{for the segment } \downarrow, \\
d_0(n,m) &= (p(n)-s(m))^2 \\
&\quad \text{for the segment } 0, \\
d_\uparrow(n,m) &= \tfrac{1}{2}[(p(n-1)-s(m))^2 + (p(n)-s(m))^2] \\
&\quad \text{for the segment } \uparrow .
\end{aligned} \quad (2.12)$$

Then the best synchronization of two profiles is obtained for the path $P = n(t), m(t))$ that minimizes the total difference defined as the sum of local differences 2.12 along the path:

$$\tilde{\mathcal{D}}(P) = \sum_t d_i(n(t), m(t)) \text{ with } i = \downarrow, 0 \text{ or } \uparrow . \quad (2.13)$$

Let $\mathcal{A}$ be the space of paths (strings) that are composed only of the segments ↓, 0 and, ↑ (figure 2.6). The path which gives the minimum of the sum 2.12, satisfies the general principle of optimality (Bellman and Dreyfus, 1962) [50]: *an optimal policy has the property that whatever the initial state and initial decision are, the remainig decisions must constitute an optimal policy with*

Magnetic Anomalies

regard to the state resulting from the first decision. In our case, this is equivalent to two basic statements (Myers et al., 1980) [373]:

1. the globally optimal path is also locally optimal;

2. the optimal path to the grid point (i,j) only depends on values of (i',j') such that $i' \leq i$, $j' \leq j$.

That is why the search for the minimum over the functional space $\mathcal{A}$ may be performed using the following two-pass dynamic programming algorithm (introduced by Gvishiani et al., in 1991 [?]) with $\mathcal{O}(\mathcal{N}^2)$ complexity (we suppose that $\mathcal{N} \approx \mathcal{M}$). The algorithm uses two matrices; a real-valued matrix $\mathcal{D}(i,j)$ and a character-valued matrix $\mathcal{D}(i,j)$, to store, respectively, intermediate sums of local differences and symbols of the last segment of the optimal path to every point of the grid. As input the algorithm receives the profiles i', j' and $s(m)$.

Forward pass:

- *Initialization*: $\mathbf{D}(-\mathcal{N}, -\mathcal{M}) = 0$, $\mathbf{T}(-\mathcal{N}, -\mathcal{M})$ *is undefined.*

- *Recursion*: $\mathbf{D}(i,j)$ *is computed for* $-\mathcal{N} \leq i \leq \mathcal{N}$, $-\mathcal{M} \leq j \leq \mathcal{M}$ *using the formula*

$$\mathbf{D}(i,j) = \min \begin{cases} \mathbf{D}(i-2, j-1) & +d_{\downarrow}(i,j) \\ \mathbf{D}(i-1, j-1) & +d_0(i,j) \\ \mathbf{D}(i-1, j-2) & +d_{\uparrow}(i,j) \end{cases} \quad (2.14)$$

  and the segment symbol, which gives the minimum in 2.14, is stored in $\mathbf{T}(i,j)$.

- *Termination: the result is a minimized accumulated difference* (2.11) *between the profiles* $p(n)$ *and* $s(m)$:

$$\mathcal{D}(p,s) = \min_{\mathcal{P} \in \mathcal{A}} \tilde{\mathcal{D}}(\mathcal{P}) = \mathbf{D}(\mathcal{N}, \mathcal{M}). \quad (2.15)$$

Now to reconstruct the optimal path $\mathcal{P} = \arg\min_{\mathcal{P} \in \mathcal{A}} \tilde{\mathcal{D}}(\mathcal{P})$ it is sufficient to make a 'backward pass' over the symbolic matrix $\mathbf{T}(i,j)$:

Backward pass:

- *Initialization: at the point* (N, M), *where the forward pass has terminated, set* $\mathbf{P} = T(N, M)$.

- *Recursion: staying at the grid point (i,j), we compose the path one segment backward (left composition). The location of the preceding point may be $(i-2, j-1)$, $(i-1, j-1)$, or $(i-1, j-2)$, depending on the segment chosen to arrive at in the recursion step of the 'forward pass':*

$$\mathcal{P} = \begin{cases} \mathbf{T}(i-2, j-1) \bigcirc \mathcal{P}, & \text{if } \mathbf{T}(i,j) = \downarrow \\ \mathbf{T}(i-1, j-1) \bigcirc \mathcal{P}, & \text{if } \mathbf{T}(i,j) = 0 \\ \mathbf{T}(i-1, j-2) \bigcirc \mathcal{P}, & \text{if } \mathbf{T}(i,j) = \uparrow \end{cases} \quad (2.16)$$

- *Termination: the recursion stops when the path reaches the point $(-N, -M)$.*

The output of the algorithm contains: 1) a global measure of the difference $\min_{\mathcal{P} \in \mathcal{A}} \widetilde{\mathcal{D}}(\mathcal{P})$ between the theoretical model and the observed profile; 2) a qualitative and, after some smoothing, a quantitative history $\mathcal{P} = (n(t), m(t))$ of oceanic crustal accretion; 3) continuous dates along the observed profile as the inverse function $n^{-1}(t)$.

The amplitude uncertainty of the theoretical anomaly profile model does not dramatically decrease the quality of the output. The only critical assumption here is that the local extremes of the profile are not missing and are correctly dated.

2.3.2 Data, results and interpretation

We show here a study based on magnetic data from the SARA (Segmentation Ancienne de la Ride Atlantique) survey on the flanks of the Mid-Atlantic Ridge between 28° and 29°N. This survey was carried out aboard the N/O Jean Charcot in May 1990 [469] to bring an off-axis complement to an on-axis survey carried out in 1988 and 1989 [454] on the CONRAD 2912 expedition between the Atlantis and Kane transform faults.

The high-resolution data cover an area of $100 \times 350 km^2$, centered on the ridge and extending to an age of 10 ma on both sides of the ridge. The data were collected along 37 profiles having a direction N101° and spaced every 3 km, with the profiles parallel to the spreading direction. The magnetic data were collected on both surveys with a towed proton magnetometer. The magnetic anomaly was calculated by subtracting the global magnetic field model IGRF90 from the magnetic records. The rectangular grid reconstructed from the profiles is shown in figure 2.7

A magnetic anomaly model for the SARA area was calculated for each profile [469] based on the calibrated magnetic inversion time scale developed

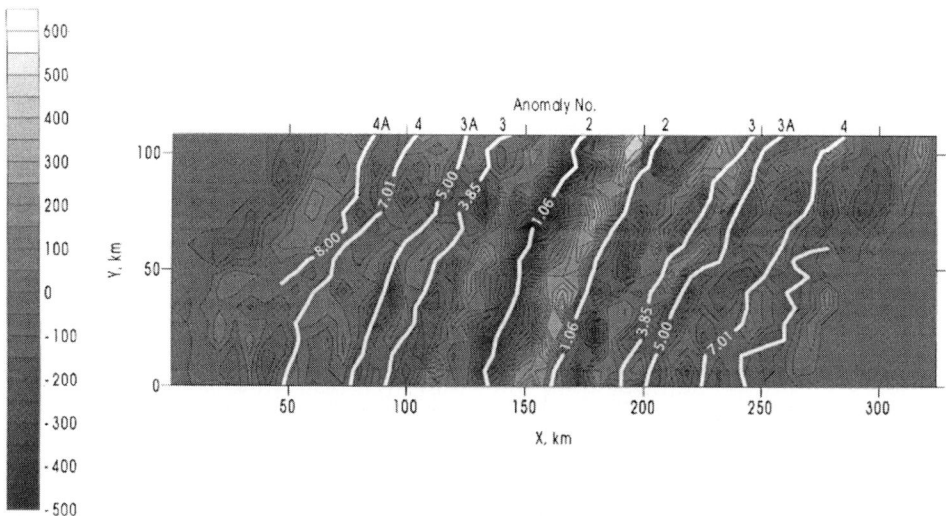

Figure 2.7: *Magnetic anomalies in the* SARA *area.* The negative anomalies are represented in dark and the positive in light. The anomalies are in the range 650-850 nT. White lines show the identification of magnetic anomalies using our method for the fixed spreading rate model and the observed profiles.

by Patriat, 1987) [401], a constant spreading rate of 12.5 mm/yr, and a magnetized layer thickness of 400 m which follows the seafloor topography.

The result of applying the non-linear alignment method to this model and measured profile No. 17 from the SARA survey is presented in figure 2.8

Lines joining the two profiles are drawn for every diagonal step in the alignment path reconstructed after the 'backward pass' of our algorithm. It follows from the method that these lines join points on both profiles with the same crustal dates. As one can see, the algorithm identifies a high percentage of the intermediate observations. After linear interpolation we estimate the crustal age for the whole observed profile using linear interpolation. In the worst case, if the spreading rate difference between the observed profile and model is two times higher (or lower), every second point of the profile would be joined. This gives a rough estimate of the precision of the algorithm as two times the sampling interval divided by the spreading rate of the model.

In some parts of the survey, magnetic anomalies are smoothed because of discontinuities between segments, and misalignment can occur in these parts. The computer program allows the user to assign higher anomaly values for the pairs of matching points after visual inspection of their alignment and

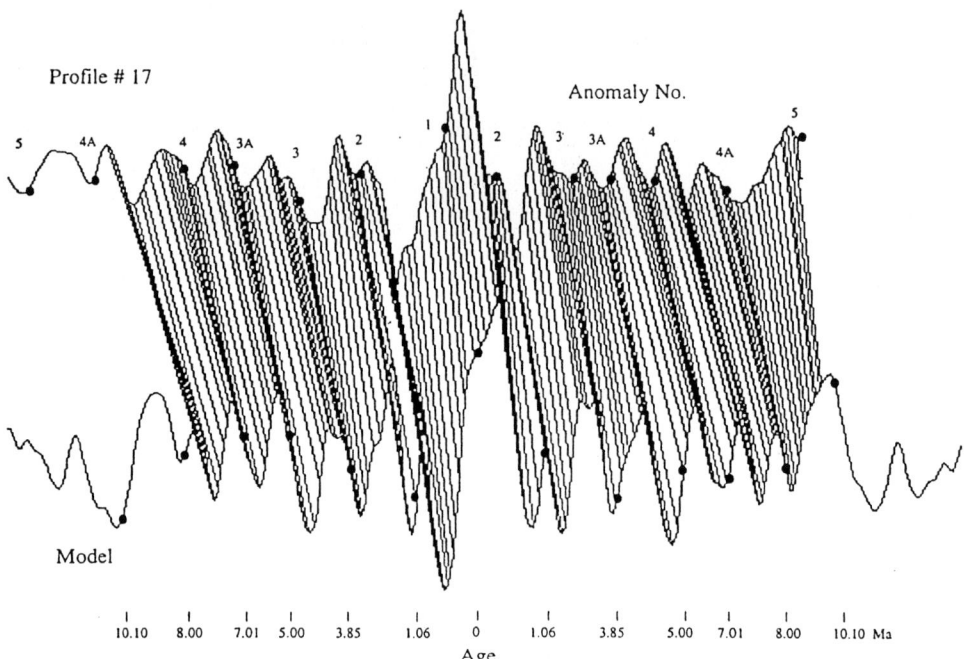

Figure 2.8: *Automatic non-linear alignment*. The result is presented for measured profile No. 17 from the SARA survey (top) with the model (bottom). The lines obtained using our method join parts of both profiles having the same age. Lines also connect the black points resulting from a manual identification by Sloan and Patriat (1993) .

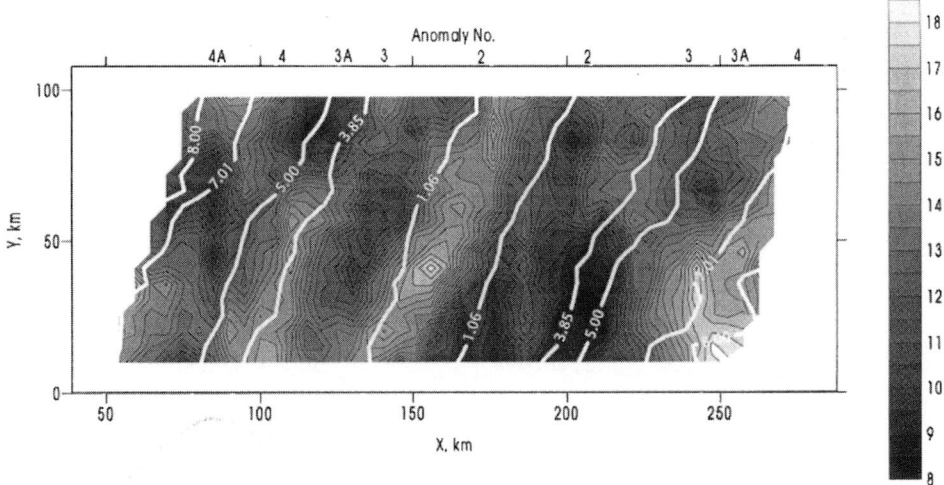

Figure 2.9: *Smoothed first derivative.* The smoothed first derivative of age in the east-west direction as an approximation of the instantaneous spreading rate in mm/Ma. It is overlaid by five magnetic anomalies to the age of 8 Ma shown by white lines.

to run automatic identification for the remaining parts. For the set of 37 profiles, manual corrections were done 7 times for the eastern flank of the ridge in the central and northern parts and 4 times for the western flank in the central part of the survey. After these corrections an age grid for the SARA survey was obtained using our algorithm. The result of the identification of the five magnetic anomalies to the age of 8 Ma on both sides of the ridge in figure 2.7 is drawn over the input magnetic variation grid. The local zigzags of the anomalies are due to the fact that the inversion of the field takes some time and is 'recorded' in several kilometers of the crust rather than as an abrupt linear boundary.

The smoothed first derivative of age in the east-west direction, as an approximation of the instantaneous spreading rate, is shown in figure 2.9.

It is overlaid by five magnetic anomalies to the age of 8 Ma. One can see the anti-symmetric variations of the spreading rate profile with respect to the accretion center: fast spreading on one flank implies slow spreading on the other. We also observe here a correlation between the extremes of the spreading rate and the inversions of the magnetic field.

For a small rectangular zone of highly detailed magnetic anomaly data in the vicinity of a stepwise linear oceanic ridge with known model profile, the crust dates can be identified using non-linear alignment of the profiles with the model. The obtained crustal surface dates as well as the instantaneous spreading rates may be used for the reconstruction of isochrons in the tectonic history of the zone.

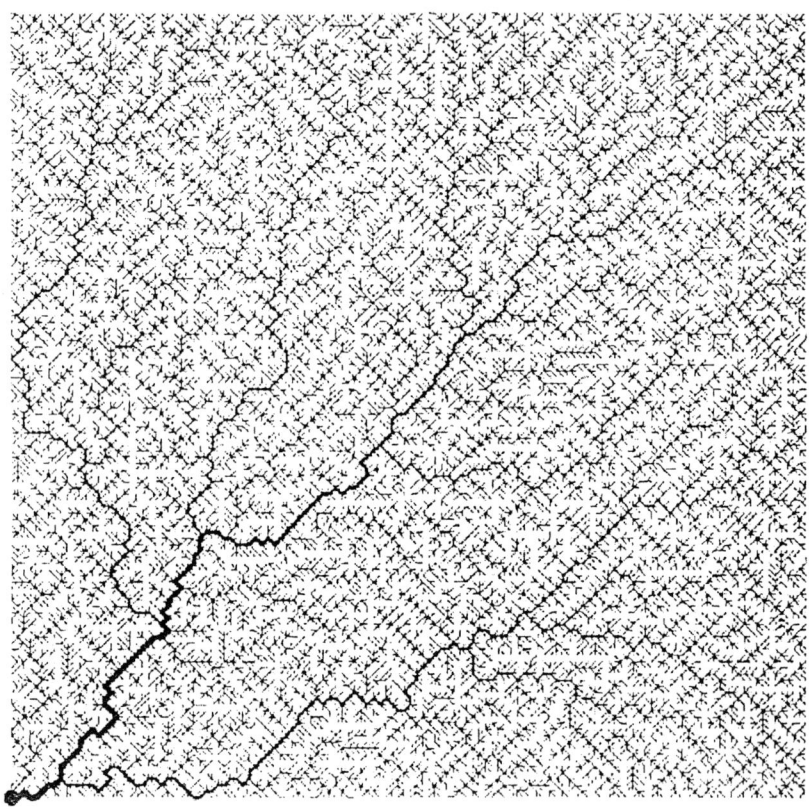

Synthetical river network generated by a self organized critical process (after Rodriguez-Iturbe and Rinaldo, 1997).

2.4 Clustering Analysis for Magnetic Field Studies

2.4.1 Euler deconvolution

Euler deconvolution technique, well known in exploration geophysics, deals with determination of the shape of causative bodies from potential field data. It is based on approximating the measured anomalous gravity or magnetic field in a running window by the field of a single elementary source of uniform density or magnetization. The method allows us to characterize the position and depth of the nearest or largest causative source in the vicinity of the window. The method in its present form is the result of the long development history of the Euler deconvolution. Hood (1965) [248] first used this method for aeromagnetic data interpretation, and demonstrated that the method is valid for point poles and point dipoles. Euler deconvolution technique was further developed by Thompson (1982) and Reid et al. (1990) [505, 430]. Thompson elaborated the 2D approach and derived structural indices for several elementary bodies. Reid et al. extended Thompson's approach to 3D and showed its applicability to gravity anomalies of finite step and magnetic anomalies of a thin dike and of a sloping contact. Keating (1998) [275] applied the Euler approach for irregular 3D grids, using weights proportional to station accuracy and interstation distance. Zhang et al. (2000) [546] applied this method to interpretation of gravity gradient tensor measurements.

Euler deconvolution appears to be a powerful method for gravitational and magnetic field interpretation, providing preliminary information on the position, shape, and depth of causative bodies. It is essentially effective for isolated compact bodies restricted by vertical side boundaries. In that case, Euler solutions cluster around contours of the bodies in the horizontal plane and provides some estimates on their depth. When anomalies of potential field result from more than one source, Euler solutions can form wide clouds, rather than dense clusters, making it difficult to outline side boundaries of the causative sources. Interpretations can often be improved by rejecting solutions with low tolerance [505], large dispersion of depth estimates or with feature located too shallow or too deep. Although efficient for isolated anomalies, these criteria are sometimes inefficient in complicated areas, where their efficiency strongly depends upon noise and the effects of shallow sources and neighboring bodies.

Herein, we present the results of the application of the clustering technique introduced in 1.3 to select best Euler solutions. Our goal is the automatic (or semi-automatic) determination of clusters of Euler solutions which outline causative bodies and provide more reliable estimates on their depth.

Euler deconvolution provides geometrical parameters estimated for elementary causative bodies using values of the anomalous potential field (gravity or magnetic) and its horizontal and vertical derivatives (measured and calculated). The method assumes that the anomaly is a homogeneous function of degree n of three co-ordinates:

$$f(tx,\ ty,\ yz) = t^n f(x,\ y,\ z).\forall t \in \Re \qquad (2.17)$$

Strictly speaking, the method is valid for magnetic or gravity anomalies caused by bodies whose position in space can be characterized by a single point (x_0, z_0) in 2D or (x_0, y_0, z_0) in 3D, where (x, y, z) represent Cartesian coordinates with the Oz axis directed downwards and the Ox axis directed to the North. In the 2D case, Ox is directed along the profile. Herein and henceforth we discuss the magnetic field case. Similar results can be obtained for gravity fields. The bodies under consideration can be point poles and/or point dipoles, as well as lines of poles and/or of dipoles. Several elementary bodies obey the Euler equation under specific conditions: for exemple a dike (vertical or inclined) when its thickness is considerably smaller than its depth. For all these bodies the Euler equation can be represented by:

$$(x-x_0)\partial f/\partial x + (y-y_0)\partial f/\partial y + (z-z_0)\partial f/\partial z = N(A - f(x,y,z)), \qquad (2.18)$$

where x_0, y_0, z_0 - are the coordinates of a point, that characterizes the position of the elementary source (further referred to as the Euler solution); x, y, z -are the coordinates of the data where potential field and its derivatives were measured or calculated; N is the structural index, which depends on the shape of the body $(N = -n)$ (structural indices for these bodies were listed in [505] and [430]; and A is a constant to be determined.

It is worthwhile to mention that A reflects a constant level in the measured field. At the same time, according to Reid et al. (1990) [430], the total magnetic field anomaly for a sloping contact obeys the equation:

$$(x - x_0)\partial f/\partial x + (z - z_0)\partial f/\partial z = A.$$

Thus, this constant should be used in equation (2.18) even for isolated magnetic anomalies having the correct zero level.

Euler deconvolution consists in determining the four unknown parameters x_0, y_0, z_0 and A in a running window (of size greater than four field points) by solving a system of linear equations. Such a system consists of the equations (2.18) written out for every window point. The system is usualy tackled by methods like singular value decomposition (SVD). By this approach dispersion of unknown parameters can be also obtained.

The structural index is assigned *a priori*, using additional information on the shape of the causative bodies. Another way is to carry out calculations for several indices choosing the solution which best fits the known superficial and/or borehole structure, seismic data, etc., or which has good clustering properties (for possible way to estimate the structural index, see Slack et al. 1967; Steenland, 1968 and Barbosa et al., 1999 [468, 485])

Let us consider a synthetic example computed for a single elementary body, where the anomalous field and its derivatives contain no errors. Even is this case, not all Euler solutions cluster around the contour of the causative body in the xOy plane. That is why, in standard Euler deconvolution [505], additional criteria have to be used to select the solution. Since the depth z_0 usually has the smallest singular value (and, consequently, the highest dispersion) this is the criteria usually used (e.g.[505]). A solution can be rejected because of its low tolerance $z_0/N \cdot \sigma < TOL$ (Thompson, 1982 [505]) or because its dispersion is higher than a given value $\sigma_{\max}$, or because it is too shallow ($z < z_{\min}$) or too deep ($z > z_{\max}$).

An important characteristic is also the distance between an Euler solution and the center of the window from which the solution has been obtained. We can reject solutions that are located at a distance several times larger than the window radius (depending on the structure of the anomalous field and the size of the window, this ratio ranges between 2 and 10).

The method gives good results for a wide range of elongate bodies. Synthetic calculations for parallel dikes having a width up to 25 times less than their length with window size equal to or less than dike width, show that Euler solutions cluster well along the four side boundaries of the dikes. For this test we used parameters of similar to those used to study the magnetic field of Saint Malo region, which we will discuss later in 2.4.3: in particular, the difference between the angle between the dike stretching and the declination of the magnetization vector was as small as 5 degrees. When the window size is smaller than the dike's width, Euler solutions outline both long sides of the dikes. When the window becomes wider, solutions cluster along their central line.

Applying the Euler approach to real data, one approximates causative sources by elementary bodies of given shapes (according to the chosen structural index). Therefore, positions of Euler solutions respective to the real body depend on the shape of this body, on the non-uniform distribution of magnetization (or density), and on the ratio size of the window versus size of the body.

Let us consider the example of a rectangular prism to illustrate the role of the window size. We consider a point located outside the contour of the prism in the vicinity of one of the vertical side boundaries. Analytical descriptions of the magnetic or gravity effect of the prism can be subdivised in two groups. One describes the effect of the nearest vertical boundary. The other one deals with the opposite boundary. Suppose that the window size is several times smaller than the width of the prism. Within the window, the effect of the opposite side boundary and its derivatives are nearly constant and considerably smaller than the effect of the nearest side boundary. As a result, when the window size is small compared to the prism dimension, the Euler deconvolution provides solutions which concentrate in a vicinity of the nearest boundary. On the contrary, when the window size is larger than the prism dimension, the effect of the elementary body approximates the entire anomaly and Euler solutions cluster in the center of the prism. The image of a prism in Euler solutions thus depends on the relation between the size of the window and the prism dimension.

The practical implementation of Euler deconvolution pose additional questions. In large areas, the size of causative bodies may differ considerably. The window size can then be too small for large bodies (providing small singular values and as a consequence large dispersion of solutions) and too large for small bodies (failing to outline or even to locate them). Because of interference between the anomalies, Euler solutions cannot cluster sharply around contours of the causative bodies. This interference and the high frequency noise in data also hamper the use of criteria based on singular value or dispersion of z_0. Indeed they increase singular values and decrease the dispersion of the estimated parameters. Thus, the solution with good tolerance may in fact be ill posed.

In order to overcome these difficulties in the selection of "good" Euler solutions, we applied the clustering technique described in 1.3.1 - 1.3.3.

The results of our numerical calculations for quite sophisticated synthetic examples (one of which is discussed in 2.4.2) show that, even if Euler solutions form broad clouds, the density of these clouds is not uniform, and more dense nuclei tend to outline contours of the causative bodies. We apply the RODIN algorithm to extract these nuclei.

Magnetic Data Clustering

Definition 2.6.1.1.

The function $f(x,y,z)$ is called a homogeneous function of degree n if and only if

$$\forall t, \quad f(tx, ty, tz) = t^n f(x, y, z). \tag{2.19}$$

If we differentiate by "t" the equality (2.19) and set $t = 1$, we obtain the Euler equation

$$x\frac{\partial f}{\partial x} + y\frac{\partial f}{\partial y} + z\frac{\partial f}{\partial z} = nf \tag{2.20}$$

The basic example is:

$$f(x,y,z) = \frac{G}{\nu^N}, \quad \nu = (x^2 + y^2 + z^2)^{1/2} \tag{2.21}$$

Here $n = -N$, and G is a constant that will be interpreted as a density. The function of type (2.21) defines different kinds of magnetic sources depending on "n". Let a source be located in the point (x_0, y_0, z_0), then the full magnetic intensity ΔT of the source (x_0, y_0, z_0) at an arbitrary point (x, y, z) is :

$$\Delta T(x - x_0, y - y_0, z - z_0) = f(x - x_0, y - y_0, z - z_0)$$

$$f_{x_0, y_0, z_0}(x, y, z) = \frac{G(x_0, y_0, z_0)}{((x - x_0)^2 + (y - y_0)^2 + (z - z_0)^2)^{N/2}}. \tag{2.22}$$

If we plug (2.22) into (2.20) we obtain :

$$(x - x_0)\frac{\partial f}{\partial x} + (y - y_0)\frac{\partial F}{\partial y} + (z - z_0)\frac{\partial f}{\partial z} = -Nf. \tag{2.23}$$

We assume that the measurements of the magnetic field ΔT in the region $\mathcal{R}$ are done on the surface.

Therefore, the gradient of the field ΔT can be calculated as follows :

$$\mathrm{grad}\Delta T(x_i, y_j) = \left(\frac{\partial \Delta T}{\partial x}(x_i, y_j, 0), \frac{\partial \Delta T}{\partial y}(x_i, y_j, 0), \frac{\partial \Delta T}{\partial z}(x_i, y_j, 0)\right) \tag{2.24}$$

We move a dynamic window of dimension 3×3, with center at (x_i, y_j) along the grid. The window is defined by the following formula:

$$\mathcal{W}_{ij} = \mathcal{W}(x_i, y_j) = (x_{i+s}, y_{j+k}), \quad k, s = 0, \pm 1. \tag{2.25}$$

We know the derivatives on the window $\dfrac{\partial \Delta T}{\partial x}, \dfrac{\partial \Delta T}{\partial y}, \dfrac{\partial \Delta T}{\partial z}$, and denote them by $a_{i,j}(s, k)$, $b_{i,j}(s, k)$, $c_{i,j}(s, k)$:

$$\begin{aligned} a_{i,j}(s, k) &= \frac{\partial \Delta T}{\partial x}(x_{i+s}, y_{j+k}) \\ b_{i,j}(s, k) &= \frac{\partial \Delta T}{\partial x}(x_{i+s}, y_{j+k}) \\ c_{i,j}(s, k) &= \frac{\partial \Delta T}{\partial x}(x_{i+s}, y_{j+k}) \end{aligned} \tag{2.26}$$

Further, we shall use the following
Agreement: The field ΔT under the window $\mathcal{W}_{ij}$ satisfies condition (2.22) with $N = 1.5$.
we plug the apparent appearence (2.26) of $\dfrac{\partial f}{\partial x}, \dfrac{\partial f}{\partial y}, \dfrac{\partial f}{\partial z}$ into (2.23), obtaining a system of 9 equations with 4 variables. Each equation is indexed by a pair of (s, k) $(s, k = 0, \pm 1)$. If we substitute f_0 for G_0 (i.e. $(x_0, y_0, z_0, G_0) \to (x_0, y_0, z_0, f_0)$) and consider that $z = 0$ on the surface, we obtain the linear system

$$(x_{i+s} - x_0) a_{ij}(s, k) + (y_{j+k} - y_0) b_{ij}(s, k) - z_0 c_{ij}(s, k) = -1.5 f_0 \tag{2.27}$$

or

$$a_{ij}(s, k) x_0 + b_{ij}(s, k) y_0 + c_{ij}(s, k) z_0 - 1.5 f_0 = -a_{ij}(s, k) x_{i+s} - b_{ij}(s, k) y_{j+k}, \tag{2.28}$$

where $s, k = 0, \pm 1$.
System (2.28) can be resolved by the method of minimal squares. Due to the system excess, a set of solutions appears $(X_0, Y_0, Z_0, G_0)(i, j)$.
The center of gravity is these solutions is the desired solution.
These solutions form the first three components of the file that will be clustered in (2.20). If $x \to \bar{x}$ is the averaging operation, then

$$(x_0, y_0, z_0)(i, j) = (\bar{X}_0, \bar{Y}_0, \bar{Z}_0)(i, j). \tag{2.29}$$

The quality of the above procedure is characterized by σ_{z_0} and tl_{z_0}, the fourth and fifth components of the file studied in (2.6.2).

The value σ_{z_0} is the dispersion of the set $Z_0(i,j)$ with respect to its average value $z_0(i,j)$.

The value tl_{z_0} is defined by the relation $\dfrac{z_0}{N\sigma_{z_0}} = \dfrac{z_0}{1.5\sigma_{z_0}}$.

We can conclude that the "geographical" set of solutions $(x_0, y_0, z_0)(i,j)$, which corresponds to the union of all the windows $\mathcal{W}_{ij}$, can be clusterized by the quality (i.e., with respect to σ_z and tl_z) and/or by the density (i.e., with respect to G, the 4th component of system (2.28)).

2.4.2 Magnetic Data Clustering

The idea to classify Euler solutions for magnetic anomalies studies using a clustering technique has been formulated by A. Gvishiani in late nineties. In 2000-2001, V Mikhailov, A. Gvishiani, A. Galdeano, M. Diament, S Agayan and Sh Bogoutdinov introduced a RODIN algorithm (see 1.3) and applied it to the magnetic anomalies studies in the Gulf of Saint Malo and in French Guyana.

Herein, we search for r-clusters of Euler solutions in a synthetic case. The idea of r-clustering deals with calculating the average distance from a given point x to other points of the cluster $\mathcal{A}$. Doing that, we take into account only those points of a cluster $\mathcal{A}$, whose distance from x is less than a given radius r. Most clustering algorithms tend to construct compact clusters of isometric shapes, but the r-algorithm enables us to find clusters of complicated irregular shapes (e.g. elongate or toroidal with empty center).

Following 1.3.2, the RODIN algorithm has two free parameters : α - which controls the density of clusters and their minimal average illumination and r, which controls the shape of the clusters. Therefore, r-clustering enables us to find clusters which satisfy *a priori* information on their properties (e.g. their structure, size, position, etc.).

It is worthwhile to mention that RODIN allows us to find dense clusters using projection of the points in the xOy plane (which can be done visually) and the density of their depth distribution. Another advantage of the RODIN approach is that it separates Euler solutions originating from different causative bodies or from their parts (if the causative bodies are large enough compared to the window size). RODIN also enables us to analyse separate clusters by determining their average depth. In addition, the anomalous field points that produce particular clusters also often form dense clusters. Such field clusters outline areas where the influence of a particular causative body (or of part of it if the body is large enough) is more prominent. It enables us to analyse such field areas separately, carrying out calculations with different window sizes.

We will illustrate these statements using a synthetic example. We assume that an anomaly of the total magnetic field ΔT is caused by four rectangular prisms, whose position on the xOy plane is shown on figure 2.10 and whose parameters are listed in Table 2.3. The prisms have different sizes, depths and magnetizations. The largest prism, $\mathcal{A}$, has an induced magnetization (the total intensity of the Earth magnetic field was 50000 nT, the magnetic susceptibility 10^{-2}, for other parameters see Table 2.3). The other prisms possess remanent magnetization with different inclinations. The resulting total field anomaly (Figure 2.10) shows that prisms $\mathcal{C}$ and $\mathcal{B}$ do not produce strong anomalies. They are smaller in size and have deeper upper boundaries than prisms $\mathcal{A}$ and $\mathcal{D}$. Figure 2.10 also shows that the anomalies from prisms $\mathcal{B}$ and $\mathcal{C}$ are disturbed by the anomalies from prisms $\mathcal{A}$ and $\mathcal{D}$.

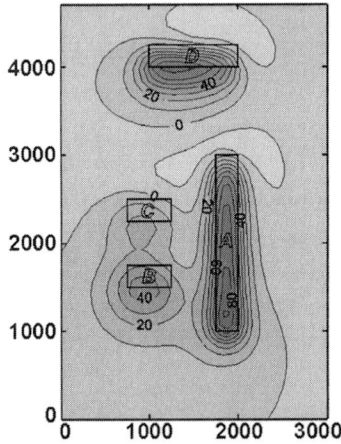

Figure 2.10: *Rodin synthetic example.* The case of the four rectangular prisms. Position on the xOy plane. Computed total field anomalies.

We computed the total field anomaly and its derivatives on a regular grid of size $6 \times 3 km^2$, with 50 points along each axis. Thus, the step Δx equals 122.4 m and Δy is $\Delta x/2$ (61.2 m). Euler solutions were calculated using running windows of size 5*5 points, assigning the structural index

Prism symbol	x_0, x_1 km	y_0, y_1 km	z_0 m	Type f magn.	$\|J\|$ A/m	I degree	D degree
$\mathcal{A}$	1.0 - 3.0	1.75 - 2.0	150	Induced	see text	70°	5°
$\mathcal{B}$	1.5 - 1.75	0.75 - 1.25	400	Rem.	1.0	30°	55°
$\mathcal{C}$	2.25 - 2.5	0.75 - 1.25	450	Rem.	1.0	50°	55°
$\mathcal{D}$	4.0 - 4.25	1.0 - 2.0	200	Rem.	0.5	50°	55°

Table 2.3: Parameters of the four prisms used for a synthetic example. Notations: (x_0, x_1)-coordinates along $0x$ axis, (y_0, y_1)-the same along Oy axis, z_0-depth to the top boundary (depth of the bottom for all the prisms was 1000 m), $\mid J \mid$-modulus of magnetization vector, I-inclination, D-declination.

$N = 3$ (point dipole). It is worth mentioning that, in comparison to smaller structural indices, the value $N = 3$ usually provides less dense clusters of points (Fairhead et al., 1994, [144]), grouping mainly at the depth between the top boundary and the center of a prism. The given grid generated $46 \times 46 = 2116$ Euler solutions which reduced to 985, by rejecting solutions with very low singular values (less than 0.01). Horizontal (xOy) locations of these solutions are shown on figure 2.11a.

For the prism, $\mathcal{A}$, having induced magnetization directed nearly to the North ($D = 5°$), Euler solutions cluster around the northern and southern vertical side boundaries. Being parallel to the direction of magnetization, eastern and western side boundaries do not cause magnetic anomalies. Euler solutions outline the southern boundary of prism $\mathcal{A}$ and are shifted to the north from its northern boundary due to the influence of prism $\mathcal{D}$. Shallow points ($z_0 < 600m$) better coincide with the northern and southern boundaries. If prism $\mathcal{D}$ was an isolated body, the calculations with the window size being two times larger than Ox prism dimension would provide elongate cluster stretching along the west-east prism axis. Here, neighboring bodies disturb this regular picture, shifting Euler solutions of prism $\mathcal{D}$ southward. Figure 2.11a also shows that the distance of shifting is different: "robust" solutions located close to the eastern and western boundaries are less shifted than "weak" solutions, which are located far from the prism axis. The distribution of the solutions in the vicinity of prisms $\mathcal{B}$ and $\mathcal{C}$ gives no way to outline these bodies.

We applied classical criteria to select these solutions. We improved the resolution in the vicinity of the southern and northern boundaries of prism $\mathcal{A}$ and the western and eastern boundaries of prism $\mathcal{D}$, but, all the points

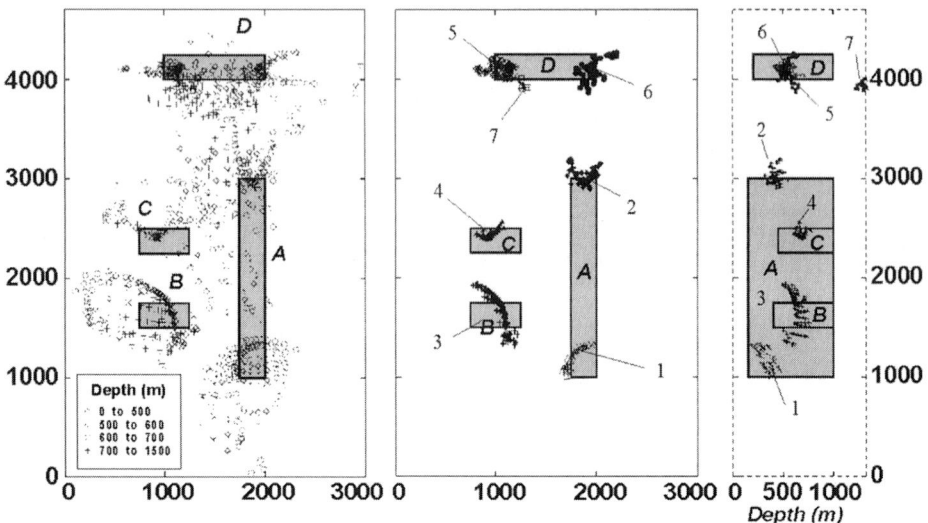

Figure 2.11: *Horizontal location of solutions.* Clusters obtained by Rodin algorithm in the case of the synthetic example. Position on the xOy plane.

around the bodies $\mathcal{B}$ and $\mathcal{C}$ were rejected because of their higher dispersion and lower tolerance, and, as a result, these bodies disappeared.

Figure 2.11b shows the result we obtained by applying the r-cluster RODIN algorithm dealing with three co-ordinates (x_0, Y_0, z_0) with $\alpha = 0.02$ and $r = 0.4$.

Seven clusters were identified (figure 2.11b). Clusters 1 and 2 clearly outline northern and southern boundaries of the body $\mathcal{A}$, while clusters 5 and 6 group densely around the eastern and western boundaries of prism $\mathcal{D}$. Cluster 4 marks prism $\mathcal{C}$ and is shifted slightly north of its center. Cluster 3 is less dense, and marks an area larger than prism $\mathcal{B}$. Nevertheless, this cluster provides some information on the position of prism $\mathcal{B}$. Cluster 7 appears to be an artifact. It exhibits low tolerance in comparison to other clusters, and we may thus reject it applying convenient criteria or taking into account that every solution of cluster 7 is situated far from the center of the windows which produced these solutions.

Figure 2.11c shows the depth distribution of the dipoles.

To construct this figure all the points were projected on the xOy plane (view from the east). One can see that the clusters provide a correct relative depth distribution of the four causative bodies. Clusters 1 and 2 clearly show that prism $\mathcal{A}$ is shallower than the other prisms. Prism $\mathcal{D}$ also appears to be shallower than the bodies $\mathcal{B}$ and $\mathcal{C}$ (clusters 3 and 4 respectively).

As we mentionned above, cluster analysis enables to separete solutions linked with different bodies, providing estimates of the average depth of each cluster. Clusters 5 and 6 have an average depth $z_{av} = 320$ and 420 m, that is between the depth to the top of the center of prism $\mathcal{A}$ ($z_1 = 150m$ and $z_c = 575m$ respectively).

Clusters 5 and 6 show that prism $\mathcal{D}$ is deeper than prism $\mathcal{A}$ ($z_{av} = 523m$ and $545m$, when $z_1 = 200m$ and $z_c = 600m$ respectively). The depth of the clusters 5 and 6 are noticeably larger. This is due to the fact that, when the prism size is comparable to the size of the running window, Euler solutions cluster closer to the prism center.

2.4.3 Euler solutions by clusterization in the gulf of Saint Malo

We applied the method described in 2.4.1 - 2.4.2 to a fragment of the aeromagnetic map for the Armorican Massif, France (Galdeano et al., 2000, [166]). The area chosen includes the eastern part of the gulf of St-Malo and the inland area south of it. Figure 2.12 shows the region's main tectonic units (Chantraine et al., 2000, [96]).

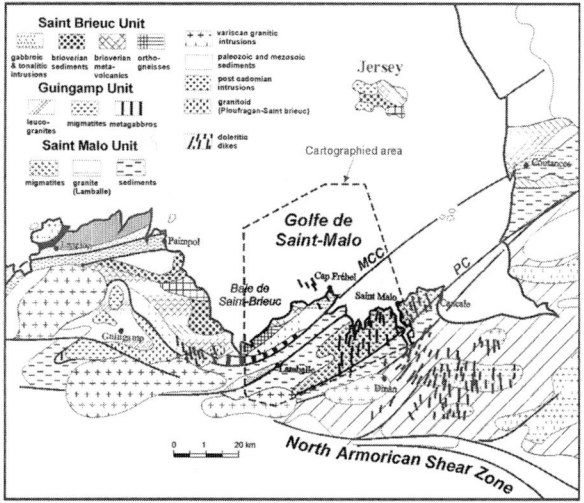

Figure 2.12: *Tectonic units of the gulf of St Malo.* Area of the gulf of Saint Malo under study.

The main tectonic structures are the Guingamp and St Malo units. They form a high temperature belt bounded by the Main Cadomian contact (labeled as MCC on Figure 2.12) to the north and by the Cancale-Plouer fault (PC on Figure 2.12) to the south (Chantraine et al, 1988, Brun and Balé, 1990, [?, 77]). To the southeast of these units is the Fougères unit. This unit is bound in the south by the North Armorican shear zone (NASZ)- an Upper Palaeozoic dextral strike-slip fault exhibiting a total shear displacement of about 30 km (Bitri et al., 1997, [61]). Thrusting at the MCC and CP fault was accompanied by left lateral movements. According to seismic data [61], the MCC, CP and NASZ faults deepen to the north. According to the interpretation of Bitri et al. (1997) [61] all north dipping reflectors in the region are covered at depth by a slightly inclined surface, which could correspond to a zone of crustal detachment.

The evolution of this region had been mainly controlled by subduction during closure of the Celtic ocean. The Guinguamp-St Malo high temperature belt was formed 600-540 Ma ago during a late Pre-Cambrian tectonic event named the Cadomian orogeny, which resulted in obduction of a back arc basin over a continental margin (Balé and Brun, 1989, [36]). Later this belt was moderately affected by hercynian deformations and metamorphism. The belt is composed of micaschists, paragneisses and migmatites. In the St Malo unit, migmatites occupy the core of an asymmetric metamorphic dome (Brun and Balé, 1990) on which anatectic granite have been dated as 541±5 Ma (Peucat, 1986) [77].

A distinctive feature of the geological structure of the region is a swarm of doleritic dikes. The dikes were intruded in the Precambrian basement and were later cut off and metamorphosed by Hercynian granites (Vidal, 1980, [526]). According to Lahaye et al., (1995) [293] St Malo dikes are typical continental tholeites, formed as a result of magma mixing, generated by partial melting of a heterogeneous mantle. Emplacement of dikes corresponds to an extensional (pull-apart) tectonic phase. Paleomagnetic and K/Ar data give an age of the dikes as 330±10 Ma i.e. of the Lower Carboniferous age (Perroud et al., 1986, [407]). Onshore dikes are thin (5 m on average) and their stretching ranges from N-S to NNE-SSW. Linear magnetic anomalies in this area are considerably wider, indicating that the dikes probably fan out from wider deep magnetic bodies.

Aeromagnetic measurements were carried out at 350 m flight altitude with small distance between profiles (Galdeano et al., 2000, [166]), enabling a grid interpolation of 250.250 m in both directions. This map demonstrates a good correlation with known geological structures: the long wavelength anomalies in the Saint-malo region are mainly associated with dikes that are exposed onshore and continue offshore. In the Saint Malo region, magnetic anomalies stretch from N(NE) to S(SW) where as in the Gulf of Saint Malo they change direction to NNW-SSE. These anomalies exhibit quasi-linear structure. Their relationship with the Guinguamp-St Malo belt is not clear, since the structure of the magnetic anomalies close to their intersection is complicated. The main purpose of the application of Euler deconvolution and cluster analysis was to investigate the structure of quasi-linear magnetic anomalies onshore and offshore and their relationship with the Guinguamp-St Malo high temperature belt.

Figure 2.13a shows the total magnetic anomaly for a small part of the region using shaded relief and isolines, as well as Euler solutions obtained with a window size of 9×9 points (2×2 km). About 40 % of the Euler solutions were immediately rejected because of their extremely small singular values. We first selected Euler solutions using different "standard" criteria. Solutions with low tolerance, situated at depth lerger than 2km and located at a distance five times larger than the window size from the center of the window were rejected. Results are shown on figure 2.13b. In comparison with the previous figure the Euler solutions now outline isometric and linear bodies. However, the position of possible causative sources remains unclear.

Figure 2.13c shows the result of clusterization. The RODIN algirithm was applied to the original set of Euler solutions. The choosen free parameters values were $\alpha = 0.8$ and $r = 0.3$. The result obtained is stable in the sense of possible changes of these parameters for α between 0.78 and 0.82 and for r between 0.25 and 0.35. The algorithm found dense clusters which more clearly outline possible causative sources. Indeed, isometric clustering is clear in the northern part of the map, as well as linear clusters stretching from SW to NE in the lower half of the map marking the Cadomian high temperature belt.

Figure 2.14 presents the results of clustering Euler solutions for the St Malo region with the same parameters as in figure 2.13c.

The initial set of Euler solutions included 34 500 points. The clustering algorithm rejected almost 7 000 points, and found dense clusters which outline isometric and linear regional structures. In particular, isometric clusters outline contours of possible causative bodies in the central parts of isometric anomalies marked on the figure 2.14 by the numbers 1, 2 and 3. The linear cluster 4 in the center of the map can be considered as the castward continuation of the Main Cadomian fault (MCC on figure 2.12), since tracing the eastward prolongation of the Cancale-Plouer (CP) fault is problematic.

The distinctive feature of clustering Euler solutions is linear clusters stretching N-S in the southern part of the map and NW-SE in its northern part.

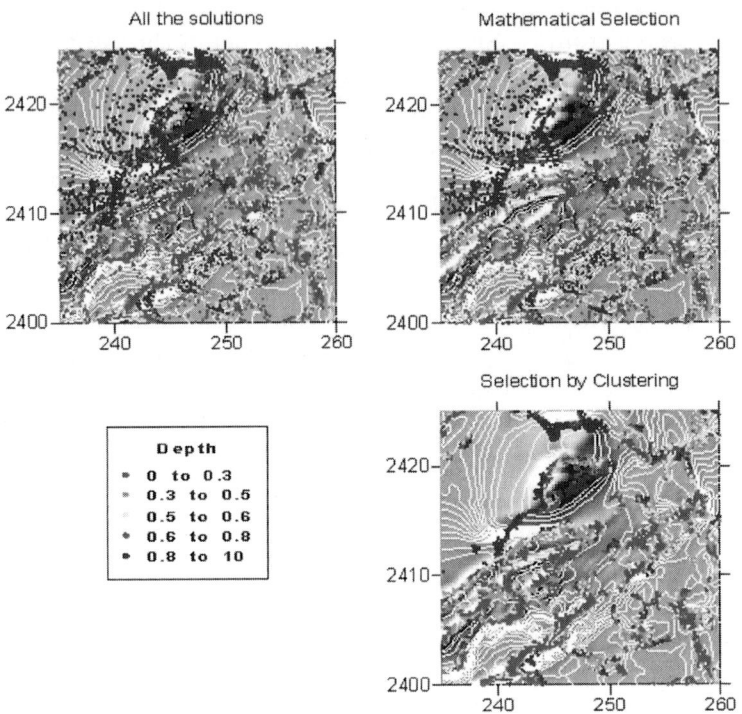

Figure 2.13: *Magnetic anomaly and Euler solutions in St Malo area.* a) All the solutions b) Solutions selected using "standard" criteria c) Solutions selected by Rodin clustering algorithm.

Inland, they coincide with the stretching of doleritic dikes. Clustering solutions in the southern part of the map shows that besides N-S trending dikes, there is probably another dike swarm stretching approximately N-S or NW-SE. Using Euler solutions, the latter dikes can be followed northward to their intersection with the Cadomian belt. Further north they are correlated with an offshore dike swarm in the NW-SE direction. The structure at the intersection of the dikes with the belt is not very clear, but linear clusters crossing the belt in N-S direction show that the dikes are younger than the belt. Comparing figure 2.13b with figure 2.13c shows that, on real data, our clustering technique appears to be more efficient than "standard" approaches for selection of Euler solutions.

Results of Euler deconvolution strongly depend on the quality of the solution selection. The clustering technique based on topological analysis of

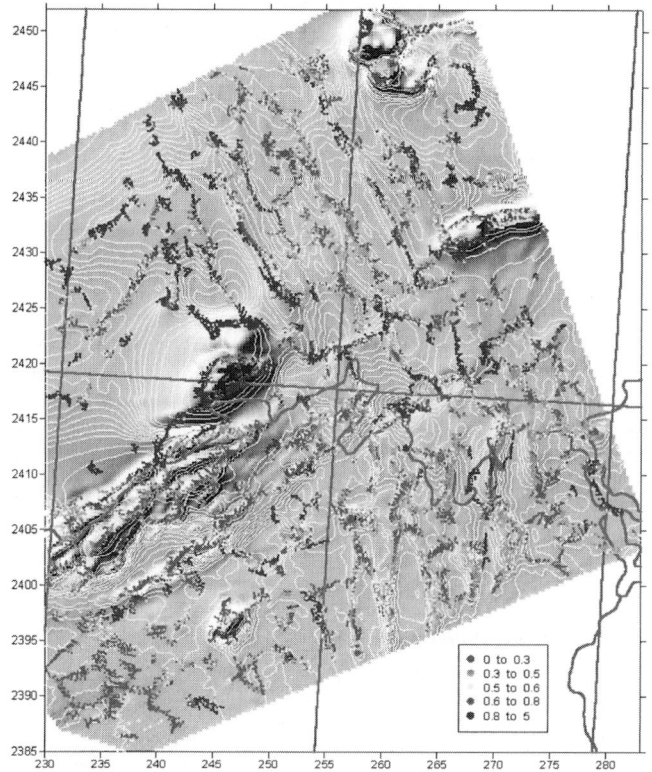

Figure 2.14: *Euler solutions clustering.* The solutions selected using Rodin clustering algorithm. In the bassin of the gulf the system of dikes turns to North-West.

the solution concentration efficiently extends methods routinely used for this purpose. The synthetic calculations (in section 2.4.2) for sophisticated causative sources distributions showed that, when Euler solutions do not group densely around the contours of the bodies, the density of their distribution appears to be higher in the vicinity of the causative bodies. That is why the application of clustering technique is so promising.

The RODIN algorithm is based on a formal definition of clustering, which enables us to construct an effective clustering algorithm (2.4.1). Thus, we proceed from informal to formal selection of Euler solutions, an important step toward automatization of the selection procedure.

On both synthetic and real Euler solutions data, the RODIN clustering technique appeared to successfully outline causative bodies. There are two other advantages of this clustering method application:

1. The found clusters provide solutions associated with particular bodies or with their parts, allowing separate analysis of different Euler solution clusters. This allows us, for example, to compute average parameters for individual causative bodies.

2. Data points of anomalous fields, which produce different clusters, usually also form rather dense clusters. Thus, the clustering technique enables us to outline areas where the influence of different causative sources is more prominent, in allowing us to reinterpret data for these areas using different window size, structural indices and so on.

2.5 Linear and Circular Clustering of Bathymetric Data in the Wharton Basin

The Wharton Basin is situated in the north-eastern part of the Indian Ocean, east of the Ninety-east Ridge. It is limited at the east by the Investigator aseismic Ridge. An area of high seismic activity, this region was investigated by multibeam bathymetry, gravity,magnetic and six-channel seismic reflection profiling data in December 1995 (Deplus et al., 1998 [119]). These data have shown active deformations in the central part of the Wharton basin, what support the interpretation of this area as a diffuse plate boundary between India and Australia (Wiens et al., 1985, Gordon et al., 1990, De Mets et al. 1994 [118, 192, 539]).

The lithosphere underlying the Wharton Basin is between 45 and 85 M.y. old. It was formed at the roughly east-west mid-oceanic Wharton Ridge on which accretion stopped ca. 45 My (Liu et al., 1983; Patriat and Ségoufin, 1988; Royer and Sandwell, 1989 [310, 401, 441]), leaving a fossil axis left-laterally offset by fossil transform faults. The principal bathymetric structures in the Wharton Basin are east-west features coinciding with the segments of the extinct ridge and north-south features linked to the old fracture zones. Most of these topographic features are buried under the sedimentary cover of the Nicobar fan.

In November-December 1995, the laboratory of gravimetry and geodynamics of the Paris Institute of Earth Physics carried out a marine geophysical survey in the Wharton Basin. During the Samudra cruise onboard the French R/V L'Atalante, multibeam bathymetry, sea-floor imagery, 3.5 kHz acoustics, gravity, magnetic, and six channel seismic reflection profiling data were gathered to investigate the Wharton fossil spreading ridge and possible present interplate deformations. Two segments of the Wharton fossil ridge axis and associated transforms were surveyed. Close to the three surveyed fracture zones, multibeam bathymetry and imagery revealed active deformation [119].

The methods of differential operations in the GIS environment (derivatives) described in (1.4.1) were applied to Wharton Basin bathymetry data[2]. The goal of the application was to test various processing techniques in order to develop an automated procedure for recognition of circular and linear structures. This approach can be expanded to analyse circle and linear structures in many other types of geophysical data, such as topography, gravity and magnetics.

A multibeam bathymetry data set selected for this study described a rectangular region located between latitudes $-4°20'$ and $-5°50'$, and longitudes $91°50'$ E and $93°10'$ E, (155 km × 174 km). The measurements were stored in the nodes of a uniform two-dimensional grid with a 100 m increment. The initial image is presented in figure (2.15).

To estimate the change in bathymetry, the first partial derivatives of the input image were calculated by deriving two images, which represent the gradients in the x and the y directions, respectively (Figure 2.16).

To find edge pixels in the input image gradient, we applied the Laplacian of Gaussian (LoG) and Canny edge detectors.

The response of the gradient edge detector was estimated using Roberts, Sobel and Cubic Spline operators (Figure 2.17, a). Edge detection was performed by thresholding the gradient magnitude image (Figure 2.17, b). The final determination of the edges would be more complex, requiring additional post-processing thinning and thresholding stages.

Edge pixels defined by the LoG edge detector usually produce closed contours. However, when applied to the bathymetry data the connectivity of edge pixels was rather poor (Figure 2.18). Also, the use of non-directional derivatives meant that the edge response was always measured parallel or perpendicular to an edge, reducing the signal-to-noise ratio. Finally, the LoG edge detector gave no indication of edge direction, which is important for circular and linear structure recognition.

The reponse of the Canny edge detector is determined by three parameters: the size, (*i.e.* the approximate standard deviation) σ, of the Gaussian kernel

---

[2]This study has been implemented in the framework of the collaboration program between Institute of Physics of the Earth in Paris (IPGP) and Schmidt united Institute of Physics of the Earth in Moscow by A. Beriozko, A. Gvishiani (USPE), M. Diament, C. Deplus and H. Hébert.

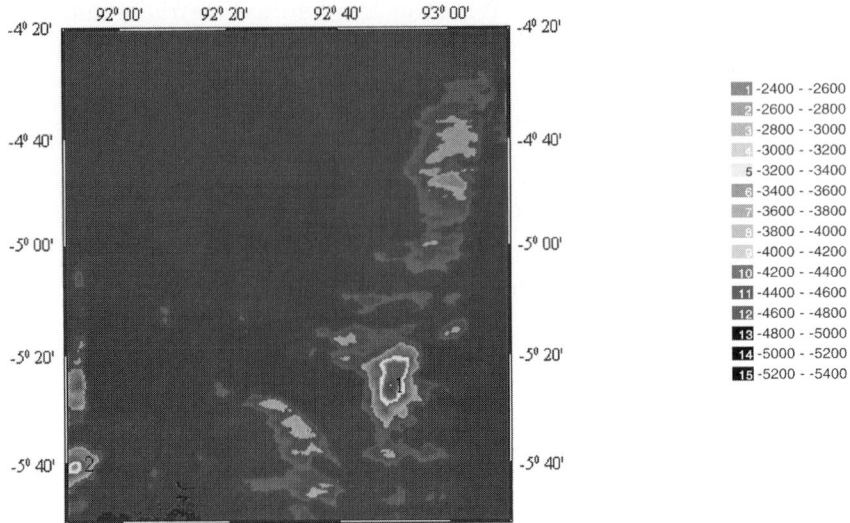

Figure 2.15: *A bathymetry image.* The Wharton Basin Region. Initial data set.

used in the smoothing stage, and the upper and the lower gradient magnitude thresholds, $T1$ and $T2$, used by the tracker. The gaussian smoothing in the Canny edge detector fulfills two purposes: to control the amount of detail appearing in the edge image and to suppress noise. Increasing the size of the Gaussian kernel reduces the detector's sensitivity to noise, at the expense of loosing some finer details and increasing the localization error of the detected edges. Setting the upper threshold too low increases the number of spurious and undesirable edge fragments appearing in the output. Setting the lower threshold too high splits noisy edges into fragments. We obtained the best result using a Gaussian kernel with standard deviation $\sigma = 1.0$, $T1 = 0.2$ and $T2 = 0.02$ (Figure 2.19, a). Most of the major edges were detected and many details were picked out, though this would be too much detail for subsequent processing. This result can be compared with the edge image obtained using a Gaussian kernel of the larger size with $\sigma = 1.5$, $T1$ lowered to 0.1 and $T2$ increased to 0.05 (Figure 2.19, b). The edges are more broken up and many details are no longer detected.

In order to improve the connectivity of the detected edge pixels and to produce the descriptions of edges as lines in the edge linking phase, we used several images derived from the edge detection phase. This took full advan-

Bathymetry in Wharton Basin

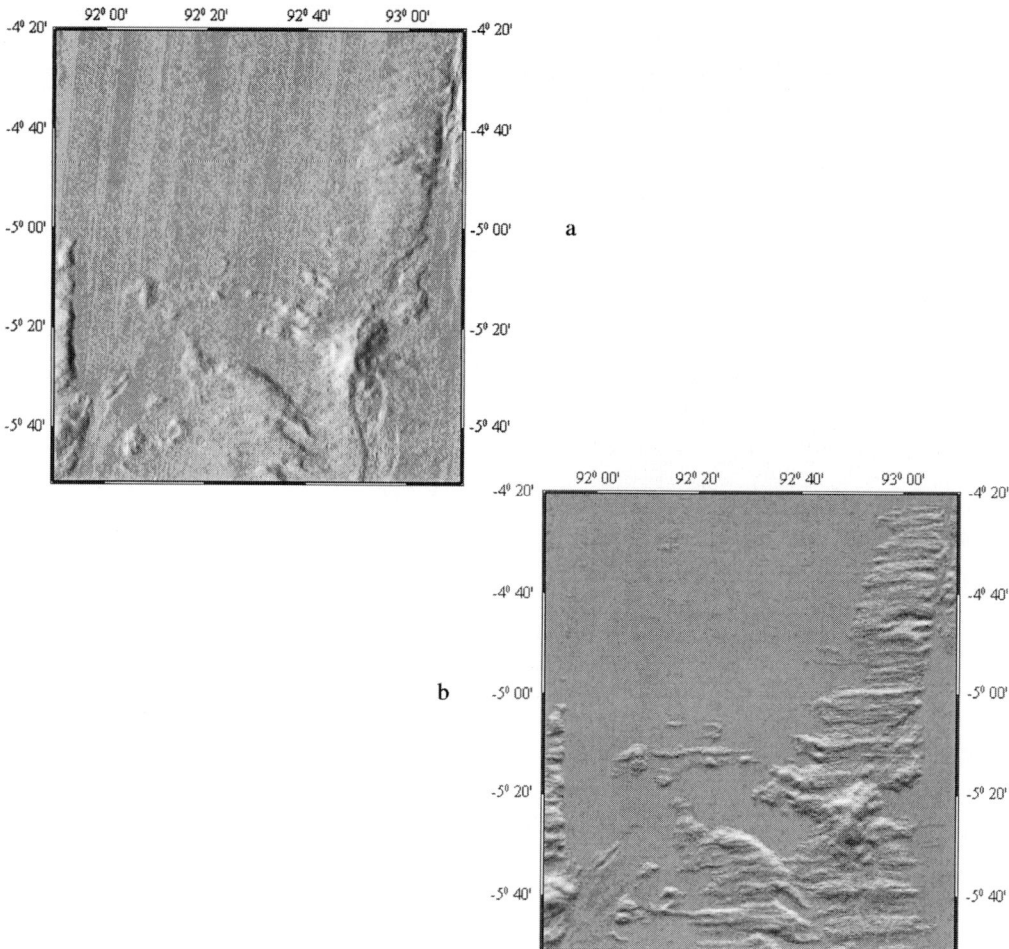

Figure 2.16: *The first partial derivatives.* a) The x gradient image; b) the y gradient image.

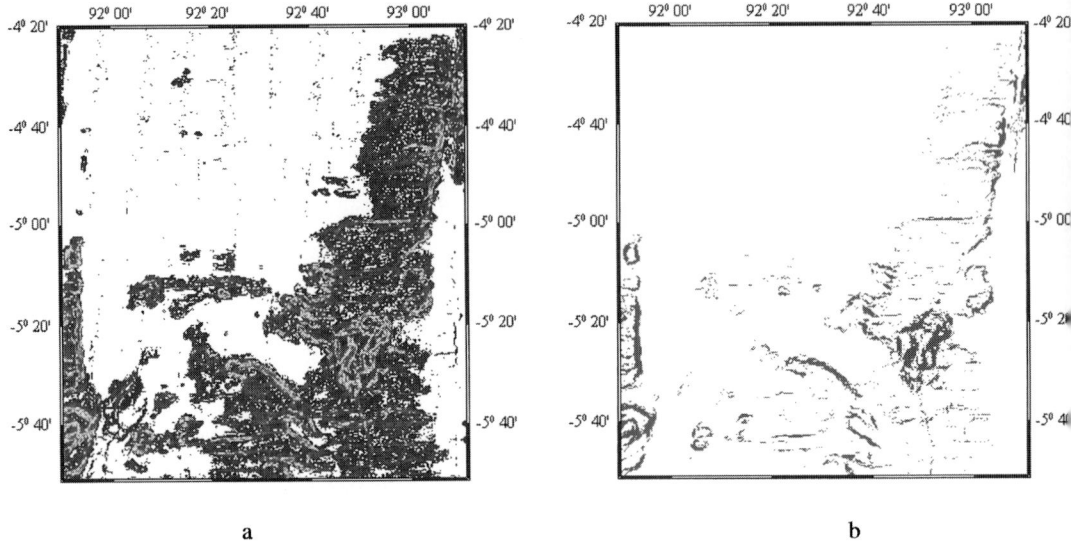

Figure 2.17: *Gradient edge detector.* a) The gradient magnitude image, Sobel operator; b) the thresholded gradient magnitude image, $T = 0.2$.

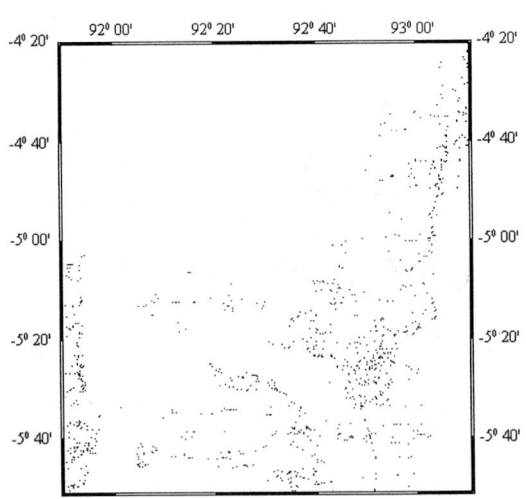

Figure 2.18: *LoG edge detector.* Zero crossings in the LoG image, $\sigma = 1.0$.

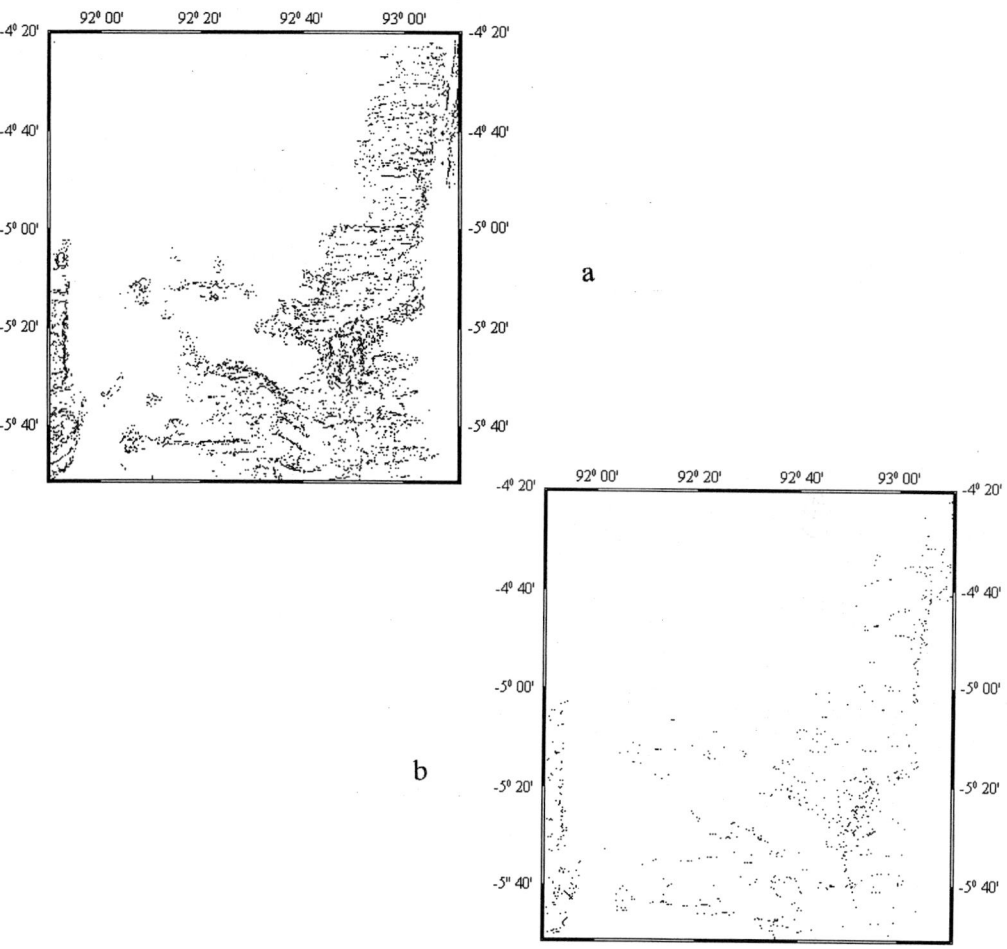

Figure 2.19: *Canny edge detector.* a) $\sigma = 1.0, T1 = 0.2, T2 = 0.02$ b) $\sigma = 1.5, T1 = 0.1, T2 = 0.05$.

tage of the used GIS environment ATTILA [58] which can analyse multiple data layers.

We first considered the information about the local edge gradient magnitude $g(x, y)$ and the direction $\theta(x, y)$ at edge pixels determined by a gradient edge detector. Edge linking started at an edge pixel with a high enough gradient magnitude value and pixels in its local neighbourhood were analysed for similarity of edge directions. A gradient magnitude threshold T was set up to identify which edges were strong enough to track. Edge pixels in the neighbourhood with similar gradient directions and gradient magnitude values above the threshold were determined to lay on the same edge.

Then, the information from the LoG and Canny edge detectors was considered. If edge pixels in a local neighbourhood satisfied the edge direction similarity constraints and had a response on both the LoG and/or Canny edge detectors in the same pixel positions then they were added to the edge set. The connected edges were described in ATTILA vector format, which presents them as separate or overlaid lines on raster images (Figure 2.20). The results suggested that this context-based edge linking process could give cleaner output than the simpler and more common one-level thresholding based on gradient magnitude.

The results of applications deriving bathymetry data in the GIS environment can be summarized as follows:

Gradient, Laplacian of Gaussian (LoG) and Canny edge detectors were applied to find edge pixels in the bathymetry image of the Wharton Basin region.

The capability of the GIS ATTILA to operate with multiple data was used for edge linking. A number of edge images were simultaneously considered in order to improve the connectivity of the detected edge pixels and to produce the description of edges as lines and circular structures. The results suggest that this context-based edge linking process could give cleaner and clearer output than the more usual one level thresholding based on gradient magnitude.

GIS-based data processing allowed the identification of sediment-covered and basalt ("crustal") areas of the Wharton Basin (Figure 2.20). The Styx volcano was also well traced. The first partial derivative images distinguished the two main directions of crustal structures in the Wharton Basin; especially the West-East direction which was not clearly observed in the initial bathymetry data (Figure 2.16).

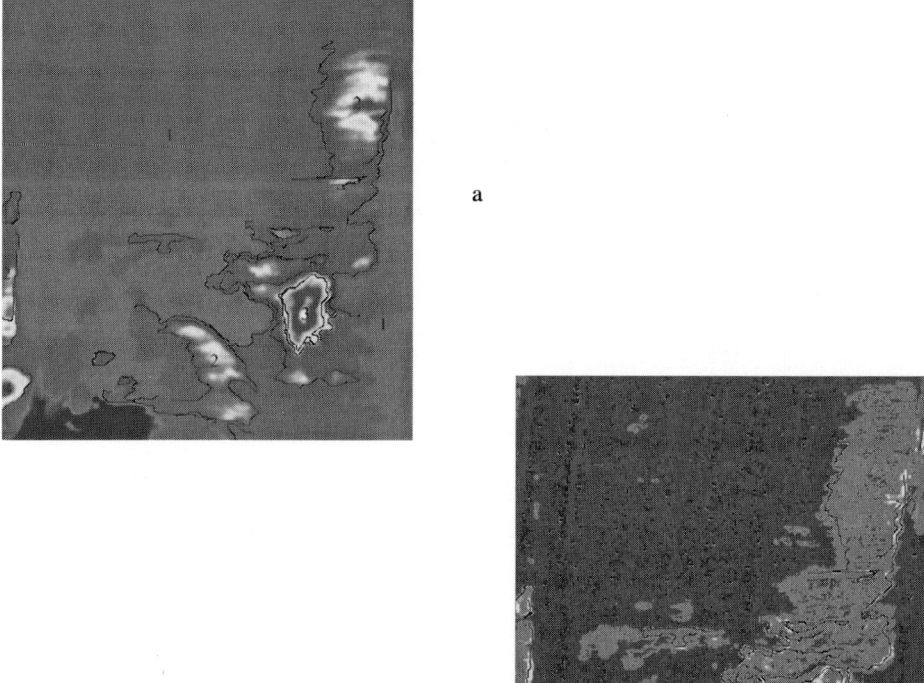

Figure 2.20: *Vector edges.* $T = 0.01$, overlayed on: a) the bathymetry image; b) the gradient magnitude image; the red spot in the South-Eastern part of the image is the Styx volcano.

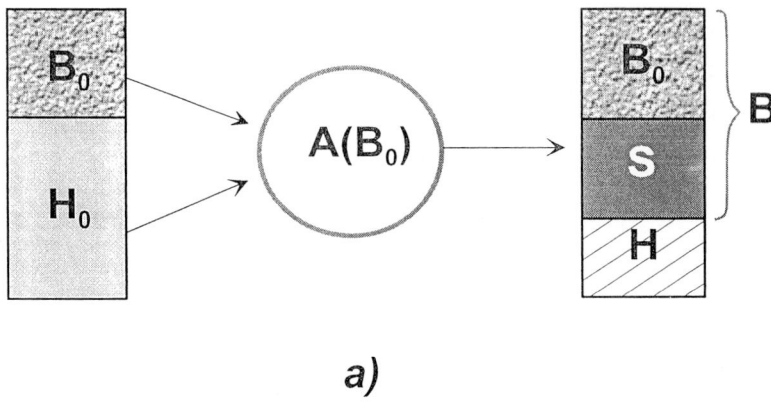

a)

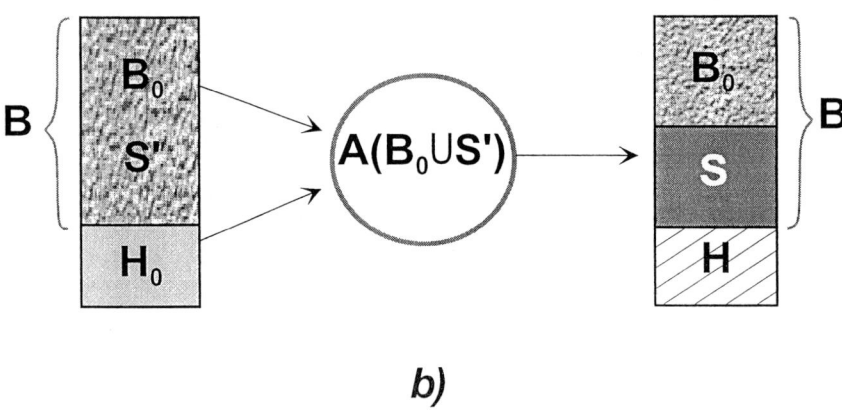

b)

Stability condition for dynamic pattern recognition problem - $\mathcal{A}(\mathcal{B}_0)$- a VSF classification algorithm with learning material a) $\mathcal{W} = \mathcal{B}_0 \sqcup \mathcal{H}_0$- initial decomposition (learning material) b) $\mathcal{S}'$ any subset $\mathcal{S}' \subseteq \mathcal{S}$, $\mathcal{W} = [\mathcal{B}_0 \cup \mathcal{S}] \sqcup \mathcal{H} = \mathcal{B} \sqcup \mathcal{H}$ - final classification (prediction).

Chapter 3

Recognition of Earthquake-Prone Areas and Seismic Hazard Assessment

3.1 Earthquake-prone Areas in the Western Alps

These results are part of the French-Russian project of investigations into earthquake-prone areas in the three regions of moderate seismicity: Alps, Pyrénées and Caucasus. The results of this section were originally published by A. Cisternas, A. Gvishiani et al. (1985) [100].
The topic was the focus of a French-Soviet project developed from 1983-1986 between the Institutes of Earth Physics in Paris, Strasbourg and Moscow. Later on, in 2000-2001, the clustering techniques described in chapter 1, was applied to cluster the solutions previously obtained dynamic pattern recognition algorithms.
The goal of the study was to establish zones in the Western Alps where earthquakes with magnitude $M > 5.0$ may occur. Two independent classification techniques were used to study the problem. The first method defined the objects of classification as segments on active faults in the Alps. The "Expert Communication" pattern recognition algorithm designed by J. Sallantin [100] was applied to classify these objects. The second method dealt with objects established using a scheme of lineaments of the Western Alps obtained by morphostructural zoning (Alexeevskaya et al., 1977), [4]. "Voting by a set of features" algorithms were used to classify these objects (see Dubois and Gvishiani, 1998, [134]).

The methods of evaluating the classification reliability were also different, including changes in the set of parameters, algorithms, free parameters in the algorithms, and comparisons of classifications obtained with real and random learning materials.

The two independent approaches gave similar results. The joint interpretation of the results allowed the classification of areas that can be considered as prone to earthquakes with magnitudes larger than 5.0 (see [172, 175, 205]).

Objects of recognition

Neotectonic scheme Here we describe the first method of selection of the recognition objects: $w \in \mathcal{W}$. The basis from which to select the objects is a tectonic-oriented map of the main faults of the Western Alps compiled from various geological and tectonic maps [21, 482, 483, 404] as well as neotectonic data [153, 530] and the geodynamic situation [409, 410]. The map was constructed by C. Weber, P. Godefroy and M. Lambert (BRGM, France) (1983) [410]. The character of the most important faults has been well defined by recent neotectonic studies in the field by H. Philip [410]. Thrust and shear-type faults dominate.

Figure 3.1 shows the neotectonic scheme of the Western Alps, which includes all large faults within 20 km of historical earthquakes with estimated magnitude greater than 5.

The scheme shows major NE-SW strike-slip faults that traverse the external crystalline massifs (Gothard, Aar, Mont Blanc, Belledonne) such as the Cévennes faults, the Durance fault, the Daluis-Guillaumes and the Mont Borel faults, NW-SE faults like the Vuache fault and the Briançonnais faults and reverse thrusts such as Clarée fault, Serenne fault, Ruburent fault, Bersezio fault. The overthrusting fronts of Subalpine chains and the contact between the Po plain and the internal zones are also included.

The lineaments presented on figure 3.1 constitute the main features of the recent and present tectonics of the region in connection with our problem [186]. For example, the Durance fault is responsable for several historical and recorded earthquakes. The "ophiolitic belt", corresponding to the suture between European and Adriatic plates, is an intensevly tectonized zone comprising faults that traverse the whole crust.

The belt, schematically shown as a continuous thick line, is consistent with a steep Moho slope separating two domains with different crustal structure, as deduced from seismic reflection profiles [243]. Some faults were detected by seismic reflection methods and confirmed by other techniques. They have been added to the scheme even though their recent activity is not

Earthquake-prone Areas in the Western Alps

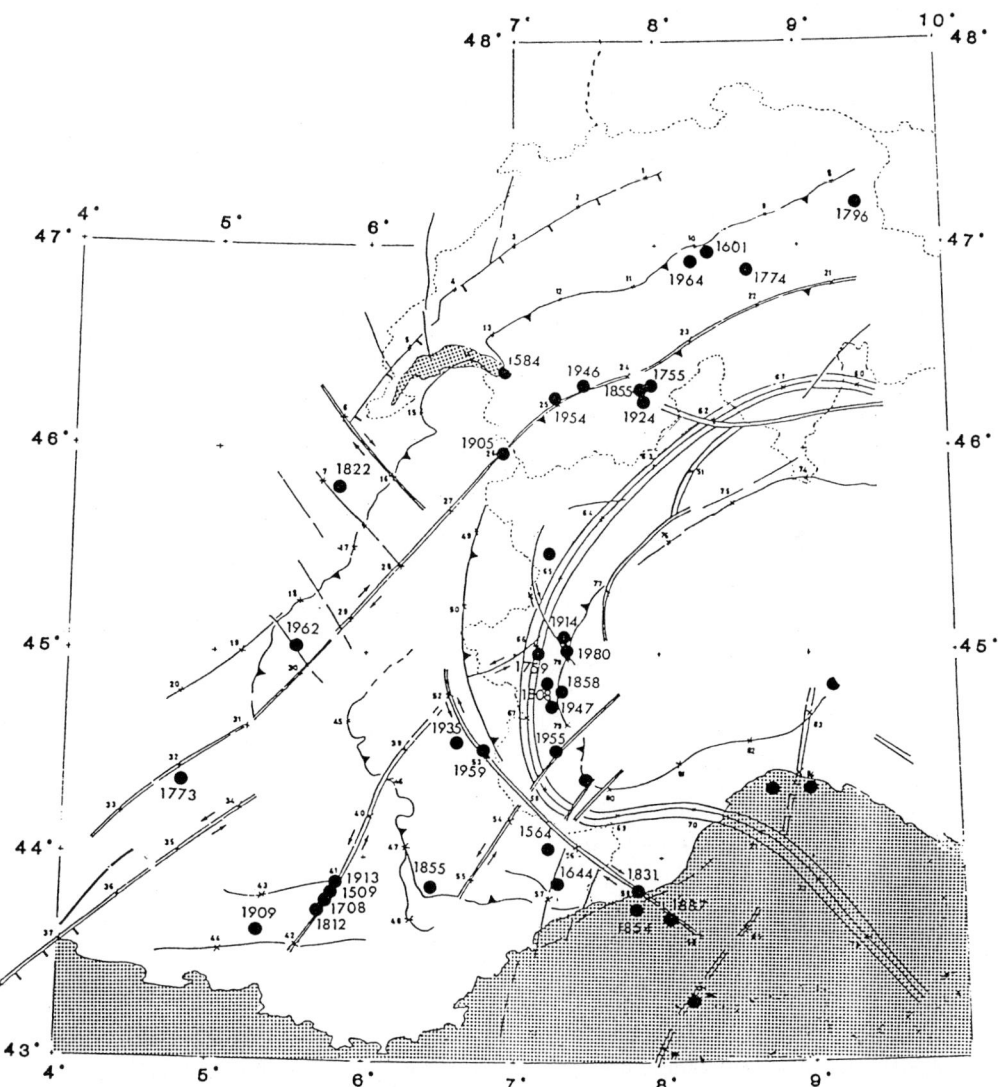

Figure 3.1: *Neotectonic scheme of the Western Alps* This scheme shows epicenters and active faults. Arrows indicate shear, dark triangles correspond to overthrusting and short segments to normal faulting. The ophiolitic belt is shown as a triple line. Small numbers give the order of more than 80 objects. Data is taken from the BRGM seismotectonic files (Cisternas et al., 1985).

firmly established due to the absence of surface breaks. However, one of them is close to a known major epicenter: Lambesc, 1909. Finally, some normal faulting is associated with Pliocene and Quaternary extension at the periphery of the region to the north and to the south-west.

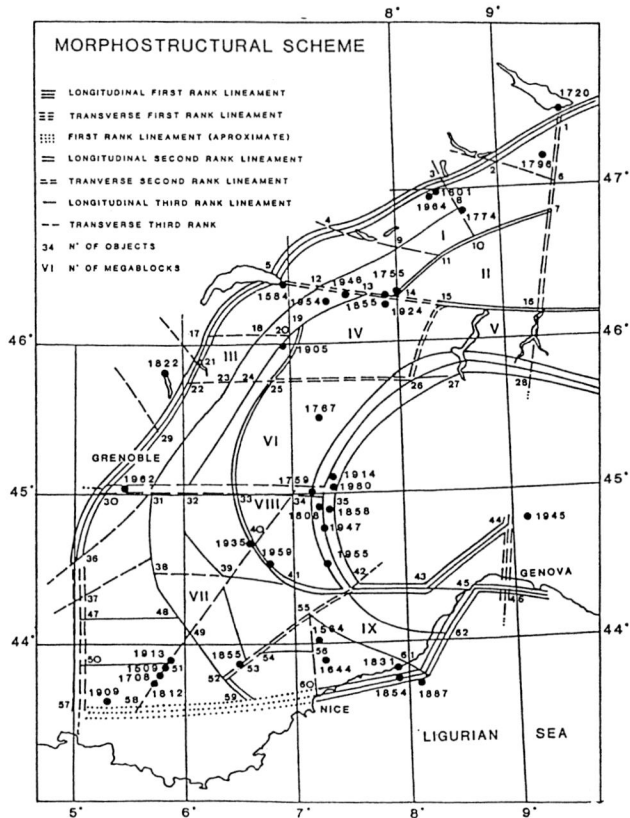

Figure 3.2: *Morphostructural zoning scheme of the Western Alps.* Morphostructural scheme showing the epicenters and the lineaments with their ranks. 62 objects are defined as intersections of lineaments. Megablocks are indicated by roman numbers (Cisternas *et al.*, 1985).

Different criteria were used to select objects: $w \in \mathcal{W}$. No rank was assigned to the different faults. Objects of recognition were selected along the linear faults at intervals of 40 km. Each object was defined as a 50 km long segment along the fault. If the center of an object was near an intersection with another fault then the objects were defined to be the union of segments. In this way 87 objects were defined, but 7 of them that lay in the sea were rejected because of insufficient information. The objects were described by a set of parameters (see Dubois and Gvishiani, 1998, [134]). In some cases the parameter value was computed using the information from within a circle 25 km in radius, centered at the middle point of the segment defining the object. Even though the parameters are usually continuous, it is possible to discretize them and, then, give them a binary form. In other words, all parameters $\mathcal{X}:\mathcal{W} \to \Re$ are interpreted as binary parameters $\mathcal{X}:\mathcal{W} \to \{0,1\}$.

Morphostructural zoning scheme The second method used to define objects is morphostructural zoning [4, 426]. A result of morphostructural zoning is a scheme of ranked lineaments. The scheme of lineaments of the Western Alps (figure 3.2) was obtained by joint analysis of topographic, geological, and tectonic maps, satellite images, and publications [16, 99, 314, 444]. It was constructed by Rantsman and Gorshkov (1988) with important contributions by C. Weber, A. Cisternas and M. Philip [208, 100].

The reader can find a detailed description of morphostructural zoning techniques in [1]. The scheme consists of two stages. First, the areas considered as neotectonics entities are singled out by their average features. Then, at the contact between these areas, the zones of lineaments are outlined by another set of features. The areas are divided into three ranks : mountain countries (I-st rank), megablocks (II-nd rank) and blocks (III-d rank). The rank of a lineament as a boundary between the blocks is the highest rank of the areas which it divides. Two types of lineaments are distinguished: longitudinal and transverse. Peculiar structures (morphostructural knots) are formed at the intersection of lineaments. The knots are determined by the intersections, but the establishment of their boundaries requires field studies.

The first-rank lineaments separate the organic belt of the Alps from the European platform, the Po depression and the deep marine basin of the Ligurian sea.

Second-rank lineaments are the boundaries between megablocks. The area of study was subdivided into 9 megablocks. Each of them has its own dominant direction of linear structures, typical for its ridges and valleys. The megablocks differ in their deep structure as well. Second-rank transverse lineaments were established in the places of sharp bending in major mophostructure [445, 355].

Third-rank lineaments are the boundaries of blocks distinguished by topographic altitude and ridge orientation. These lineaments correspond to steep edges in the surface basement (figure 3.2).

The comparison with epicenter distributions shows that all the epicenters are located in the vincinity of lineament intersections. This observation allows us to consider the problem of determinating earthquake-prone areas in the Western Alps by dynamic pattern recognition techniques introduced by Dubois and Gvishiani, 1998.
Territories inside circles of 25 km radius with centers at the lineament intersections are considered as objects of recognition $w \in \mathcal{W}$. All the objects were described by a list of parameters connected with the regional seismicity. The parameter values were measured using topographic, geological, and gravity maps as well as the lineament scheme itself.

Classification Algorithms

The classification algorithms used in this study divide binary vectors representing the objects $w \in \mathcal{W}$ into two classes, using information from the learning material $\mathcal{O}$:

$$\mathcal{A}(\mathcal{O}, \mathcal{P}) : \mathcal{W} = \mathcal{B} \bigsqcup \mathcal{H}. \tag{3.1}$$

In this study, the classification is done in two different ways. The first one, the "Expert Communication process" (E.C.), consists of constructing a topology. The learning material is used to define the "set of dangerous objects" and new "dangerous objects" are obtained by "assimilation" to that set. The EC algorithm was introduced by J. Sallantin specifically to study earthquake-prone areas in the Western Alps. It should be emphasized that the EC algorithm does not belong to the family of VSF algorithms introduced in Dubois and Gvishiani (1998) [134], which is why we describe it in detail below.
The second algorithm used for independent classification is CORA, which belongs to the family of VSF algorithms. It is described in detail in [134] We denote:

- $\mathcal{W}$ to be the finite set of objects of recognition $w \in \mathcal{W}$.

- $\mathcal{L} = (\mathcal{D}_0, \mathcal{N}_0)$ to be the learning material. $\mathcal{D}_0, \mathcal{N}_0 \subset \mathcal{W}$, $\mathcal{D}_0$ corresponds to dangerous and $\mathcal{N}_0$ to non-dangerous objects.

- $\Pi : \mathcal{W} = \mathcal{D} \bigsqcup \mathcal{N}$ to be a classifiction of $\mathcal{W}$ with an algorithm Π i.e. $\Pi : \mathcal{W} = \mathcal{D} \bigsqcup \mathcal{N}$ to be a division of W into two subsets $\mathcal{D}$ and $\mathcal{N}$ such that $\mathcal{D} \bigcup \mathcal{N} = \mathcal{W}$ and $\mathcal{D} \bigcap \mathcal{N} = \emptyset$.

- $\Omega_p = \{0,1\}^p$ to be the binary space of p dimensions.

- $d: \mathcal{W} \to \Omega_p$ to be the binary description of objects, for each $w \in \mathcal{W}$ $d(w) = \omega = (\omega_1, \cdots, \omega_i, \cdots, \omega_p)$. $\Lambda = d(\mathcal{W}) \subset \Omega_p$.
- $|\mathcal{X}|$ to be the number of objects in the finite set $\mathcal{X}$.

Process and stability of communication between experts (E.C.) In the E.C. method $\mathcal{L} = (\mathcal{D}_0, \emptyset) \equiv \mathcal{D}_0$. That is we only use learning material from the dangerous class.

To build the E.C. process we first need to define a pattern recognition operator that we will call an expert Exp_i, which depends on a set of parameters $\mathcal{U}$. This operator either transforms a point of Ω_p into another, or it doesn't give an answer.

$$Exp_i: \Omega_p \times \mathcal{U} \to \{\Omega_p, \emptyset\}.$$

Let us define the parameter set $\mathcal{U}: u \in \mathcal{U}$ and $u = (\mathcal{D}_0, a, b, \Delta)$ where a, b, Δ are thresholds to be defined below.

Let us consider the following partition of the learning set $\mathcal{D}_0$,:

$$\mathcal{D}_{01}^i = \{w \in \mathcal{D}_0 : \omega_i = 1\}$$
$$\mathcal{D}_{00}^i = \{w \in \mathcal{D}_0 : \omega_i = 0\}.$$

The operator Exp_i constructs a system of logical rules to discriminate $\mathcal{D}_{01}^i$ from $\mathcal{D}_0^i$. These rules satisfy three conditions that allow only those parameters that contain the most information:

- A) Each rule must individually verify some conditions related to their information measure (parameter a). Rules that are too general are eliminated (low information content)
- B) A system of rules should also verify some minimal information criteria (parameter b).
    - i) The maximum number of rules is less than a given number.
    - ii) The sum of the information measure of the rules is higher than a given threshold.
- C) Decisions are made by voting using the learned rules which estimate for an object of $\mathcal{L}$ whether the set of rules is in favour of $\omega_i = 1$ or of $\omega_i = 0$; Δ is the threshold if the minimal information necessary to make a decision.

    Three outcomes are possible:

- i) a logical justification:
$$Exp_i(\omega; u) = \omega$$

- ii) a logical contestation:
$$Exp_i(\omega; u) = (\omega_1, \cdots, \omega_{i-1}, 1 - \omega_i, \omega_{i+1}, \cdots, \omega_p).$$

- iii) a logical silence:
$$Exp_i(\omega; u) = \emptyset.$$

The parameter u is selected in such a way that:
$$Exp_i(\mathcal{D}_0; u) \subset \mathcal{D}_0.$$

Exp_i verifies:
if $y = Exp_i(x; u)$ then $Exp_i(y; u) = y$.

Communication between experts

Let us consider the set of experts $\{Exp_i, i \in \mathcal{I}\}$.
We can build a recursive mathematical filter with ω in its kernel:

$$\mathcal{V}_1(\omega, u) = \{\omega\} \cup \bigcup_{i \in \mathcal{I}} Exp_i(\omega, u) \tag{3.2}$$

$\mathcal{V}_1$ contains ω and all the points generated by the experts.
It is clear that by considering the inductive process in the same way we will obtain the basic system of neighbourhoods of $w \in \mathcal{W}$ which generates a topology

$$\mathcal{V}^p = \{\omega\} \cup \left(\bigcup_{i \in \mathcal{I}} Exp_i(\mathcal{V}^{p-1}, u) \right). \tag{3.3}$$

In this study we consider only $\mathcal{V}_1(\omega)$. Let us denote by $\mathcal{C}_1(\omega)$ the number of experts $Exp_i(i \in \mathcal{I})$ which contest ω and by $\mathcal{J}_1(\omega)$ the number of experts that justify ω. We would like to have many experts justifying and few contesting ω. Let us introduce two thresholds α_1, α_2. If we want to define a classification from $EC(\mathcal{D}_0, \alpha_1, \alpha_2, a, b, \Delta)$:

$$\mathcal{D} \equiv gen(\mathcal{D}_0 = \{w \in \mathcal{W} : (\mathcal{J}_1(\omega) \geq \alpha_1 \wedge \mathcal{C}_1(\omega) \leq \alpha_2)\}. \tag{3.4}$$

We emphasize that only $\mathcal{D}$ was established by the EC algorithm and that we have to build a subset $\mathcal{N}$ which is a consistent logical negation of $\mathcal{D}$: $\mathcal{N} = \mathcal{W} \setminus \mathcal{D}$.

Logical stability of the communication between experts

Now that we have described the E.C. method, we analyse its stability. Different types of logical justification are possible

- a) Logical consistency of "$\mathcal{D}_0$ as a concept of danger". The set $gen\dot{g}en(\mathcal{D}_0)$ should be rather similar to $gen(\mathcal{D}_0)$.

- b) No contradiction of the decision (contraposition principle)

  Let us consider a set $\mathcal{D}_1$ of undecided objects such that:

$$\mathcal{D}_1 = \{\omega \in \mathcal{L}: \mathcal{J}_1(\omega) \leq \alpha_3\} \tag{3.5}$$

  and $\mathcal{D}_1 \cap \mathcal{D}_0 = \emptyset$.

  To study the non-contradiction of the decision we compare:

$$gen(\mathcal{D}_0) \quad \text{and} \quad gen(\mathcal{D}_1)$$
$$gen(gen(\mathcal{D}_0)) \quad \text{and} \quad gen(\mathcal{D}_1)$$

  $\exists w \in gen(\mathcal{D}_0) \cap gen(\mathcal{D}_1)$ then there exists a logical contradiction for the decision (non contradiction principle).

  If $w \in gen(\mathcal{D}_0 \cup gen(\mathcal{D}_\infty)$ the logical contradiction corresponds to a non verification of the excluded middle principle;

- c) Coherency of the generalization Let us consider the set of new points obtained from the generalization:

$$\mathcal{D}_3 = \{w \in \mathcal{W}: w \in gen(\mathcal{D}_0) \text{and} w \in \mathcal{D}_0\}, \tag{3.6}$$

  $gen(\mathcal{D}_3) \cap \mathcal{D}_0$ gives the objects of the learning set which can be learned from the extension of the generalization. Thus the small number $|\mathcal{D}_0 \setminus (gen(\mathcal{D}_3) \cap \mathcal{D}_0)|$ indicates that the classification has good quality.

CORA algorithm

CORA is the other algorithm which we used for alternative, independent recognition of earthquake-prone areas in the western Alps. This algorithm is a member of the VSF family, introduced by the authors in the first volume of this monograph [134]. That volume gives [134] a detailed description of the CORA algorithm, as its correspondence to other algorithms of the VSF family and associated control experiments.

However, for the reader's convenience we give herein a short schematic description of CORA as it was applied to the Western Alps.

We define a feature as a matrix

$$\tau = \begin{pmatrix} i_1 & i_2 & i_3 \\ \alpha_1 & \alpha_2 & \alpha_3 \end{pmatrix} \qquad (3.7)$$

where $i_r = 1, \ldots, n;\ r = 1, 2, 3;\ \alpha_r = 0$ or $\alpha_r = 1$.

An object $\omega \in \mathcal{W}$ possesses the feature τ, if $\omega_{i_r} = \alpha_r$ for its description $d(\omega) = (\omega_1, \cdots, \omega_p)$ in the binary space Ω_p. Let $\mathcal{D}_0$ and $\mathcal{N}_0$ be the subsets of *a priori* dangerous and non-dangerous objects (learning material), $Y = \mathcal{D}_0$ or $\mathcal{N}_0$. We denote by $\mathcal{K}_Y(\tau)$ the number of objects $\omega \in Y$, which possesses the feature τ and $\mathcal{K}_\mathcal{D}$, $\widetilde{\mathcal{K}_\mathcal{D}}$, $\mathcal{K}_\mathcal{N}$, $\widetilde{\mathcal{K}_\mathcal{N}}$ be some fixed free integers. A feature τ is considered to be a characteristic $\mathcal{D}$-feature if $\mathcal{K}_{\mathcal{D}_0}(\tau) \geq \mathcal{K}_\mathcal{D}$ and $\mathcal{K}_{\mathcal{N}_0}(\tau) \leq \widetilde{\mathcal{K}_\mathcal{D}}$.

It is a characteristic $\mathcal{N}$-feature if $\mathcal{K}_{\mathcal{N}_0}(\tau) \geq \mathcal{K}_\mathcal{N}$ and $\mathcal{K}_{\mathcal{D}_0}(\tau) \leq \widetilde{\mathcal{K}_\mathcal{N}}$.

Let $X = \mathcal{D}$ if $Y = \mathcal{D}_0$ and $X = \mathcal{N}$ if $Y = \mathcal{N}_0$. Two characteristic X-features τ and ν are equivalent if they appear on the same objects from Y, and τ is subordinate to ν if every object $\omega \in Y$ which possesses τ also possesses ν. We denote $\mathcal{R}_{\mathcal{K}_X, \widetilde{\mathcal{K}_X}}(Y)$ as the maximal subset of the set of all characteristic X-features, that does not have equivalent or subordinate features.

For every $\omega \in \mathcal{W}$ we define $\Delta(\mathcal{W} = \mathcal{N}_\mathcal{D}(\mathcal{W}) - \mathcal{N}_\mathcal{N}(\mathcal{W})$, where $\mathcal{N}_X(\mathcal{W})$ is the number of features from $\mathcal{R}_{\mathcal{K}_X, \widetilde{\mathcal{K}_X}}(Y)$ which appear on the object ω.

Let δ be an additional free parameter. The CORA algorithm gives the classification $\mathcal{W} = \mathcal{D} \bigcup \mathcal{N}$, defined as:

$$\mathcal{D} = \{\omega \in \mathcal{W}: \Delta(\omega) \geq \delta\},\ \mathcal{N} = \{\omega \in \mathcal{W}: \Delta(\omega) < \delta\}.$$

Control experiments for CORA algorithm

Generally speaking, the result of the recognition depends on which of the equivalent feature groups take part in the voting. This dependence can be evaluated by voting with all available equivalent features using the formulas:

$$\overline{\mathcal{N}_\mathcal{D}}(\omega) = \sum_{i=1}^{p_1} \frac{n_i}{m_i},\ \overline{\mathcal{N}_\mathcal{N}}(\omega) = \sum_{i=1}^{p_2} \frac{\nu_i}{k_i}$$

where $\overline{\mathcal{N}_\mathcal{D}}(\omega)$, $\overline{\mathcal{N}_\mathcal{N}}(\omega)$ are the votings of the object, p_1, p_2 are the numbers of selected characteristic features for the $\mathcal{D}$ and $\mathcal{N}$ classes, m_i, k_i are the numbers of features in the group of these equivalent to the i-th feature, and

Earthquake-prone Areas in the Western Alps

n_i, ν_i are the numbers of features belonging to the object from the respective group of equivalent features.

A large group of control experiments follows from the application of stability theory (see [134]) for dynamic pattern recognition problems to the case of CORA algorithm.

One such experiment is called "seismic future". It justifies the possibility of refinding the classification $\mathcal{W} = \mathcal{D} \sqcup \mathcal{N}$ using new learning material $\mathcal{D}'_0 = \mathcal{D}$, $\mathcal{N}'_0 = \mathcal{N}$.

Another type of control experiment deals with randomized data. One can find the formal description of these experiment in [134]. This group of experiments is based on comparing the quality of classifications obtained using real and random learning material, taking into account that the latter is randomized among the real objects of recognition.

Seismicity

Epicenters Because the Alps-Provence region has only moderate seismicity we found only 40 epicenters corresponding to historical or instrumentally recorded earthquakes with magnitude greater than 5 or intensity higher than VII (see figures 3.1 and 3.2).

This catalog corresponds to a synthesis made by the "Bureau de Recherches Géologiques et Minières (BRGM)", that includes previous regional catalogs, in particular that of Vogt (1979) [529] for France and that of the ENEL (Caputo, 1981) [88].

Special care was taken to include in our list only those epicenters for which the coordinates were known to within less than 50 km, even for historical earthquakes [529]. The selected epicenters cover the region of study fairly well (see figure 3.2). The list of coordinates and magnitudes (or intensities) of these events is given in table 3.1. These are crustal shocks with depth less than 30 km that are very likely associated to surface faulting.

The maximum magnitude is 6.1 and corresponds to the 1946 shock with epicenter near the Pennic frontal thrust. The maximum estimated intensity is X and was observed during the 1564 earthquake in the Nice region and also the 1887 Ligure earthquake. The Lambesc 1909 earthquake had maximum intensity IX.

Selection of the learning set .

- a) *Selection of the learning set for the neotectonic scheme*

Date	Intensity	Magnitude	Coordinates		remark
20 apr. 1796	VIII-IX		47°12'	9°25'	Macro
13 mar. 1964	VII	5.3	46°55'	8°16'	Instr.
18 sept. 1601	IX		46°50'	8°20'	Macro
12 sept. 1774	VIII		46°50'	8°30'	Macro
19 may 1954	VII	5.1	46°16'	7°17'	Instr.
25 jan. 1946	VIII	6.1	46°19'	7°30'	Instr.
15 apr. 1924	VII	5.1	46°15'	7°55'	Instr.
25 jul. 1855	IX		46°10'	7°50'	Macro
09 dec. 1755	VIII-IX		46°10'	7°50'	Macro
11 mar. 1584	VIII		46°20'	6°50'	Macro
18 feb. 1822	VIII		45°40'	5°40'	Macro
29 apr. 1905	VIII		45°50'	6°50'	Macro
25 apr. 1962	VII-VIII	4.8	45°00'	5°30'	Macro
05 jan. 1980	VII	5.3	45°01'	7°24'	Instr.
26 oct. 1914	VII	5.2	45°05'	7°20'	Instr.
26 may 1767	VIII		45°30'	7°10'	Macro
30 mar. 1759	IX		45°00'	7°10'	Macro
02 apr. 1808	VIII-IX		44°50'	7°10'	Macro
23 jan. 1773	VII-VIII		44°20'	4°40'	Macro
19 mar. 1935	VII-VIII		44°30'	6°30'	Macro
05 apr. 1959	VII-VIII	5.5	44°32'	6°47'	Instr.
20 jul. 1564	X		44°00'	7°10'	Macro
20 jun. 1955	VII	5.0	44°32'	7°18'	Instr.
17 feb. 1947	VII-VIII	5.5	44°45'	7°16'	Instr.
01 nov. 1858	VIII		44°50'	7°20'	Macro
11 jun. 1909	IX		43°30'	5°10'	Macro
14 may. 1913	VII-VIII		43°50'	5°50'	Macro
13 dec. 1509	VIII-IX		43°50'	5°40'	Macro
14 aug. 1708	VIII-IX		43°40'	5°40'	Macro
20 mar. 1812	VIII		43°40'	5°40'	Macro
15 feb. 1644	IX		43°50'	7°10'	Macro
26 may. 1831	IX		43°50'	7°50'	Macro
29 dec. 1854	IX		43°40'	7°50'	Macro
23 feb. 1887	X		43°40'	8°00'	Macro
12 dec. 1855	VIII		43°50'	6°20'	Macro

Table 3.1: List of epicenters with magnitude larger or equal to 5 or maximum intensity larger or equal to VII in the western Alps. Historical and instrumentally recorded earthquakes are included (after Cisternas et al., 1985).

Earthquake-prone Areas in the Western Alps

The set of *a priori* dangerous objects for the neotectonic scheme was chosen as follows: for each object we measured the distance of the nearby epicenters to their projections on the object. If this distance was smaller than 25 km we incorporated the object in the learning set. Epicenters may thus be related to several segments. We obtained a set of 27 dangerous objects.

- b) *Selection of the learning set for the morphostructural scheme*

Instrumentally recorded epicenters (for 1900-1980) with $M > 5.$ are situated near the following objects: NN 3, 12, 13, 14, 20, 30, 31, 35, 40, 41, 42, 44, 51, 57.

These objects are referred as *a priori* dangerous and they form the learning set for the "CORA-algorithm". The epicenters are sometimes located in the middle of two objects (for example the 1962 epicenter between N 30 and N 31). In such case both objects were included in the learning set. Historical epicenters prior to 1900 with intensity higher than VII are located near objects 1, 5, 6, 8, 53, 56 and 61. These objects were excluded from the learning set and were used as a control on the recognition result. Objects NN 18 and 19 were also used as controls, as the 1905 epicenter is located in their vicinity. The whole set of objects was thus divided into three classes : a) 14 objects of the learning set (or the first class); b) 37 objects of the second class; c) 11 objects of the third class for control.

3.1.1 E.C. application to the neotectonic scheme

Our objects are divided at the beginning into two classes : dangerous and unknown (Figure 3.1). We seek for a final clessification in which the unknown objects are either assimilated to the dangerous and non dangerous classes or remain unknown. For this purpose we use the E.C. method described above. As we have seen E.C. works by learning some assimilation procedure from a learning set that corresponds to only one class (the "dangerous" one in our case).

The descriptors we considered are related to tectonics, geology, seismicity, gravimetry and topography. They may be grouped into two classes : those related to the geometric characteristics of sources and those associated with the characteristics of recent deformations (vertical movements, stress field, etc...).

Seismicity parameters are only occasionnally used; in any case they do not appear to bias the classification. The descriptors are either quantitative or qualitative, while E.C. algorithm needs binary data. For qualitative data the conversion is easy. For quantitative data we define a number of thresholds and relate each threshold to a binary descriptor. We obtain a rather large and redundant set of binary descriptors. Then, in a selection stage, we retain a subset $\mathcal{S}$ of the original descriptions which contains only one or two binary descriptors for each qualitative descriptor and which does not contain unsignificant descriptors (this is done by using the different collective behaviours of the descriptors over the set of dangerous and unknown objects). In the case where there are still too many binary descriptors left relative to the size of the learning set, we may divide the set of descriptors to work separately with each part or, alternatively, to make another selection using the knowledge about the correlation between descriptors.

In the latter case we build several groups of correlated descriptors, then we may choose one descriptor in each group and obtain a set of poorly correlated descriptors (decorrelation). In the former case, on the contrary, we may keep all the descriptors and divide them into several subsets, each of them constituted of several clusters (correlation).

By applying the E.C. method to the neotectonic scheme we obtained, after the selection step, 32 binary descriptors, we clustered these descriptors and devided the set of descriptors into three overlapping subsets (16 descriptors each) according to the clustering order. Each subset contained several clusters of correlated descriptors. The procedure used on them is the following: for a given learning set we obtain three independent outputs of the E.C. and we compare for each object the total number of justifying experts and the total number of contesting experts (these numbers constitute the vote).

The classification proposed here is related to the well known Jack-knife test: from the basic learning set $\mathcal{D}_0$ (27 objects), one is put aside at a time, reducing the learning set to 26 objects voted by the E.C.. This gives us 27 learning sets and therefore 27 votes. The final vote is obtained by averaging the number of justifying and opposing experts.

Let x and y be respectively the numbers of experts justifying and opposing. The straight lines with a slope-1 ($y+x = b$) represent "boundaries of silence": beneath these boundaries less than b experts have spoken. In a similar way the straight lines $y = C_1$ represent contestation boundaries and $x = C_2$ represent justification boundaries. Several decision rules may be used, for example, in working with lineaments: we assimilate an object to $\mathcal{D}_0$ if $y > C_1$

Earthquake-prone Areas in the Western Alps 137

and $x > C_2$. Another possibility is to assimilate the object to $\mathcal{D}_0$ if it is outside some silent zone and some contestation zone. Here we merely use the difference between the number of objects justifying and the number of objects opposing: $x - y > d$ (straight lines with slope 1). Four such lines can be established on the map and we obtain five corresponding decisions:

1. $x - y > 20$ strong assimilation to $\mathcal{D}_0$
2. $20 \geq x - y > 14$ weak assimilation to $\mathcal{D}_0$
3. $14 \geq x - y > 10$ uncertain
4. $10 \geq x - y > 4$ weak exclusion
5. $x - y < 4$ strong exclusion.

The objects with decisions 1 and 2 are considered as dangerous.
Results of the application of E.C. to neotectonic scheme are shown in figure 3.3. Among the 27 objects of $\mathcal{D}_0$, two are not seen as dangerous in this experiment ($n°32$ and $n°56$).
The first one is related to a rather weak historical epicenter whose magnitude may be less than 5. The second one is the Nice earthquake of 1564.
We will see in the paragraph about the logical tests that our results show poor logical confidence in the southern Alps. Among the 53 objects of $\mathcal{D}_0$, 15 are recognized as dangerous ans 28 are rejected (weakly or strongly).

3.1.2 CORA algorithm classification results

As a result of applying the CORA algorithm to the morphostructural scheme of the Western Alps, 11 characteristic features of the class $\mathcal{D}$ and 8 characteristic features of the class $\mathcal{N}$ were obtained. The following parameters are included in these features:

- $\mathcal{Q}$-the percentage of soft (quaternary) deposits;

- n_i-the number of lineaments forming the intersection (objects of recognition);

- N_i-the number of lineaments in the circle of $R = 25$ km centered in the object;

- ρ_1-the distance to the nearest I-st rank lineament in km;

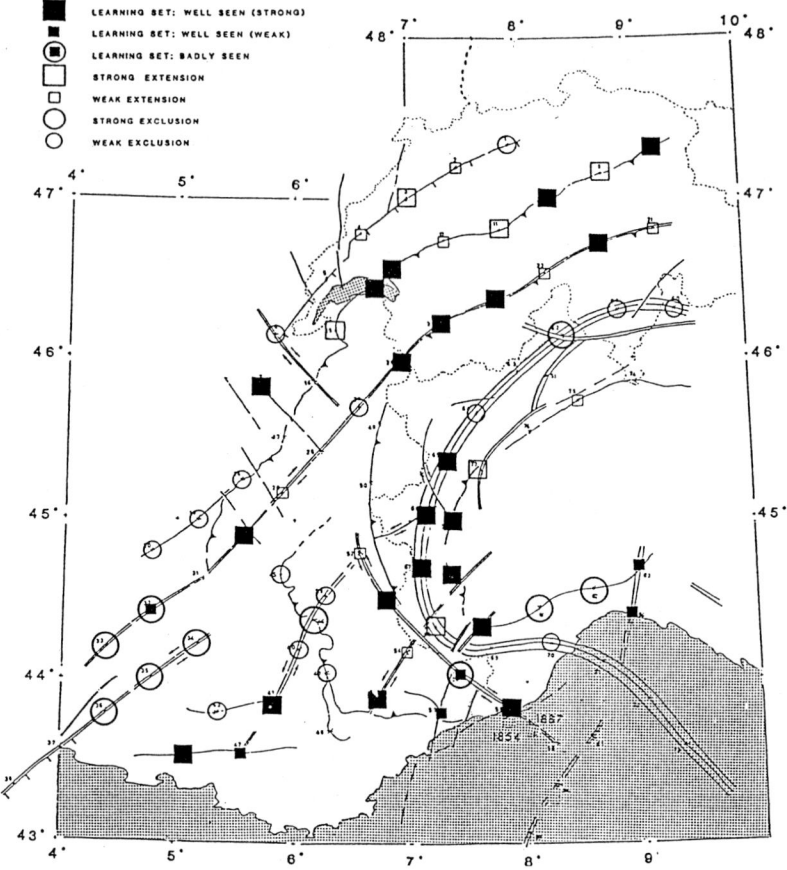

Figure 3.3: *The second and third rank lineaments.* Results of "Experts Communication process" application to neotectonic scheme of western Alps

- ρ_2-the distance to the nearest II-nd rank lineament in km;

- $\Delta B = (B_{\max} - B_{\min})$, where $B_{\max}$ ($B_{\min}$) is the maximum (minimum) value of the Bouguer anomaly in mGal;

- $(\nabla B)^{-1}$-the minimum distance between two Bouguer isolines spaced by 10 mGal, in km/mGal.

Let $\Delta(\omega) = N_{\mathcal{D}}(\omega) - N_{\mathcal{N}}(\omega) \geq 0$, where n is the number of votes for the class $\mathcal{D}$ and v the number of votes for the class $\mathcal{N}$. The objects for which $n - v = \Delta(\omega) \geq 0$ are referred to as the set $\mathcal{D}$. The results are presented in figure 3.4 in such a way that there are 34 objects in $\mathcal{D}$ and 28 objects in the set $\mathcal{N}$.

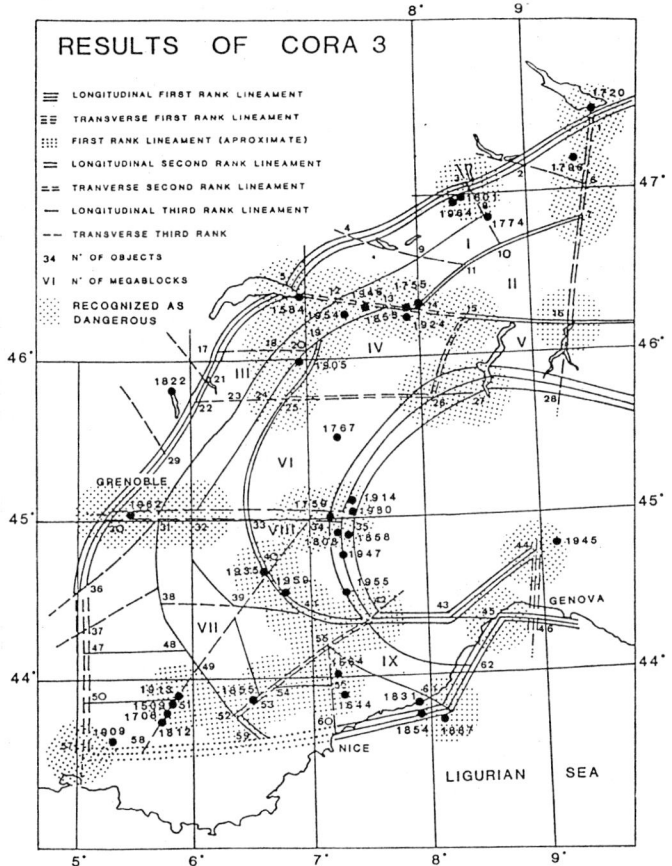

Figure 3.4: CORA *algorithm application*. The results of CORA algorithm application to the scheme of morphostructural zoning of western Alps

The learning sets $\mathcal{D}_0$ and $\mathcal{N}_0$ have respectively 14 and 36 objects and hence 12 objects were examined. All 14 objects from $\mathcal{D}_0$, 11 objects from $\mathcal{N}_0$ and 9 objects examined were referred to $\mathcal{D}$. Therefore, the classification obtained does not contradict the suggestion that the percentage of $\mathcal{D}$-objects in the set $\mathcal{N}_0$ is rather small.

The stability of the obtained classification was estimated by performing new classifications with successive exclusion of one of the above-listed parameters. Each of these classifications differs from the main result by no more than 15% of the objects.

Furthermore, the exclusion of the parameter $\mathcal{Q}$ doesn't change the result at all. The biggest differences (5 to 9 objects) take place when parameters n_1, $(\nabla B)^{-1}$ and ΔB are excluded. This fact shows the essential role of gravity parameters in the problem.

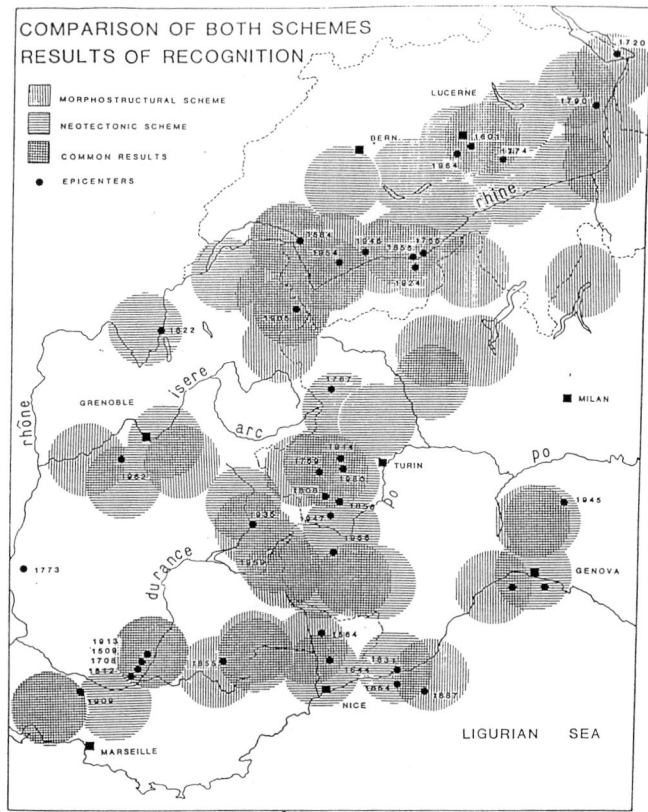

Figure 3.5: *Comparison of the results obtained by E.C. and* CORA *algorithm.* The largest differences are found to the SE of the city of Bern. The coincidence of non-dangerous areas (white area) should also be considered as an indicator of the success of common recognition.

By applying the CORA algorithm to the (figure 3.4) formal morphostructural scheme, 34 objects out of 62 were recognized as dangerous. The recognized objects from several extensive dangerous zones along transverse lineaments of the second rank (lineaments (1-16), (5-16), (30-35) and (42-52)). All the known epicenters (table 3.1) are located within dangerous objects.

The most dangerous objects (22 objects out of 34) are located in the intramountain part of the western Alps. The 18 intramountain objects have lineaments of the second rank, established in the places of sharp bending of major morphostructures. The importance of the second rank lineaments for seismicity are emphasized by the dangerous classification.

The seismicity features selected by "CORA-algorithm" consist of three parameter types:

1. Parameters related to the lineaments scheme: n_{l_i}, N_{l_i}, ρ_I, ρ_{II}. These parameters characterize tectonic parcelling level for the study area.

2. Parameters connected with Bouguer gravity anomalies: ΔB and $(\nabla B)^{-1}$. These parameters indirectly characterize the level of deep crustal heterogeneity.

3. Parameter Q is the fraction of the soft (quaternary) deposits with respect to older rocks. The area of the soft deposits indirectly characterizes the contrast in vertical movement.

It is easy to see that the dangerous objects have features of high tectonic parcelling level (large values n_l and N_l). The dangerous objects are also characterized by the presence or proximity to first and second lineament ranks. ($\rho_{II} = 0$; $\rho_I \leq 32$km) in combination with clear deep heterogeneity parameters ($\Delta B \leq 65$ mGal; $(\nabla B)^{-1} < 2$).

Nondangerous objects are characterized by features of low tectonical parcelling level ($n_l = 2$, $N_l \leq 3$) and by remoteness from second rank lineaments ($\rho_{II} > 40$ km). These objects are also characterized by deep homogeneity of the crust (small value ΔB). The remoteness of the nondangerous objects from the second rank lineaments is an important feature for their identification. The five features in the nondangerous class (see table 2) include $\rho_{II} > 40$ km parameter.

Each dangerous object has one or several selected features. The distribution of them into dangerous objects shows that these objects are grouped in seismic feature sets. The dangerous objects which belong to the first rank lineaments have large values of n_l and N_l parameters. The intra-mountain dangerous objects have another set of seismic features. These objects, NN 18, 51, 56, well recognized by only one feature $n = 2$, $\rho_I > 32$km, $\Delta B > 45$ mGal. It is clear that the seismicity of these objects is determined by the deep heterogeneity of the crust. This conclusion also applies to objects N 16 and N 41 recognized by only one feature (N 9 from table 2). All dangerous objects are characterized by features including parameters with the presence or proximity to first and second rank lineaments and the parameters of the high tectonic parcelling of the crust.

The analysis of the distribution of dangerous objects thus shows that the most dangerous objects are the second rank transverse lineaments. This result stresses the very important role of transverse lineaments in the formation of the recent morphostructure of the western Alps.

3.1.3 Comparison of CORA-algorithm results with E.C.-algorithm results

As a result of the recognition we can now consider the territories recognized as dangerous by both algorithms (figure 3.5).
We first compare the territories along the western Alps boundaries, then the intra-mountain areas.
The major inconsistencies are located along the NW Alps boundary (between the Leman lake and Bodensee). According to the E.C.-algorithm, this boundary is dangerous while, according to the CORA-algorithm, only three places are dangerous.
Both algorithms give the western Alps boundary as a non-dangerous zone (except the areas in the vicinity of Grenoble and the lower Durance river). The southern boundary between the Alps and Provence is dangerous in isolated places according to the E.C.-algorithm. The southern part of the boundary between the Alps and the Po depression was found as the most dangerous zone in both cases.
The extensive territories in the intramountain part of the Alps along large faults and along the second rank lineaments have been considered dangerous. The territories along the boundary of the Penninic zone and the Helvetic zone was found to be dangerous by the E.C.-algorithm. The CORA-algorithm defines as seismic, areas which are located along the second rank transverse lineaments (5-16).
Let us stress that the region between the Aar massif and the Mont Blanc massif is considered dangerous by both algorithms.
The areas between megablocks (formal morphostructural scheme) and between large faults (neotectonic scheme) were recognized as non-dangerous, indicating that the territories adjacent to the boundaries of large tectonic units are most dangerous.
The most successful coincidence between the two algorithms is in the southern part of the western Alps.

3.1.4 Control experiments

Numerous control experiments have been executed to evaluate the reliability of the classification represented in figure 3.5. For the "Expert Communication" algorithm, logical consistency, coherency and stability were verified.

As for logical consistency, $gen(\mathcal{D}_0)$ and $gen(gen(\mathcal{D}_0))$ have been compared. Among the 80 objects only 8 changes appear and they are all non-dangerous objects becoming dangerous. This does not significantly change the map of the results and the control experiment should be considered as successful.
By performing learning on $(gen(\mathcal{D}_0) - \mathcal{D}_0)$, we could control what was learned. This experiment also gave quite good results and we conclude that the recognition result shown in figure 3.5 is sufficiently coherent. Furthermore, verification of the logical stability of the classification (fig.3.3, fig. 3.5) also gives satisfactory results. In other words $gen(\mathcal{D}_0)$ and $gen(\mathcal{D}_1)$ are in correct correspondence.
We conclude that the results of recognition of earthquake-prone areas in western Alps obtained by the E.C.-algorithm applied to the neotectonic scheme (fig. 3.3) is sufficiently stable towards the whole set of the control experiences under consideration.
The classification obtained by CORA-algorithm voting using all equivalent features is the same as the studied recognition result (fig. 3.4).
The seismic experiment with $\widetilde{K}_\mathcal{D} = \widetilde{K}_\mathcal{N} = 0$ gives the same classification as the recognition result for $K_\mathcal{D} = 12$ and $K_\mathcal{N} = 10$. With $\widetilde{K}_\mathcal{D} = \widetilde{K}_\mathcal{N} = 1$ it gives the classication which differs from the main result for 3 objects: (the objects 33, 43 and 59 are transferred into $\mathcal{D}$-objects. When $\widetilde{K}_\mathcal{D} = 2$, and $\widetilde{K}_\mathcal{N} = 1$, the experiment classifies the 3 objects differently (objects 29, 33 and 43 become $\mathcal{D}$-objects).
Two stability experiments were also performed. In the first experiment, sets $\mathcal{D}'_0$ and $\mathcal{N}'_0$ consisted of objects from $\mathcal{D}$ and $\mathcal{N}$ respectively with numbers from 1 to 10, 21 to 30 and 41 to 50. In the second experiment, sets $\mathcal{D}'_0$ and $\mathcal{N}'_0$ consisted of objects with numbers from 11 to 20, 31 to 40 and 51 to 62. In the first experiment the new classification differs from the main result for 6 objects (objects 33, 39 and 59 become $\mathcal{D}$-objects and objects 18, 51, and 56 become $\mathcal{N}$-objects). In the second experiment the new classification differs for 4 objects (object 29 becomes a $\mathcal{D}$-object and object 1, 3 and 30 become $\mathcal{N}$-objects).
These results suggest that the recognition result is stable (fig. 3.5).
Additional control experiments involve learning on random material. We take the same vector of descriptors and free parameters $K_\mathcal{D}$, $\widetilde{K}_\mathcal{D}$, $K_\mathcal{N}$, $\widetilde{K}_\mathcal{N}$ as in the main result of CORA 3 recognition, and generate 100 random transpositions $\{\sigma\}_1^{100}$ which define 100 random problems $\widehat{\mathcal{P}} = \{(\widetilde{\mathcal{L}}, \widetilde{\mathcal{D}}_0)\}_1^{100}$. For each $(\widetilde{\mathcal{L}}, \widetilde{\mathcal{D}}_0)$ we define:

$$\Delta = \min\{t \in \mathcal{R} : |\{\omega \in \mathcal{W} : \Delta(\omega) \geq t\}| < 36\} \tag{3.8}$$

and the classification

$$\mathcal{C}(\mathcal{L}; K_\mathcal{D}, \widetilde{K}_\mathcal{D}, K_\mathcal{N}, \widetilde{K}_\mathcal{N}, \Delta): \quad \mathcal{W} = \widetilde{\mathcal{D}} \bigsqcup \widetilde{\mathcal{N}}. \tag{3.9}$$

For 17 jumbled problems out of 100 we have $\mathcal{D}_0 \subset \mathcal{D}$ as in the "problem of earthquake-prone intersections of morphostructural lineaments". So $\widehat{P}\{\nu(g(\widetilde{\mathcal{L}}), \widetilde{\mathcal{D}}_0) \leq 0\} = 0.17$. If we restrict the conditions (exclude parameters $\mathcal{Q}$ from the description and change the threshold 36 for 35 in the definition of Δ) the estimate becomes 0.11.

The estimate of the upper bound for the probability of misclassification $P_g(\mathcal{L})$ gives the following values: 0.3932 for the main recognition of CORA 3, and 0.3881 for the restricted version. Note that these values are in fact larger than $P_g(\mathcal{L})$ which is about 15% in both case.

Conclusions

1. The work of two independent teams, French and Russian, using different methods for the definition and description of objects and different algorithms of recognition, gives similar interpretations, a further argument in favor of dynamic pattern recognition applications [134] to seismic hazard and risk studies in particular in regions of moderate seismicity like the Alps.

2. The experiment of selecting objects of recognition in two different ways was fruitful for the Alps, and can be useful for other regions.

3. It is reasonable to apply pattern recognition algorithms that have one learning class to investigations of earthquake-prone areas because the class $\mathcal{N}_0$ includes dangerous objects. Expert Communications is one example of such algorithms, but there are numerous other dynamic pattern recognition algorithms [134].

4. The comparison of the results shows (fig. 3.5) that E.C. reveals more dangerous areas than CORA. The most important inconsistencies are located on the N.W. Alps border. The strongest agreements are in the southern part of western Alps.

5. The characteristics of dangerous objects obtained by the CORA algorithm are physically reasonable: high degree of tectonic parcelling and strong heterogeneity in the vicinity of the intersection of lineaments.

6. The largest number of dangerous objects are located in zones of bending of the morphostructural orientation. This indicates that transverse

lineaments plays an active role in the formation of the block structure of the western Alps.

7. The results obtained in the region common to both this study and that of Caputo *et al.* (1980) [89] are not contradictory.

3.2 Seismically Dangerous Zones in the Pyrénées

A similar pattern recognition study was performed by a French-Russian research group led by A. Cisternas from the French side and A. Gvishiani from the Russian side. The main result of this study was published by Gvishiani *et al.*, (1987) [218].

On some special aspects of morphostructural zoning

The principles of morphostructural zoning are given in detail elsewere [4, 426, 134]. The formal procedure is strictly applied here. A region is subdivided into a system of hierarchically ordered territorial and linear morphostructures. The territorial morphostructures are: mountain countries (first rank), megablocks (second rank) and blocks (third rank). Linear morphostructures are the boundaries of the territorial ones. We call them morphostructural lineaments or, simply, lineaments. The rank of a lineament is the maximum rank of the adjacent territories. In practice, lineaments are the expression of tectonic activity during the more recent orogenic period.

Our scheme (figure 3.6) was obtained from the joint analysis of topographic, geological and tectonic maps, and from satellite photos. It may be surprising for people familiar with Pyrenean tectonics that the lineament (13-36), corresponding to the North Pyrenean Fault is shown only as a third rank lineament in our scheme. It is important to stress that the meaning of rank is formal. Rank is determined, first of all, by the geomorphological difference between the territories on both sides of the lineament. For example, first order lineaments separate orogenic belts from plains or platforms. It may happen that an active fault that does not correspond to important relief changes, is the boundary between Mesozoic and paleozoic rocks. This last feature is associated to third rank lineaments [4].

The most important features of the morphostructural zoning scheme constructed for the Pyrénées by E. Rantsman, A. Gorshkov and Philip (fig.3.6) can be formulated as follows:

1. First rank lineaments separate the Pyrénées from the west European platform, the Ebro depression and from deep marine basins. These lineaments include long segments of the north and south Pyrenean thrusts which are expressed in the relief by steep ledges. In the Mediterranean and the Atlantic they follow the border of the continental slope.

2. Megablocks: The Pyrénées are divided into three megablocks related to abrupt changes in the altitude of the crest axis. This subdivision

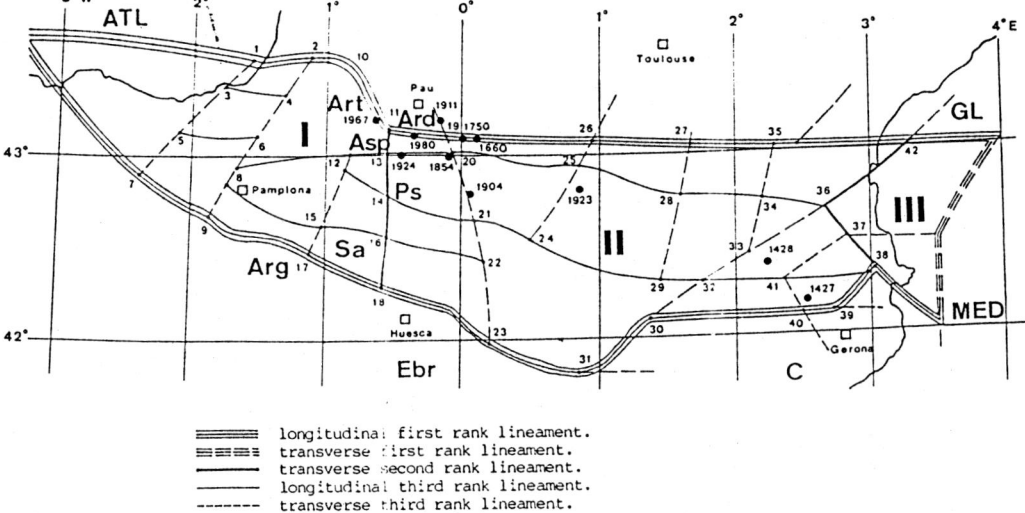

Figure 3.6: *Scheme of morphostructural zoning of the Pyrénées.* Cities are represented by open squares. ATL: Atlantic Ocean, MED: Mediterranean Sea, GL: Gulf of Lyon, Ebr: Ebro Depression, Asp: Aspe Valley, Arg. Aragon Valley, Ps: Pass of Somport, Sa: Sierra de Aragon, Art: Arette, Ard: Arudy, C: Catalonia.

is therefore determined by the geometry of the chain that is characterized by a continuous, huge and extended ridge with a stable E-W orientation.

Megablock I (western) exhibits a Pyrenean low altitude axis (1400-1500 m) and landforms with moderate slopes, the northern one being wider. It is comprises a Mesozoic cover countaining some isolated Paleozoic massifs in the Basque country.

Megablock II (central) includes the most elevated part of the Pyrénées (3000-3400 m). Slopes within the megablock vary considerably. The northern slope is characterized by a steplike cross section with an altitude difference of about 1000 m between steps. The southern slope is wider and has a smoother cross section without sharp flexures. A considerable part of the megablock is composed of Precambrian and Paleozoic rocks.

Magablock III (eastern) is a region where the Pyrenean ridge disappears, but geological structures continue into the Gulf of Lyon.

3. Transverse second-rank lineaments are boundaries between megablocks. Lineament (11-18) is traced along sub-meridional segments of the Aspe and Aragon valleys. The height of the Pyrenean ridge decreases sharply across the lineament zone (Pass of Somport). Lineaments (36-38) and (36-42) are boundaries between mountains and plains, and they include fault segments. Moving towards the SW, lineament (36-42) changes to third-rank (30-36) and then to first-rank (30-31) according to the differences in relief on both sides of it.

4. Longitudinal third rank lineaments separate blocks composed by Precambrian and Paleozoic rocks from blocks with Mesozoic and Cenozoic cover. An exception is lineament (8-22) which is the northern boundary of a range (Sierra de Aragon) and of a long segment of a longitudinal valley. These lineaments correspond to faults.

5. Transverse third rank lineaments are associated with places where the altitude of the Pyrenean axis changes by more than 300 m, in agreement with the formal rules [4]. Transverse lineaments are typefied by discontinuities in the relief having a common strike and that are clearly observed in satellite photos. Lineament (24-26) is the continuation of the Toulouse fault according to geomorphologic evidence and geophysical data [114].

Seismicity and selection of the learning set

The seismicity of the Pyrenean region is moderate and the hypocenters are located in the crust. A map of instrumental seismicity for the period 1965-1985 from the files of the LDG (Laboratoire de Détection et de Géophysique, Commissariat à l'énergie Atomique de France) was used to form the learning material (figure 3.7). The dominant feature is the sequence of aftershocks of the Arette (1967) and the Arudy (1980) earthquakes. A comparison of various catalogs (ISC, NOAA, BCSF, Mezcua and Martinez, 1983; Karnik, 1969 and 1971; Vogt, 1979 [357, 272, 273, 529]) permitted selection of the parameters of the most important earthquakes. In case of conflict we gave preference to local network data and to the information of J. Vogt for historical seismicity. We chose 11 epicenters corresponding to historical and instrumentally recorded earthquakes with magnitude greater than 5 or intensity higher than VII (see table 3.2).
We can clearly see that the events listed in table 3.2 are close to lineament intersections in the morphostructural scheme of figure 3.7. Forty one intersections are considered as objects of recognition (intersections at sea are

N	Date	Coordinates	Magnitude	Intensity
1	02.03.1427	$42.2N\ 2.5E$		X
2	02.02.1428	$42.4N\ 2.2E$		X
3	21.06.1660	$43.1N\ 0.1E$		VIII-IX
4	24.05.1750	$43.1N\ 0.0E$		VIII-IX
5	20.07.1854	$43.0N\ 0.1W$		VIII-IX
6	13.07.1904	$42.8N\ 0.0E$		VIII
7	24.07.1911	$43.2N\ 0.2W$	5.0	VII
8	19.11.1923	$42.8N\ 0.8E$	5.4	VIII
9	22.02.1924	$43.2N\ 0.7W$	5.3	VIII
10	13.08.1967	$43.2N\ 0.7W$	5.3	VIII
11	29.02.1980	$43.1N\ 0.4W$	5.1	VII

Table 3.2: Strongest earthquakes in the Pyrénées with magnitude larger than 5 or maximum MKS intensity larger than VII (after Gvishiani et al., 1987).

excluded). Of these, objects 11, 13, 19, 21, 25, 33 and 40 form the learning set $\mathcal{D}_0$, which is a subset of the dangerous class $\mathcal{D}$. $\mathcal{D}_0$ is defined as the set of intersections (objects) next to earthquakes with intensity larger than VII before 1900 or to those with magnitude greater than 5. Other earthquakes are used for control. Even though the epicenters of old earthquakes have a large uncertainty, we select objects (for the learning set) that are near to these epicenters or clusters of epicenter. We do not take all of the objects near the estimated epicenters, but select those that are representative in the sense of their parametrization. Most other objects are placed in the *a priori* non-dangerous class $\mathcal{N}_0$. Object 41 was excluded from both classes and left aside for control.

Description of the objects

All objects were described by the same set of parameters used in 3.1 to study the earthquake-prone area of the western Alps [100]. We distinguish five groups of parameters (table 3.3): morphometrical, geomorphological, geological, gravimetric and those coming from the morphostructural scheme. The first three groups indirectly characterize the contrast and intensity of the vertical movements. The gravity parameters indicate deep heterogeneity of the Earth crust. Parameters from the scheme of lineaments point out

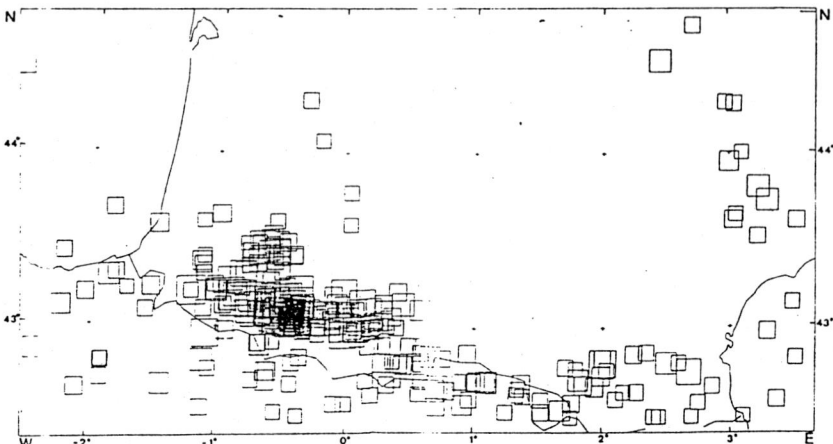

Figure 3.7: *Instrumental seismicity of the Pyrénées.* Data from 1965 to 1985 given by the LDG (Laboratoire de Détection Géophysique, Commissariat à l'Énergis Atomique).

---

the level of tectonic parcelling of the area. Thus, the selected parameters describe several tectonic attributes which may be associated to seismicity.
The values of the parameters were measured within circles of radius $R = 50$ km around each intersection of lineaments. The quantitative parameters were discretized by subdividing each one's range into three intervals, corresponding to "small", "intermediate" and large values of a given parameter. Each interval contains one third of the objects.

Results of recognition

The Pyrenean objects were classified using a few dynamic pattern recognition algorithms including the CORA-algorithm (see [134]). The procedures of CORA-algorithm are described in detail in section 3.1.
The following thresholds were used to select the characteristic features: $K_\mathcal{D} = 3$, $\overline{K}_\mathcal{D} = 5$, $K_\mathcal{N} = 20$, $\widetilde{K}_\mathcal{N} = 0$. Three characteristic features of class $\mathcal{D}$ and four of class $\mathcal{N}$ were obtained (see table 3.4). Only four parameters were used in the final recognition: ΔH, the combination of landforms, ΔB and R_{int}.
The result (table 3.5) of voting shows that $\mathcal{D}$-objects are well separated from $\mathcal{N}$-objects by the zero diagonal. $\mathcal{D}$ contains seven objets of $\mathcal{D}_0$, seven from $\mathcal{N}_0$ and one left for examination (41). This classification is called the "main

I. *Morphometric*		
	1)	H_{max} (in m). Maximum elevation above sea level
	2)	H_{min} (in m). Minimum elevation
	3)	$\Delta H = H_{max} - H_{min}$ (in m).
	4)	$\Delta H/l$. Rough slope. l is the distance between H_{max} and H_{min}.
II. *Geomorphic*		
	5)	a) Mountains ? (m/m). b) Plains ? (p). c) Mountains and Plains ? (m/p). d) Mountains and foothills ? (m/pm) e) Foothills and plains ? (pm/p).
III. *Geological*		
	6)	Q : area of quaternary deposits.
IV. *Gravity parameters*		
	7)	B_{max} : maximum value of the Bouguer anomaly in mGal
	8)	B_{min} : minimum value of the Bouguer anomaly in mGal
	9)	$\Delta B = B_{max} - B_{min}$ in mGal
	10)	$\bar{B} = (B_{max} + B_{min})/2$ in mGal
	11)	$(\nabla B)^{-1}$ = minimum distance between two Bouguer isolines separated by 10 mGals in km/mGal.
V. *Parameters related to the scheme of lineaments*		
	12)	R: highest rank of lineaments in the intersection.
	13)	n_1: number of lineaments at the intersection.
	14)	N_1: number of lineaments within the circle.
	15)	O_I: distance to the nearest I-st rank lineament in km.
	16)	O_{II}: distance to the nearest II-rank lineament in km.
	17)	R_{int}: distance to the nearest intersection in km.

Table 3.3: Parameters used to describe the objects of recognition.

Class	N°	ΔH m	m/p	Relief m/pm	m/m	ΔB mGal	R_{int} km
D	1	> 1409	yes				
	2	> 2448		no			
	3					> 47.5	< 27.5
N	1	< 2448	no				
	2		no		no		
	3	< 2448					> 22.5
	4				no		> 22.5

Table 3.4: Characteristic features selected by the CORA algorithm for the Pyrénées.

classification" (figure 3.8).

The set of dangerous objects ($\mathcal{D}$) represented in the figure 3.8 has the following properties according to the characteristic features: i) Strong topographic contrast reflecting vertical movements (large values of ΔH and combinations of different landforms). ii) Deep heterogeneities (large ΔB). iii) High level of tectonic parcelling (small values of R_{int}). The set of non dangerous objects has the opposite characteristics.

Only 15 out of 41 objects were classified as dangerous, as we might expect given the moderate seismicity of the Pyrénées. These dangerous objects are concentrated in two areas, which coincide with places where instrumental seismicity is highest and the largest intensities have been reported (BRGM, 1981). The dangerous area in the central Pyrénées is characterized by a stepwise cross section of the northern slope and by a sharp transition between the mountains and the adjacent planes. Transverse river valleys (Pau, Arudy, and Adour rivers) show an increase in width to the north of lineament (13-35), that may be connected to subsidence. Geodetic data (Ruegg, private communication) support such relative vertical motions accross lineament (13-25). Moreover, the seismic belt on the western side of the North Pyrenean Fault provides clear evidence of current tectonic activity on this lineament. Another fact to be observed is the vicinity of longitudinal lineaments (11-26) and (13-25), a factor that may be related to the importance of stresses within the region.

The second dangerous area is located in the eastern part of the Pyrénées, where NE-SW lineaments are dominant. Young depressions in the neighbourhood of lineament (32-36) and Pleistocene basalts near objects 39 and

Seismically Dangerous Zones in the Pyrénées 153

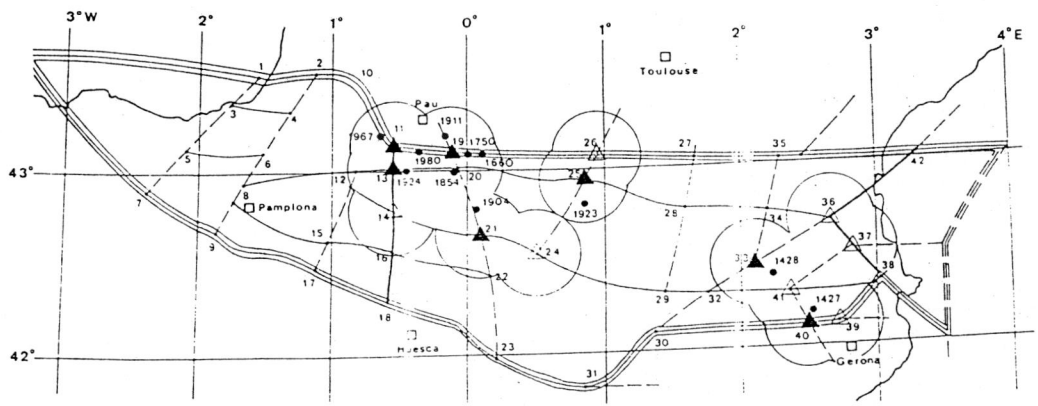

Figure 3.8: *"Main classification"* of recognition of earthquake-prone areas in the Pyrénées. Filled triangles represent the learning set, while open ones are the new objects recognized as dangerous. The region recognized as dangerous is the union of circles centered at the objects.

40 show evidence of tectonic activity. Thus, the recognition results agree with independent geological data.

Control experiments

The system of control experiments for the VSF algorithm is described in Dubois and Gvishiani (1998) [134]. In the case of the CORA-algorithm, the control experiments are also described in 3.1. Here we apply a few experiments from that system that allow us to verify the stability and non-randomness of the "main classification" of recognition:

- i) "Seismic future" experiment: This experiment tests the stability of the dangerous set $\mathcal{D}$, by using as learning set $\mathcal{D}_0$ the set of objects recognized as dangerous in the "main classification". Table 3.6 shows the characteristic features obtained in experiment for both classes. They do not differ much from those obtained for the "main classification" and have a similar meaning.

  The CORA algorithm used thresholds $K_\mathcal{D} = 8$, $\widetilde{K}_\mathcal{D} = 1$, $K_\mathcal{N} = 10$, $\widetilde{K}_\mathcal{N} = 1$ in this experiment.

  The voting is shown in table 3.7. Only one object (30) from $\mathcal{N}_0$ was classified as belonging to $\mathcal{D}_0$, while all of the 15 objects already known as dangerous remained within $\mathcal{D}$. The separation of both classes in table 3.7 is well defined. The zero diagonal contains only two objects (26 and 30).

- ii) Jack-Knife experiment: This experiment is related to the well known Jack-Knife test. The CORA algorithm is run using a learning set with one of the seven objects of the original learning set $\mathcal{D}_0$ left out. The test is repeated for all seven possible combinations.

  We thus have seven learning sets of six objects each, and seven voting results. There were no misclassifications of the excluded objects, giving a probability of misclassification less than 1/7. Five classifications out of seven coincide with the "main classification", and two have a slight difference. If objects 21 or 33 are excluded from the learning set, then we obtain a classification in which object 30 is included in $\mathcal{D}$.

- iii) Recognition by several algorithms: We used several alternative algorithms, Bayes, Hemming and weighted Hemming (Gvishiani and Kossobokov, 1981 [216]) to study the objectivity of the "main classification". The CORA result is almost identical to that obtained with the other methods. Only two differences were noticed: object 22 was assimilated to $\mathcal{D}$ when using weighted Hemming, and objects 22 and 29 were assimilated when standard Hemming was applied. This result may be expressed in terms of probabilities. Table 3.7 shows the

Number of votes for 2nd class	Number of votes for the first class			
	0	1	2	3
4	−P1 −P4 −P5 −P6 −P7 −P8 −P9 −P10 −P12 −P16 −P18 −P27 −P31 −P34 −P35	−P38		
2	−P2 −P3 −P15 −P17 −P23 −P28 −P32	−P22 −P29 −P30		
1			−P37	−P36
0	−P39	∗P11∗ −P14 −P24 P41	∗P13∗ ∗P19∗ ∗P21∗ ∗P25∗ ∗P33∗ ∗P40∗ −P20 −P26	

Table 3.5: Results of voting in the "main classification" for the Pyrénées (after Gvishiani.

| Class | N° | ΔH | Relief | | | | ΔB | R_{int} |
		m	m/p	m/pm	mp/p	m/m	mGal	km
D	1	> 2448						
	2			no			> 47.5	< 27.5
	3			no	no			< 22.5
	4			no	no		> 67.5	
N	1	< 2448	no					
	2	< 2448						> 22.5
	3		no			no		
	4		no				< 67.5	> 22.5
	5					no		> 27.5

Table 3.6: Characteristic features selected by the CORA algorithm in the experiment of "seismic future".

Bayesian *a posteriori* probability for an object belonging to class $\mathcal{D}$. From these data we are able to see a clear clustering of $\mathcal{D}$ and $\mathcal{N}$-objects of the "main classification".

- iv) Random problems: This experiment tests the non-randomness of the classification. The idea is to generate random learning sets $\tilde{\mathcal{D}}_0$ containing seven objects each, and verify that good classification is unlikely. Fifty such random problems were generated. The CORA algorithm was first run using 17 descriptors and then with only four, those that had more information according to the "main classification". Figure 3.9 shows that the estimation of non-randomness is improved with a reduction of the number of descriptors. In the first case there were 22 % errorless (for the learning set $\tilde{\mathcal{D}}_0$) random classifications. In the second case the percentage lowers to 6 %.

These control experiments allow us to classify the "main classification" as stable and non-random. It is important to stress the fact that this stable result for the Pyrénées was obtained in spite of having a very small learning set ($\mathcal{D}_0$ has only seven objects).

Number of votes for 2nd class	Number of votes for the first class					
	0	1	2	3	4	5
4	−P1 −P2 −P4 −P7 −P8 −P9 −P10 −P17 −P18 −P31 −P35					
3	−P3 −P5 −P6 −P15 −P16 −P27 −P38					
2	−P12 −P29 −P32 −P34	−P23				
1	−P22 *P26* −P28	*P40* −P30	*P39*			
0		*P21* *P33*	*P14* *P24* *P41 ?	*P25*	*P11* *P36*	*P13* *P19* *P20 *P37

Table 3.7: Results of voting for the "seismic future" experiment.

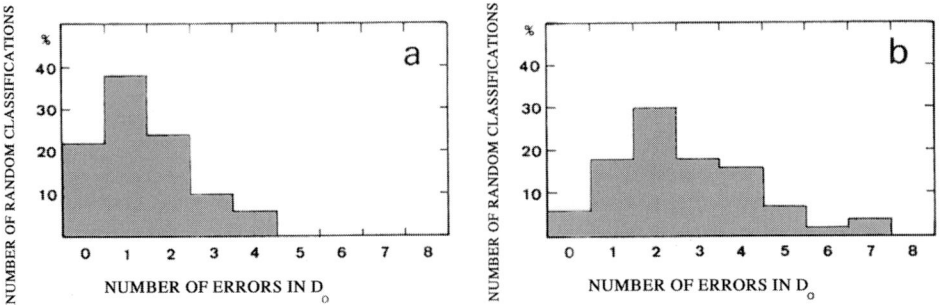

Figure 3.9: *Estimation of non-randomness*. Empirical distribution of the number of $\mathcal{D}_0$ misclassifications in the random problems; a) for 17 descriptors; b) for 4 descriptors.

3.3 Comparison between Earthquake-prone areas in the Pyrénées and the Alps

The control experiments described in 3.1-3.2 tested the mathematical and logical coherence of applying the dynamic pattern recognition technique to the given data sets. Let us try to examine the results from a more physical point of view. A natural way to do this is to compare the recognized earthquake-prone area in the Pyrénés with the area in the western Alps, since the classification criteria were different (see 3.1-3.2). Two complementary approaches are possible: the first, transfering the criteria of seismicity $M \geq 5.0$ used in the Alps to the Pyrénées and *vice versa*; and the second, treating the Alps and the Pyrénées as a joint region.

- i) Transference of criteria: The set of characteristic features obtained for the Alps may be used to obtain a classification in the Pyrénées and *vice versa*. It is obvious that each seismic region has its own structural and geodynamical properties, but it is not unreasonable to consider that these two regions belonging to the Alpine belt and located within zones of continetal deformation with moderate seismicity, may have some similarities.

    Figure 3.10 shows the dangerous places in western Alps according to the Pyrénées criteria. We compare them with the original result for Alps shown in the figures 3.4 - 3.5.

Comparison between the Pyrénées and the Alps

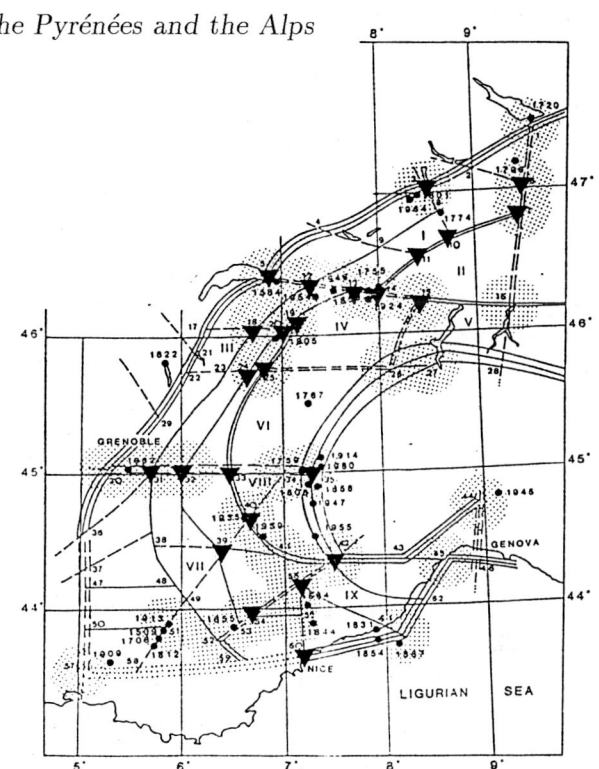

Figure 3.10: *Results of transference of the Pyrenean criteria of seismicity $M \geq 5.0$ to the western Alps.* Objects recognized as dangerous are represented by inverted triangles. Shaded areas are the dangerous regions determined by Cisternas *et al.*, 1985..

Most of the differences correspond to the southern part of the western Alps (objects 36 to 62). The result is better for the northern part of western Alps (objects 1 to 35) where all places with epicenters corresponding to $M > 5$ are recognized. The structural difference between the northern and southern parts of western Alps was already discussed in [536]. It is possible to infer that the Pyrénées are more similar to the northwestern Alps than to the southwestern Alps.

Figure 3.11, on the other hand, shows the classification for the Pyrénées according to the criteria obtained for western Alps. All of the objects of Pyrénées close to the epicenters of table 3.2 are recognized as dangerous. Nevertheless, the number of objects included in $\mathcal{D}$ is too large to be acceptable. The criteria coming from the western Alps, given the heterogeneity already mentiond, are not strict enough to be valid for the Pyrénées.

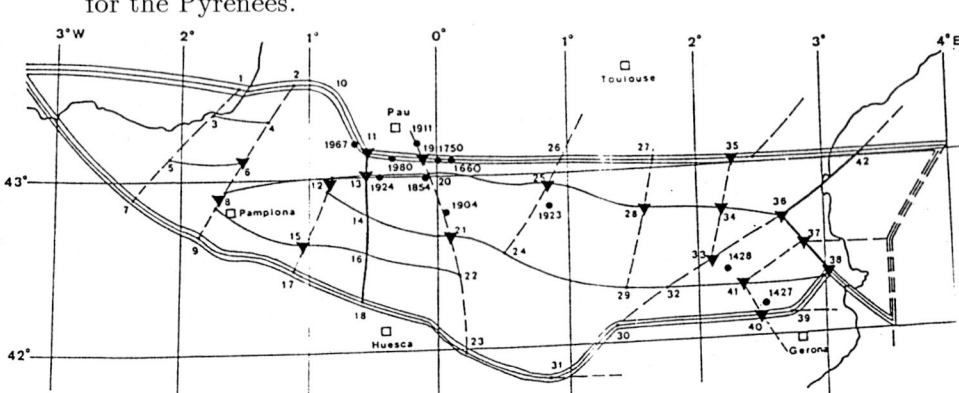

Figure 3.11: *Results of the transference of the Alpine criteria of seismicity* $M \geq 5.0$ *to the Pyrénées.* Objects recognized as dangerous are shown as inverted triangles..

- ii) Recognition for the joint region formed by the union of the sets of objects of recognition in Alps and Pyrénées: we consider the joint region (northwestern Alps and Pyrénées) as a new pattern recognition problem. This experiment has the advantage of increasing the small learning set of the Pyrénées on one hand, and of allowing for a new comparison between the Alps and the Pyrénées on the other. The new joint region contains 76 objects (41 from Pyrénées and 35 from nothwestern Alps). The new learning set is the union of those used in each region. $\mathcal{D}_0$ therefore includes 14 objects. The objects are represented in the space of the four parameters that contain the most information. As in the Pyrénées they are: ΔH, ΔB, combinations of landforms and R_{int}.

The result of the CORA algorithm is shown in figure 3.12. A total of 27 objects are classified as dangerous, 14 in the Pyrénées and 13 in the northwestern Alps. The output is very good for the Pyrénées because only one object (36) differs from the "main result". The correspondence is not as good for northwestern Alps due to the decrease of the number of $\mathcal{D}$-objects (22 objects were previously recognized as dangerous, see [100]). It is convenient to verify this result using the same control experiments already described. Using the "seismic future" test, only 3 objects (object 36 from the Pyrénées and objects 11 and 26 from the Alps) out of 76 changed from $\mathcal{N}$ to $\mathcal{D}$. In the Jack-Knife experiment only 3 objects from the Alps (3, 12 and 31) produced problems. Classification by different algorithms gives practically the same results: there are differences for less than 10 % of the objects. Finally, when 50 random problems were generated, the number of misclassifications on $\mathcal{D}_0$ was never less than 2, even when using all 17 descriptors.

The similarity between the northwestern Alps and the Pyrénées is qualified by the fact that the criteria used for the Pyrénées are stronger than those used for the Alps.. Nevertheless, we may say that the features: large values of ΔH and ΔB, contrasted landforms and small values of R_{int}, are valid for the Pyrénées and for the northwestern Alps.

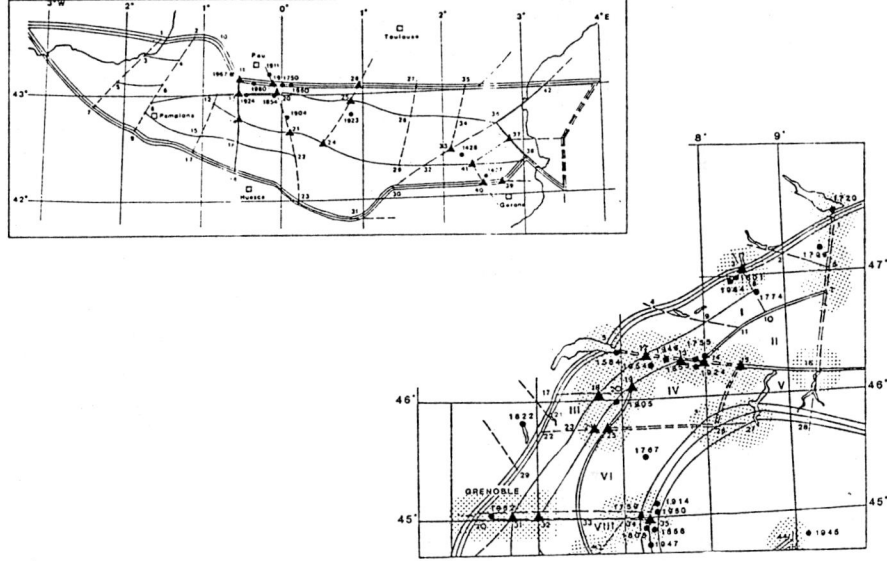

Figure 3.12: *Results of recognition for the joint region: Pyrénées union with northwestern Alps.* Objects recognized as dangerous are indicated by triangles.

Conclusions

Dynamic pattern recognition techniques introduced by the authors in [134] have sufficient capacity to produce a stable and physically consistent classification in a region of moderate seismicity such as the Pyrénées. In particular, the epicenters of strong Pyrenean earthquakes are situated close to intersections of lineaments.

The reliability of the classification obtained for the Pyrénées was confirmed by several independent control experiments such as: "seismic future", Jack-

Knife, random problems, different algorithms and transference of criteria from and to the Alps. It was shown that the characteristic features of the dangerous objects are: strong topographic contrast related to vertical movements, strong heterogeneities at depth in the crust and high level to tectonic parcelling. It was found that the objects classified as dangerous are clustered into two regions: the western end of the North Pyrenean Fault around the epicenter of the 1967 Arette earthquake, and the southeastern part of the Pyrénées in Catalonia. Both are regions of historical and present seismic activity.

A comparison between the Pyrénées and the western Alps showed that: The criteria used for the Pyrénées are useful for determining dangerous objects in the Alps, especially in the North. On the contrary, those obtained for the Alps are too broad for the strict classification of dangerous objects in the Pyrénées. The joint study of the northwestern Alps and the Pyrénées helped us to improve our evaluation of those characteristic features that can be considered as useful for the recognition of dangerous places in both regions.

3.4 Strong Earthquakes Prone-areas in the Great Caucasus

The scheme of seismic zoning of the Great Caucasus was represented at 1 : 1 000 000 scale by E. Rantsman, A. Gorshkov and M. Zhidkov in 1980-1985. Later on, A. Cisternas and H. Philip (1985-1992) provided their important contribution and the scheme has been further developed. The final version of the morphostructural scheme of the Great Caucasus is shown in figure 3.13. The scheme was constructed on the basis of the formal principles worked out and formulated in [426].

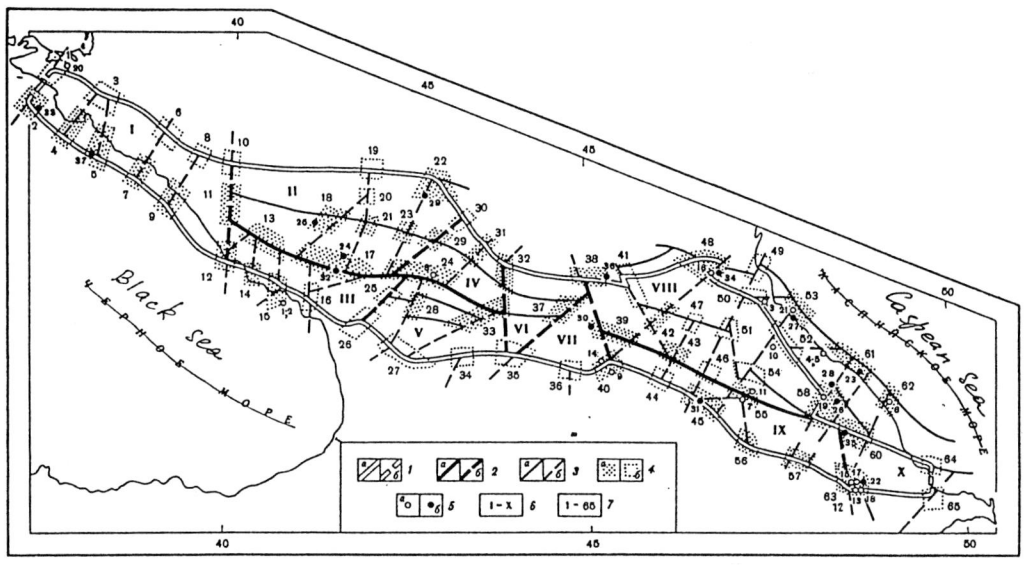

Figure 3.13: *Great Caucasus area*.Scheme of morphostructural zoning of the Great Caucasus and results of recognition of earthquake-prone areas for $M \geq 5.5$: 1-3 lineaments of first, second and third rank correspondingly ; 4-morphostructural knots ; a) prone for $M \geq 5.5$; δ) $M < 5.5$; 5-epicenters of earthquakes with $M \geq 5.5$: a) up to 1900 ; δ) after 1900 (numbers from the table 3.4.1) ; 6-megablock numbers; 7-knot numbers.

All three morphostructural schemes for the western Alps (3.1), the Pyrénées

(3.2), and the Great Caucasus were constructed homogeneously and the results obtained in these three regions can therefore be analyzed together. The borders of the knots were determined by field observations, following general priniples described in [134].

The list of crustal earthquakes of the Great Caucasus with $M \geq 5.5$ is given in table 3.8. The maximum values of magnitude ($M = 7.0$) and intensity ($I_0 = 8 \div 9$) are founded in Shemakhin earthquakes (N 12, 22 in table 3.8). The sources of most Caucasian earthquakes are located at depths greater than 70-90 km. However, in the Eastern part of the Caucasus, earthquakes with sources between 70 and 90 km exist. In particular, the hypocenter depth for earthquakes N 26 and 28 was recalculated in [380, 127]. Following [127], we included earthquakes N 26, 28 in table 3.8, but we use them only to control the reliability of the recognition.

As we see from figure 3.13, epicenters of practically all of the earthquakes from the table 3.8 are located within the limits of morphostructural knots. This corresponds well with the previously reported clustering of numerous Great Caucasus earthquakes to the zones of intersection of significant faults [10, 68]. Two exceptions are the epicenters of two historical earthquakes in the years 743 and 918 (N 4 and 5 in the table 3.8), located, according to [380], with precision of ±1.0°.

We considered as a learning set $\mathcal{D}_0$ of the "dangerous" class, the knots containing epicenters of earthquakes with $M \geq 5.5$ since 1900 (N 22-37 in the table 3.8 except N 26 and 28). Thus, the following 13 knots were selected: N 2, 5, 17, 18, 22, 38, 39, 45, 48, 53, 60, 61, 63. The learning set $\mathcal{N}_0$ of the low seismicity class includes 42 knots where no earthquakes with $M \geq 5.5$ are known. Nine knots (N 1, 15, 24, 40, 50, 52, 55, 58, 62) containing earthquake epicenters with $M \geq 5.5$ before 1900, were neither included in the learning material $\mathcal{D}_0$ nor in $\mathcal{N}_0$. They were used to evaluate the classification. The classifications under consideration were limited by the condition $\mid \mathcal{D} \mid \leq 0.6 \mid \mathcal{W} \mid$, where $\mathcal{W}$ is the whole set of objects (knots) represented in figure 3.13.

The objects of recognition (morphostructural Knots in figure 3.13) are described by parameters similar to those used in the Alps and the Pyrénées (see 3.1 and 3.2). Among them are geological, geomorphological and gravity parameters as well as parameters defined by the geometry of the zoning scheme.

The recognition of earthquake-prone areas in the Great Caucasus was performed by the CORA-algorithm using two types of objects: knots and intersections of lineaments (Gvishiani et al., 1988 [208]). Herein, we describe the result obtained for the morphostructural knots.

N	Date	Lat.N	Long.E	Magnitude	Intensity	h, km	n. knots
1	50	42.9	41.0	5.5	8	3-30	15
2	400	42.9	41.0	5.5	8	3-30	15
3	650	42.6	47.7	6.1	8	7-60	50
4	743	42.1	48.2	5.5	7	7-60	
5	918	42.1	48.2	5.5	8	7-60	
6	957	41.5	49.0	5.5	7	7-60	62
7	1250	41.6	47.2	5.7	7-8	5-50	55
8	1350	43.0	43.0	6.5	8-9	10-40	24
9	1530	42.0	45.4	5.7	8	5-50	40
10	08.06.1652	42.1	47.7	5.8	8-9	3-30	52
11	17.12.1667	41.7	47.3	6.5	8	20-45	55
12	1967	40.9	48.2	7.0	9-10	6-24	63
13	14.01.1669	40.6	48.6	6.0	9	5-20	63
14	24.07.1742	42.1	45.2	6.2	7	5-50	40
15	09.08.1828	40.7	48.4	5.7	8	5-20	63
16	09.03.1830	43.0	47.0	6.3	8-9	11-24	48
17	11.06.1859	40.7	48.5	5.9	8-9	7-15	63
18	28.01.1872	40.6	48.7	5.7	8-9	3-14	63
19	04.05.1878	41.6	48.1	5.7	7	7-60	58
20	09.10.1879	45.1	37.8	5.7	7	22	1
21	26.06.1889	42.5	48.0	5.9	6	28-63	53
22	13.02.1902	40.7	48.6	6.9	8-9	15	63
23	05.07.1903	41.8	48.7	5.5	6	40	61
24	21.10.1905	43.3	41.7	6.4	7	35	17
25	21.10.1905	43.6	41.2	5.6	6	32	18
26	20.02.1906	41.5	48.4	5.9	6	25	58
27	30.10.1909	42.4	48.0	5.8	6	40	53
28	25.03.1913	41.8	48.3	5.7	7	15	58
29	29.06.1921	43.9	42.8	5.6	7	22	22
30	15.08.1947	42.5	45.0	5.5	7	25	39
31	29.06.1948	41.6	46.4	6.1	7	48	45
32	16.07.1963	43.25	41.58	6.4	9	5	17
33	12.07.1966	44.7	37.3	5.8	7	55	2
34	14.05.1970	43.0	47.09	6.6	8-9	13	48
35	20.12.1971	41.23	48.38	5.5	7	5	60
36	28.06.1976	43.10	45.50	6.4	-	-	38
37	3.09.1978	44.38	38.03	5.5	6	25	5

Table 3.8: Catalogue of Great Caucasus earthquakes with $M \geq 5.5$.

The results of the main and additional classifications of the objects concerning the possibility of an earthquake with $M \geq 5.5$ are given in table 3.9.

In the main classification of Great Caucasus 64 knots 38 are classified as seismically dangerous and 26 as non-dangerous. The condition $\mid \mathcal{D} \mid \leq 0.6 \mid \mathcal{W} \mid$ is satisfied.

Characteristic features of the knots from the classes $\mathcal{D}$ and $\mathcal{N}$ are given in the table 3.10. They have been selected using the following thresholds of selection and contradiction: $K_\mathcal{D} = 5$, $\widetilde{K}_\mathcal{D} = 6$, $K_\mathcal{N} = 16$, $\widetilde{K}_\mathcal{N} = 1$. The selected features are sufficiently representative. Each of $\mathcal{D}$-features occure on minimum 38% of all the knots from $\mathcal{D}_0$ and each of $\mathcal{N}$-features on 38% of all the objects from $\mathcal{N}_0$. The objects for which $\Delta(\omega) = V_\mathcal{D}(\omega) - V_\mathcal{N}(\omega) \geq -2$ have been classified to the class $\mathcal{D}$. Geographical positions of the knots $\mathcal{D}$ and $\mathcal{N}$ are represented in the figure 3.13.

There are much more objects in the class $\mathcal{N}_0$ rather than in the class $\mathcal{D}_0$. Therefore, an attempt was made to the recognition, using as a learning material only half of the knots from $\mathcal{N}_0$. Therefore, we obtained two additional classifications of the knots, which we will call additional versions of recognition. The learning material for the additional versions was formed as follows. The learning set $\mathcal{D}_0$ included the same 13 knots as in the main classification. The learning set of the low seismicity class was defined by formal division of the 42 knots from $\mathcal{N}_0$ of the main classification into two half: $\mathcal{N}_1$-knots with odd numbers, $\mathcal{N}_2$-knots with even numbers. Their space distribution is shown in the figure 3.13.

Additional versions of classifications were obtained using the threshold two times smaller than in the main classification ($\widetilde{K}_\mathcal{D} = 3$, $\widetilde{K}_\mathcal{N} = 8$), what well corresponds to the fact that $\mid \mathcal{H}_i \mid = 1/2 \mid \mathcal{H} \mid$, $i = 1, 2$.

Additional versions of recognition not much differ from the main classification (see table 3.9). Furthermore, the characteristic features selected in the additional versions are quite similar to the features established in the main classification (see table 3.10). The fact that all the three classifications are really similar gives another evidence in favor of reliability of the executed classification. However, higher priority should be given to the main classification since it is more stable in control experiences on one side and is based on twice higher learning material.

Numerous control experiments have been done to evaluate the reliability of the obtained results. In the "Seismic Future" (SF) experiment the learning set $\mathcal{D}_0$ included all the knots recognized as dangerous in the main classification. The experiment was done with different values of the thresholds ($\widetilde{K}_\mathcal{D}$, $\widetilde{K}_\mathcal{N}$, chosen in the way that the number of selected features in SF is

1	2	3	4	5	1	2	3	4	5
$\mathcal{D}_0$	2	+	+	+	$\mathcal{H}_0$	28	+	+	+
	5	+	+	+		29	+	+	+
	17	+	+	+		30	−	−	−
	18	+	+	+		31	+	+	+
	22	+	+	+		32	+	+	+
	38	+	+	+		33	+	+	+
	39	+	+	+		34	−	−	−
	45	+	+	+		35	−	−	−
	48	+	+	+		36	−	−	−
	53	+	+	+		37	+	+	+
	60	+	+	+		41	−	−	−
	61	+	+	+		42	+	+	+
	63	+	+	+		43	+	+	+
$\mathcal{N}_0$	3	−	−	−		44	−	−	−
	4	+	+	+		46	−	−	−
	6	−	−	−		47	−	−	−
	7	+	+	+		49	−	−	−
	8	−	−	−		51	−	−	−
$\mathcal{H}_0$	9	+	+	+		54	−	−	−
	10	−	−	−		56	+	+	+
	11	+	+	+		57	+	−	−
	12	−	−	−		64	−	−	−
	13	+	+	+		65	−	−	−
	14	+	+	+	Examination	1	−	−	−
	16	−	+	−		15	+	+	+
	19	−	−	−		24	+	+	+
	20	−	−	−		40	+	+	+
	21	+	+	+		50	−	−	−
	23	+	+	+		52	−	−	−
	25	−	−	−		55	+	+	+
	26	−	−	−		58	+	+	+
	27	−	−	−		62	+	+	+

Table 3.9: Results of recognition for the knots of the Great Caucasus. Column 1: Class; 2 Number of the knot; 3, 4, 5 Classification (3 main classification; 4,5 additional classifications with learning: 4 $\mathcal{N}_1$, 5 $\mathcal{N}_2$) Note: (+) knots recognized as seismically dangerous; (−) knots recognized as non-dangerous ($M \geq 5.5$).

1	2	3	4	5	6	7	8	9	10	11
1						> 5				< 27
2	< 600			not					> 2100	
3						> 20			< 2100	
4				not		> 5	> 5			
5	< 600				not		> 5			
6		not				> 20	< 5			
7				not		> 20				
1						< 5			< 2100	> 27
2				not					< 2100	> 27
3				not			< 5			> 27
4			not	not						> 27
5					not			< 1050	< 2100	
6						< 20			< 2100	
7					not		< 5	< 1050		
8						< 20		< 1050		
9						< 20	< 5			

Table 3.10: Characteristic features selected by CORA-algorithm for recognition of earthquake-prone areas of the Great Caucasus. Class $\mathcal{B}$ from N 1 to 7, and Class $\mathcal{N}$ from N 1 to 9. The 11 columns are: 1, N ; 2, ; 3,4,5,6, Combination of the types of relief; 7, Q % ; 8, Number of faults ; 9 Bouguer anomaly in mGal ; 10 ; 11, Distance to lineament intersections, km.

similar to the main classification.

If $\widetilde{K}_\mathcal{D} = \widetilde{K}_\mathcal{N} = 0$ and $\widetilde{K}_\mathcal{D} = \widetilde{K}_\mathcal{N} = 1$, the result of the SF is equal to the result of the main classification. If $\widetilde{K}_\mathcal{D} = \widetilde{K}_\mathcal{N} = 2$ the SF result differs only on one object (the knot 54 is recognized as $\mathcal{D}$). Thus, the results of SF control experiment for the Great Caucasus should be considered as succesful. The SF was successful for additional versions of recognition as well. The SF classification differ from the additional versions of recognition not more than on three objects.

In "Jack-knife" experiment the knots have been excluded one by one from the learning material. Then, the excluded knots have been examined after learning on the new learning material. In our case of main classification, only 15% from the total number of the objects $\mathcal{D}$ do not pass this control (the object 2 and 22). Among the objects $\mathcal{N}$ only 10% (the knots 25, 27, 49, and 54) change their classification.

The "Jack-knife" experiment was also executed for the additional versions of recognition. In the classification withe the learning material $\mathcal{N}_1$ the knots 18 and 22 do not pass the control. For the learning material $\mathcal{N}_2$ these are the knots 2, 5 and 22. Apparently, the "Jack-knife" experiment speaks in favor of reliability of the results obtained for the Great Caucasus.

In the experiment of stability evaluation to new learning set $\mathcal{D}'_0$ was formed by the knots from $\mathcal{D}_0$, affiliated with the epicenters, which occured before 1971. The rest of knots from $\mathcal{D}_0$ (5 from the 38, see table3.8) along with all the knots, from the knots from $\mathcal{N}_0$ created the new learning set $\mathcal{N}'_0$. This kind of control experiments is usually called "seismic history". The value of Δ has been choosen in the way that $|\mathcal{D}| \leq 38$. The experiment was successful and it gave exactly the same results of recognition as the main classification.

Among numerous other control experiments which confirm the reliability of the main classification (Gvishiani et al., 1988 [208]) it is worth to mention the following:

1. The estimation of the probability to "miss the goal" gives less than 15%. In other words the probability that there are still seismically dangerous objects among those that have been classified as $\mathcal{N}$ is not more than 15%.

2. 19 objects from 42 are recognized as seismically dangerous. Therefore, a priori estimation of the probability is $p = 19/42 = 0.452$. For the examination there are more dangerous objects: $q = 6/9 = 0.666$. Thus, $q > p$. If we propose that the knot from the examinating part

are recognized with the same probability p, then the probability to recognize not less than 6 knots from 9 is sufficiently small. It is:

$$C_9^6 p^6(1-p) + C_9^7 p^7(1-p)^2 + C_9^8 p^8(1-p)^3 + C_9^9 p^9 = 0.169.$$

3. For the knots of the Great Caucasus 100 random pairs of the learning sets ($\widetilde{\mathcal{D}}_0$, $\widetilde{\mathcal{N}}_0$) was generated with the same number of objects as in the pair ($\mathcal{D}_0$, $\mathcal{N}_0$) of the main version of the recognition. While doing the recognition in these 100 "mixed" problems, only in 10 cases we obtained the inclusion $\widetilde{\mathcal{D}}_0 \subset \mathcal{D}$. Therefore, the measure of non-randomness of the main classification is 0.1. For the two additional versions of the recognition in the Great Caucasus we obtained 0.06 and 0.09 as the estimation of the measure of non-randomness. Sufficiently small values of these estimations speaks in favor of the reliability of the classification obtained.

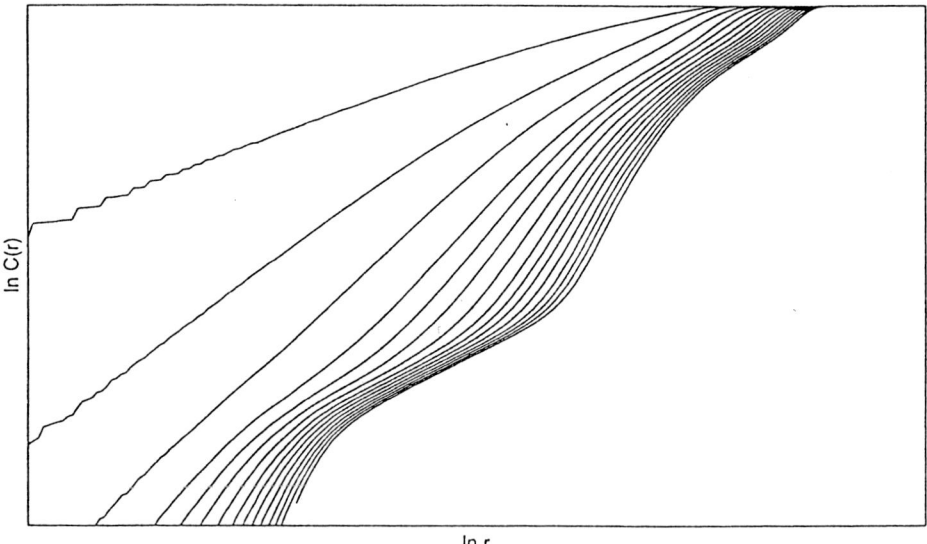

Correlation Integral Function when the time series (unfiltered daily discharge of Oubangui river) presents a strong short term correlation (after David Aubert, 2000).

Part II

Fractals and Dynamic Systems

Lorenz Attractor

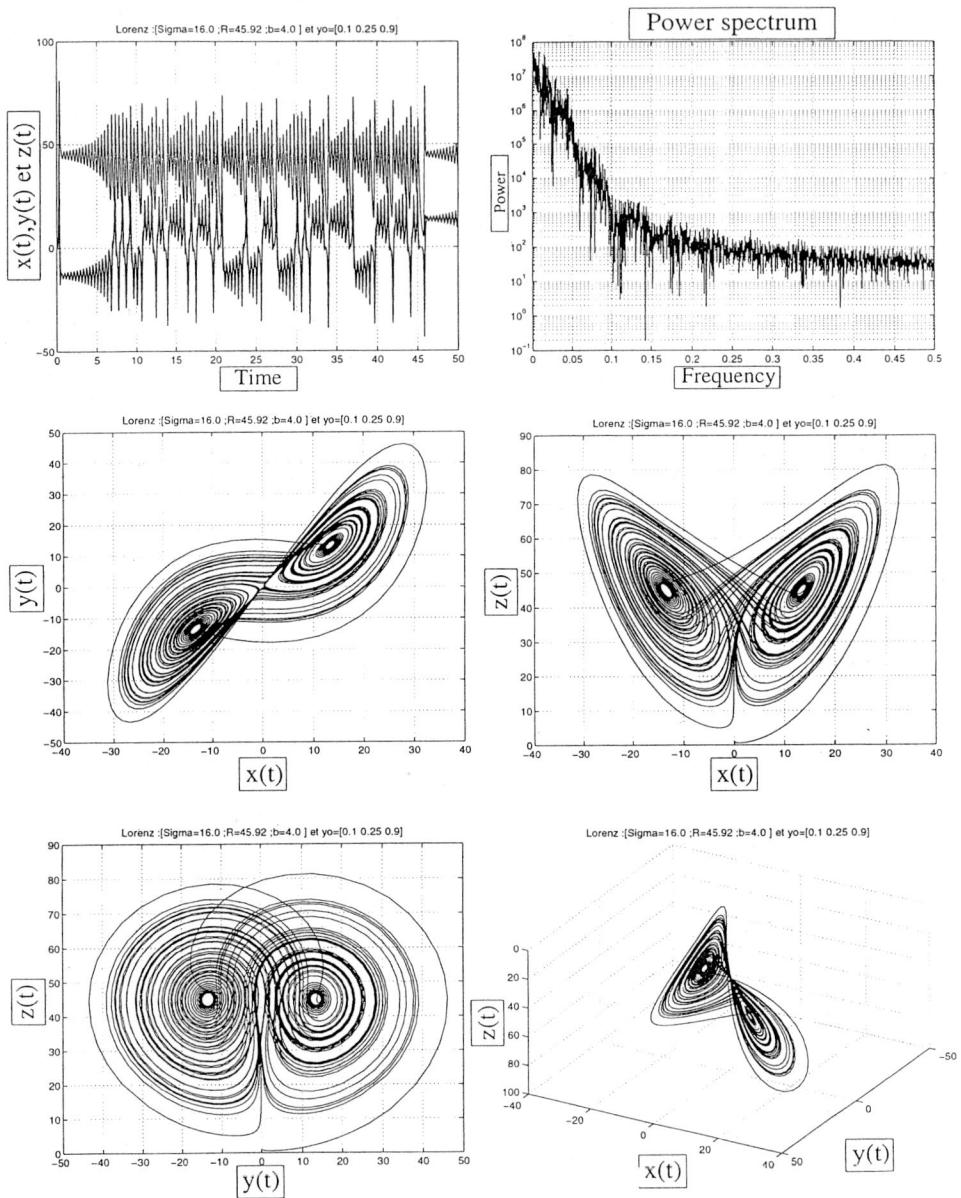

Lorentz Attractor - The Lorenz model is successively analyzed for coordinate variations as a function of time, in frequency domain, and in the phase space (after Lionel Hongre, 1997).

Chapter 4

Fractals and Multifractals

4.1 A brief Review of Fractals and Multifractal Analysis

In this section we will review some developments given in our previous book (see [134] p.195-205). Following the style of a synthesis published ([446]) by P. Sailhac (1997) we will emphasize the physical meaning of fractal and multifractal analysis.

Fractal analysis allows a good description of objects being either black or white (an image), either true or false (a test), either active or not (earthquake or not, eruption or not); it characterizes the support S of the activity (its contour), but does not describe the degrees such as the shades of greys in the image or the amplitudes of the eruptions activity in the volcano. Therefore, it only considers the support S of the activity as shown on figure 4.1.

To describe the degree of activity, one could use several different fractal analysis, each considering, for example, a prescribed level of greys separating bright and dark regions of an image, or critical seismic levels inside the volcano separating areas of high activity. Multifractal analysis does this by assuming that the levels have a physical meaning. This is done by assuming that the feeling of the degrees follows local power laws so that the sensitivity levels are the exponent characterizing how the measure is concentrated, namely the singularity strength of the Hölder exponent α.

In the case of grey levels in an image, our eyes feel the intensities of objects according to a logarithmic scale of optical magnitudes. From each elementary grey point $P(x,y)$ of size $dxdy$, the eye receives the intensity $\delta I(x,y)dxdy$; from a small disc of radius λ centered in $P(x,y)$ the eye receives the sum-intensity $I_\lambda(x,y) = \int \delta I(x,y)dxdy$ which is felt as the optical

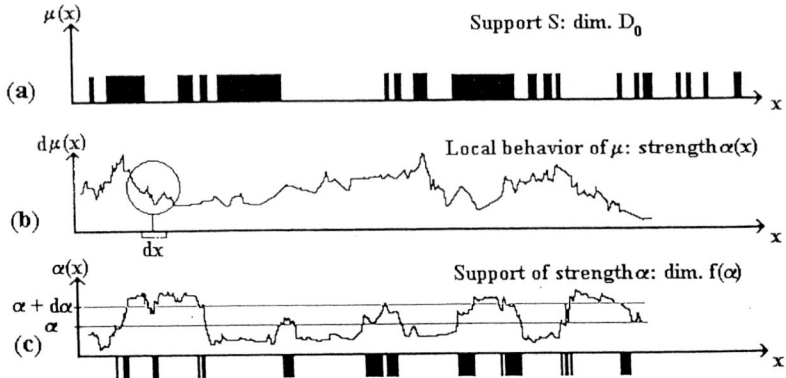

Figure 4.1: *Fractal and Multifractal analysis*. A fractal object is characterized by the box counting of its support S, the set where its measure is not null (a) ; a multifractal object has irregular variations of its measure (b) ; measure which can be characterized by the singularity strengths α ; the subset of S having singularities in the range $[\alpha, \alpha + d\alpha[$ with $d\alpha$ small is fractal and has a fractal dimension $f(\alpha)$ (c) (after Sailhac, 1997).

magnitude $M_\lambda(x,y) = a + \log I_\lambda(x,y)$, where a is a constant, and the basis of the log depends on the magnitude range: for instance, the Pogson Law for stellar objects, $M^* = M^\circ - 2.5 \log_{10}(I^*/I^\circ)$, gives the visual magnitude M^* of a star which has intensity I^*, compared with a reference star with magnitude M° and intensity I°.

To have an idea of what eyes can actually measure as the grey level in the center point $P(x,y)$ look at the asymptotic behaviour of the magnitudes as $\lambda \to 0$:

- in the simplest case where the distribution of grey point intensities is continuous and uniform, one finds

$$I_\lambda(x,y) = \pi \lambda^2 \delta I(x,y) \text{ and } M_\lambda(x,y) \sim 2 \log \lambda ; \qquad (4.1)$$

- in the case where the distribution of grey point intensities is uniform on a fractal support having a local dimension $D(x,y)$ near $P(x,y)$, one finds

$$I_\lambda(x,y) \sim \lambda^{D(x,y)} \delta I(x,y) \text{ and } M_\lambda(x,y) \sim D(x,y) \log \lambda ; \quad (4.2)$$

- in the case where the distribution of grey point intensities is not uniform, assuming the disc intensity should tend to a zero value for $\lambda \to 0$ as in the two previous cases, one can postulate the existence of a parameter $\alpha(x,y)$ such that

$$I_\lambda(x,y) \sim \lambda^{\alpha(x,y)} \delta I(x,y) \text{ and finds } M_\lambda(x,y) \sim \alpha(x,y) \log \lambda ; \quad (4.3)$$

Obviously in this last model, the magnitude does not converge as $\lambda \to 0$, and the eye should feel the local grey level in the form of the finite parameter $\alpha(x,y)$. $\alpha(x,y)$ equals the local dimension $D(x,y)$ of the point distribution if intensities are constant in the neighbourhood of the point $P(x,y)$, it is the local degree of "sparseness"; otherwise it measures the strengths of intensity variations, the local degree of "roughness": it is the local singularity strength of intensity. Note that the global grey level of the finite-frame-size image can be measured by the defined finite magnitude value $M = a + \log I$, where I is the total intensity. By the way, this global grey level gives the reference level which is used to calibrate each singularity strength by comparison to their averaged value.

When it is possible to define a sensitivity level that follows a power law (like the preceding grey level) and when the singularity strength is not constant, the process is said to be multifractal. We can then introduce special statistics for the singularities.

Let us consider a $1-D$ support so that the singularities are labelled $\alpha(x)$, and introduce the fractal dimension calculated from the fractal analysis of the subset of S supporting the singularities $\alpha(x) \in [\alpha, \alpha+d\alpha[$, as shown in figure 4.1: labelling $f(\alpha)$ this dimension transforms [134] of a fractal object into the form of a multifractal one:

$$\begin{cases} N(\mu_\lambda \geq \lambda^\alpha) \sim \lambda^{-f(\alpha)} & \text{if } \alpha \geq <\alpha> \\ N(\mu_\lambda \leq \lambda^\alpha) \sim \lambda^{-f(\alpha)} & \text{if } \alpha \leq <\alpha> \end{cases} \quad (4.4)$$

$N(\mu_\lambda \geq \lambda^\alpha)$ and $N(\mu_\lambda \leq \lambda^\alpha)$ the number of balls of size λ required to cover all points x where the strength of the measure μ_λ in a ball of size λ is respectively larger or smaller than α (i.e. $\mu_\lambda(x) \geq \lambda^\alpha$ or $\mu_\lambda(x) \leq \lambda^\alpha$), and $<>$ is the averaging operator.

The graph $f(\alpha)$ versus α is the so-called multifractal or singularity spectrum. It draws the distribution of the singularity strengths using a special kind of statistics where the probability of finding the average value is exponentially larger than the probability of finding another value. It is similar to the Large Deviation theory of statistics (Oono, 1989, Evertsz and Mandelbrot, 1992 [389, 143] who formalize statistical physics ; Tsallis, 1988 who introduced a generalization of Bolzmann-Gibbs statistics for system having a multifractal distribution of its energy states [510]).

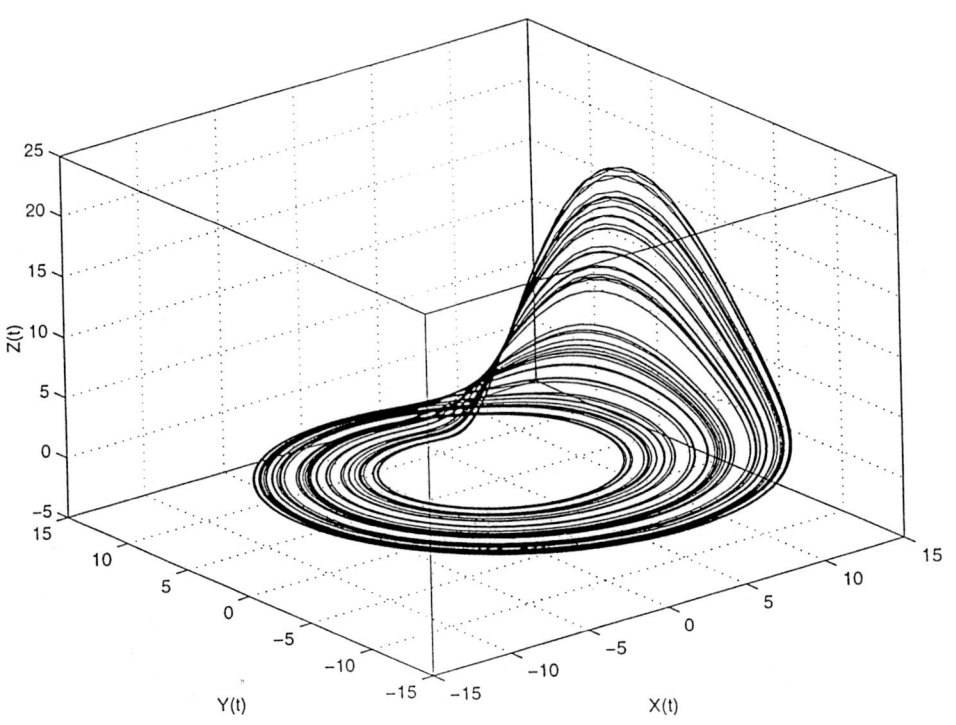

Rössler Attractor,

$$\begin{cases} \dfrac{dX}{dt} = -(Y+Z), \\ \dfrac{dY}{dt} = X + aY, \\ \dfrac{dZ}{dt} = b + XZ - cZ. \end{cases}$$

(after Dubois and Gvishiani, 1998).

4.2 Geomorphology (Continental and Marine) Hydrology

4.2.1 Continental Earth's Relief, Topography, Self-affine Fractals

Mandelbrot (1967) introduced the concept of fractals and determined the fractal dimension by measuring, (in the same way as Richardson (1961)), the length of the west coast of Great Britain [324, 433].

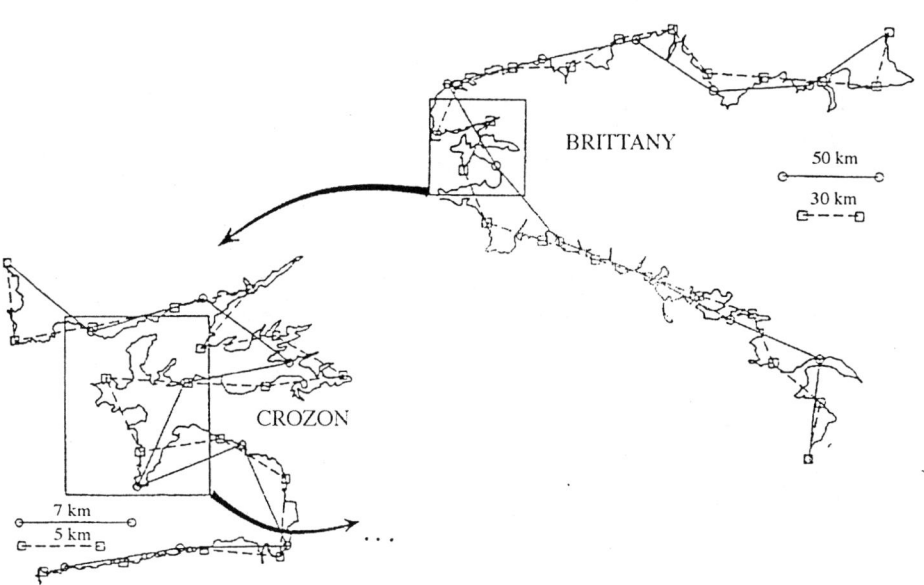

Figure 4.2: *The length of the Brittany's coast.* This map shows the application of the Richarson method to measure the length of the Brittany's coast using measuring rods of different lengths (after Dubois, 1995).

On figure 4.2 we represent the measurement technique. If N_i is the number of steps of length r_i and if the contour is fractal we get: $N_i = C/r_i^D$ where C is a constant and D is the fractal dimension of the set. For a given i the perimeter $P_i = N_i r_i = C r_i^{1-D}$. Therefore, if the set is fractal a bi-logarithmic plot of the perimeter length versus the road length will fall on a line with slope $1 - D$. We present some examples from different coastlines in figure 4.3. We obtained a fractal value of 1.25 for the Brittany's coast.

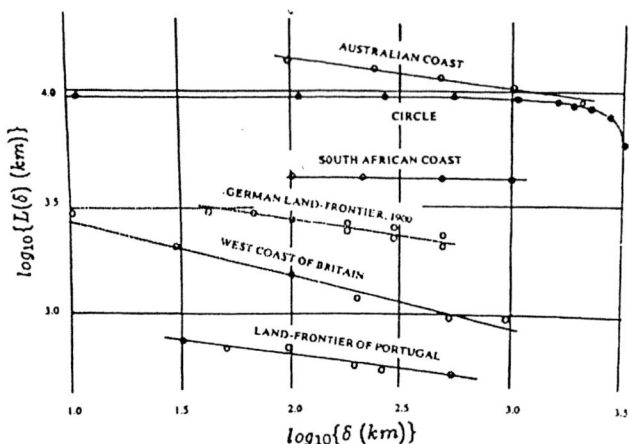

Figure 4.3: *The approximative length of some coasts, after Richardson (1939).* Plotted on a bi-logarithmic graph, the slope coast length versus measuring step, is $1 - D$ (after Mandelbrot, 1975).

The method may be also applied to contour lines of topography and bathymetry. In general, D is found to be between 1.3 and 1.5 [516]. It should be noted that not all topography is fractal (Goodchild (1980) and Turcotte (1997) [191, 516]), as we confirmed when studying contour lines of young intraplate volcanoes in the Pacific ocean.

The classification of oceanic islands was studied before the introduction of the fractal geometry. Korcak (1938) [287] proposed an empirical relation for the number of islands on with an area greater than a specified value, finding a power law of the form $N = C/r^D$. The subject was reexamined by Mandelbrot (1975) [325] who found a fractal dimension $D = 1.3$.

It was when searching for a model describing topographic variations that Mandelbrot introduced fractional Brownian walks. He proposed a family of stochastic processes generating random surfaces (actually fractal surfaces) depending on one parameter, the fractal dimension, which may be arbitrarily fixed. The resulting fractal island, is a very famous example in the fractal literature [326].

4.2.2 Bathymetry, Seafloor Roughness

In this subsection we follow a study presented by Ballu (1992) and Ballu et al. (1993) [37, 38] in which a bathymetric profile, along a flowline of the Mid-Atlantic ridge, south of Açores Island was studied using non-linear analysis (a correlation function method applied to the 300 km long profile and a moving window analysis) in addition to a classical spectral method.

Physical models explaining the creation of lithosphere along slow spreading ridges [307, 492] cannot account for the apparently chaotic short wave-length behaviours of the spreading centers, as expressed in their off-axis morphology. To address this question, Ballu (1992) [37] used a statistical and quantitative approach. The first goal was to determine whether the process which generates the lithosphere there is deterministic, as is assumed in physical models, or whether it is random, as suggested by the chaotic appearance of the detailed topography. The second purpose is to examine the temporal variability of the process within the geological context. As has been repeatedly demonstrated [154], seafloor bathymetry contains a great amount of morphological information which may reflect seafloor spreading processes.

Considering the similarity between real bathymetry profiles and synthetic models constructed using fractal geometry the fractal dimension seems an appropriate parameter to describe topography [325, 335, 319, 250]. Moreover, in Ballu's work the MAR topography sampled lies along a flowline of sea-floor growth and thus, to a first approximation, corresponds to a time series.

The fractal dimension can be considered in different ways for the purpose of analysing time variations in the accretionary process; either as a Hausdorff dimension, which can be calculated using a box-counting algorithm [325, 76], using the covariance function [188], or by methods based on the topographic energy spectrum [48, 154, 184, 183, 250]. In the study of the Sigma profile, neither the box-counting, ruler methods, nor the covariance function method were used to calculate the Haussdorff dimension because these methods cannot easily be used for self-affine fractals [76]. Topography and bathymetry have a self-affine fractal geometry, since the vertical and horizontal scales do not represent the same measure.

In their study, Ballu et al. (1993) [38] use the correlation function and spectral analysis. In order to describe the degree of organisation (random or deterministic) of the accretionary process, they consider the variations of the observable (here the digitalized values of sea-floor depth) and they compute the dimension of the attractor of the dynamic system which, if it exists, generated the series in time.

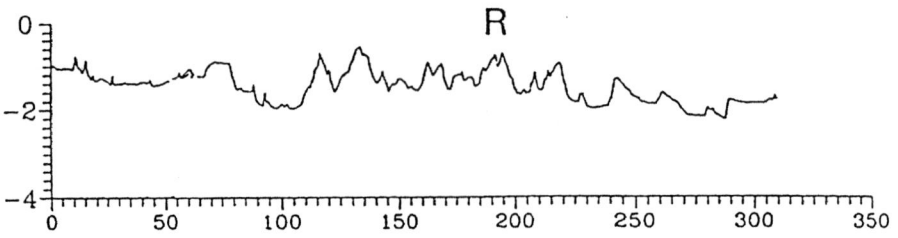

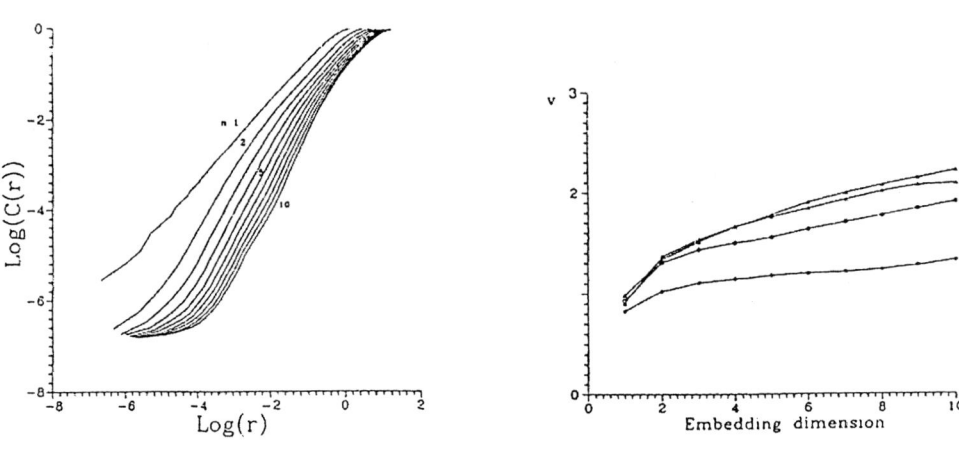

Figure 4.4: *Correlation function method applied to a bathymetric profile.* The first plot shows the bathymetric profile westward from longitude 29°20.3 W. R is the estimated position of the ridge axis. The second graph shows $\log C(r)$ vs $\log(r)$ for the whole bathymetric data series using increasing embedding dimensions. The slopes stop increasing while the embedding dimension keeps increasing. The last two graphs show the correlation dimension and similarity dimension versus the distance between the moving window's center and the beginning of the profile. The largest and smallest windows are respectively solid and empty circles. (after Ballu *et al.*, 1993 and Dubois, 1995).

Let us briefly recall the Grassberger and Procaccia (1983) method [195]. We denote the N points of such a long time series by

$$\{\vec{X_i}\}_{i=1}^N = \{\overrightarrow{X(t+i\tau)}\}_{i=1}^N. \tag{4.5}$$

where τ is an arbitrary but fixed time increment.
The correlation integral is:

$$C_r = \lim_{N \to \infty} \frac{1}{N^2} \sum_{i,j=1}^N H(r - |\vec{X_i} - \vec{X_j}|). \tag{4.6}$$

where $H(x)$ is the Heavyside function. Note that $C(r)$ behaves as a power law of r for small r, $C(r) \sim r^v$, when the attractor has a finite dimensional geometry.
Thus from the series, it is possible to compute v, the slope of the $\log C(r)$ vs $\log r$ graph.

$C(r)$ is computed for a number of r values and may be considered as proportional to the statistical average of numbers of points of the phase trajectory inside the hypersphere centered at any point of the attractor. When the attractor has a fractal structure, any value $C_i(r)$ has an assymptotic behaviour with respect to r^v. This allow a good approximation of $C(r)$ to be obtained by averaging a limited number of $C_i(r)$. On the graph $\log C(r)$ vs $\log r$ we obtain a series of plot lines (figure 4.4) for the increazing values of n (the embedding dimension). When the slope saturates for a given value, the computation is stopped. The constant value obtained is v. the correlation dimension of the set, that is, the fractal dimension of the attractor.

In the figure we see that the correlation method applied to the whole profile shows a fractal behaviour and reveals the few number of degrees of freedom of the dynamical system which generates it.

In addition to this first test, the authors applied a classical spectral analysis to study the sea-floor roughness (see [48, 179]), computing the power spectrum using the Fast Fourier Transform method.

The spectrum obtained from bathymetric profiles is linear in a bilog graph for self affine fractals. The spectral density is given by a power law which depends on the wave number k ([184, 250]):

$$S(k) = k^s \tag{4.7}$$

$S(k)$ is the spectrum energy for k and s is a negative real number.
The spectra are drawn on a bi-logarithmic graph. When they are linear, the slopes give estimates of their fractal or similarity dimension.

For fractal geometry to be applicable to bathymetric profiles it is necessery that $-3 \leq s < -1$. The relationship between D and s is given by Gilbert and Malinverno (1988) (see [184]), $D = (5 + s)/2$, so $1 \leq D < 2$.

The correlation function analysis method showed that the seafloor topography is generated by a dynamic system, itself controled by a low-dimensional attractor. The system is thus deterministic and can be described by equations of only two or three parameters. To the extent that the spatial series is a time series, we can say that the generator system is not steady state and that the correlation dimension is unstable along the profile. On the other hand, the variation is relatively small and the dimension stays smaller than 3.

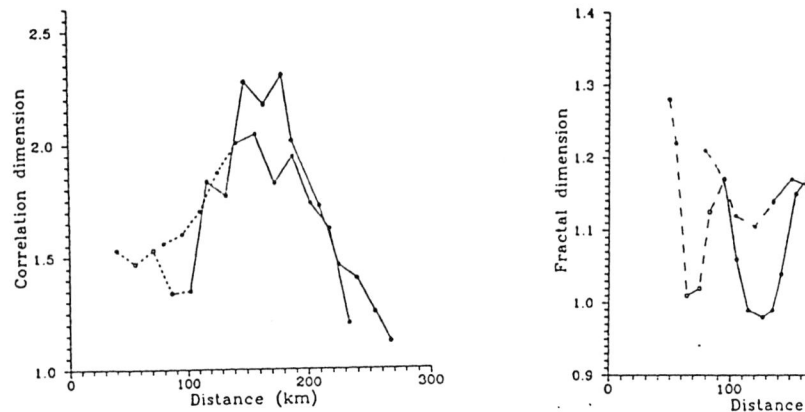

Figure 4.5: *Correlation dimension and similarity dimension.* On the two schemes are represented the correlation dimension and the similarity dimension versus the distance between the moving window's center and the beginning of the profile. Largest and smallest windows are respectively solid and empty circles. If the dimension is computed using interpolated data (to fill up the data gap), the connecting line is dashed. For interpretation see the text (after Ballu *et al.*, 1993 and Dubois, 1995).

The second result provides evidence of temporal variability in the accretion mechanism, using the fractal tool. The correlation dimension has a maximum at a zone centered about 170 km from the beginning of the profile. The fractal dimension computed from the power spectra, also shows a maximum at the same place (see figure 4.5). This is probably not a coincidence even though the two dimensions do not have exactly the same significance, the first being a function of the generator of the topography and the second a function of the bathymetry itself. Our profile is short for extrapolation, but it appears that the fractal dimension is nearly periodic along the profile. The corresponding time period is about 10 Ma, which is of the same order as life-span of some local ridge segments based on off-axis mapping [513, 402].

Much of the effort in seafloor analysis in recent years therefore has concentrated on using statistical quantification methods [51, 473] suited to investigation of complex morphology. For example, abyssal hills are created at mid-ocean ridges by extrusive volcanism and faulting and modified through time by mass wasting and sedimentation. The product of these four interacting processes is a complex, chaotic, multiscale morphology that defies simplistic (i.e. deterministic) quantitative description [189].

In a study very similar to the above-described study of Ballu, Smith and Shaw (1990) [458, 474] remarked that the seafloor is unique among Earth's topographic surfaces in that the depth sampled along a track parallel to the spreading direction approximates a spatial time series. They applied the methods of Grassberger and Procaccia to a sequence of topographic slopes calculated from Sea Beam swath bathymetry data collected along a flow line. Their results differ from Ballu's results in that the estimated values of the correlation dimension do not roll off to a constant as the embedding dimension is increased. This indicates that the system can not be described by a low order attractor. The difference between the results of the two studies lies in the choice of the parameters: the rough bathymetry in the Ballu's study, the slope of the small surface facets of the seafloor in the Smith and Shaw's study. The dynamical system which generates the second is apparently more complicated than the first which deals only with the accretionnary rate of the ridge.

4.2.3 Fractal and Multifractal analysis applied to river basins and to river flows

Before presenting a specific approach to the geomorphology of river drainage networks, we present a very simple application to river platforms that defines their scaling properties. The fractal analysis using the Richardson method (1961) as adapted by Mandelbrot (1967, 1975) [433, 324, 325], was applied by Beauvais et al. (1994) [45] to the Mbomou basin in the South-eastern Central African Republic. The rivers of Haut-Mbomou (figure 4.6) were divided into areas with different channel bed-slopes. The obtained fractal dimensions range from 1.08 to 1.32.

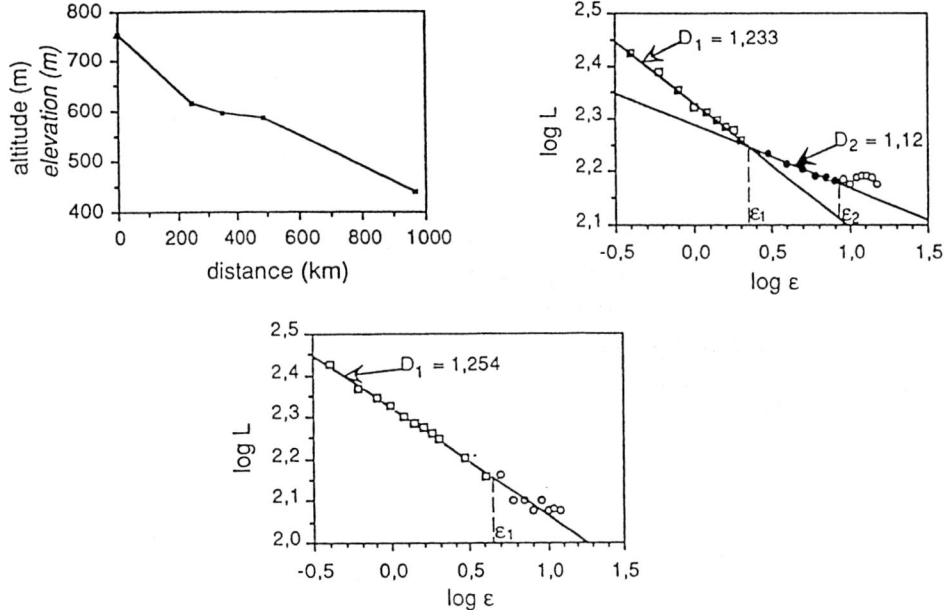

Figure 4.6: *Géomorphology of a hydrographic network*. This map shows the High Mbomou Basin, a tributary of the Oubangui and Congo rivers.

The analysis distinguished two groups of rivers, depending on channel bed-slope. The first group rivers have gradients higher than $0.5 m.km^{-1}$, and present two fractal dimensions, indicated by two regression-curve slopes (figure 4.6). The second group, with gradients lower than $0.5 m.km^{-1}$, present only one fractal dimension ranging from 1.13 to 1.29. These results were compared to environmental factors. The river platforms have two fractal dimensions where the bed slopes are high: over quartzitic and schisto-quartzitic

rocks (generally stongly fractured) and where lateritic soils are relatively thin. These regions are in the upstream parts of the basin where mean annual rainfalls are lowest and where the vegetation cover is thinnest.
In another paper Beauvais and Montgomery (1996) studyied the influence of valley type on the scaling properties of river planforms [46]. Scaling properties of 44 individual river planforms from the Cascade and Olympic Mountains of Washington State were defined using the divider method. Analysis of the standardized resuduals for least squares linear regression of Richardson plots reveals systematic deviations from simple self-similar that correlate with geomorphological context defined by valley type:

- A single fractal dimension describes rivers flowing through bedrock valleys.

- Those flowing in inherited glacial valleys exhibit two distinct fractal dimensions, with a larger fractal dimension at small scales.

- Rivers flowing in alluvial valleys are also described by two fractal dimensions, but with a larger dimension at large scales.

The authors further find that the wavelength of the largest meander defines an upper limit to scaling domain characterized by fractal geometry. These results relate scaling properties of river planforms to the geomorphological processes governing valley floor morphology [46].

Let us briefly recall the quantitative stream ordering (Horton, 1945 ; Strahler 1957) [249, 486], and models of this ordering based on Fractal trees (Tokunaga, 1978, 1984; Tarboton *et al.*, 1988 ; Peckhaam, 1995 ; Turcotte, 1997) [405, 500, 508, 509, 518].
Long before the concept of fractals was introduced, Horton (1945) and Strahler (1957) proposed a quantitative stream-ordering system (Figure 4.7).

The system calls the upstream tributories "first-order" streams as,which then branch into second-order, third-order, etc streams. A first-order stream can directly join a second-order or a third-order stream, a second-order stream can join a third-order stream or fourth-order stream, and so forth.
Horton then defined the bifurcation ratio $R_b = N_i/N_{i+1}$ and the length-order ratio $R_r = r_{i+1}/r_i$ where N_i is the number of streams of order i, (N_{i+1} the number of streams of order $I+1$) in the given drainage network, r_i is the mean length of streams of order i.
Empirically, Horton observed that the two ratios were constant for a range of stream orders in a given drainage network. That is called Horton's law.

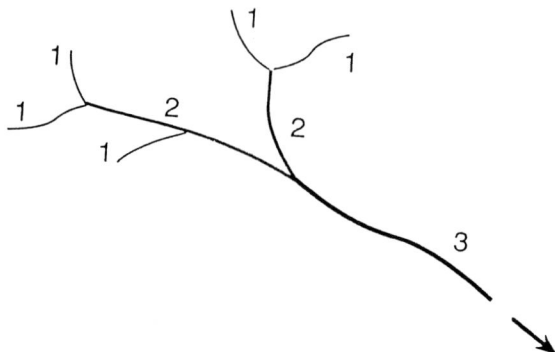

Figure 4.7: *Strahler stream ordering system.* The different order tributaries in a drainage network are numbered, for the upstreams tributaries, first-order streams, $i = 1$, then combination of two first-order streams give a second-order stream $i = 2$ and so on... .

Coming back to our fractal approach we may compute a fractal dimension of a drainage network by writing :

$$D = \frac{\ln(N_i/N_{i+1})}{\ln(r_{i+1}/r_i)} = \frac{\ln R_b}{\ln R_r}. \qquad (4.8)$$

These observations imply that the standard stream-ordering parameters are directly related to the fractal dimension of the network.

Turcotte (1997) observe that the number-length statistics for the United States are coherent, with a fractal dimension of 1.83 [516].

DLA technique (Diffusion-limited aggregation)

. The DLA technique was used by Masek and Turcotte (1993) [345] to generate realistic drainage networks. Many studies have been devoted to DLA since Van Damme *et al.* (1986) presented their interpretation of fractal viscous fingering in clay slurries [519]. The principle consists of randomly introducing an accreting cell on a "launching circle". The accreting cell then follows a random path on the grid until it accretes to the growing cluster of cells or wanders accross the "killing circle" (see figure 4.8). In both cases the random path is stopped and a new cell is introduced in the launching circle.

Geomorphology

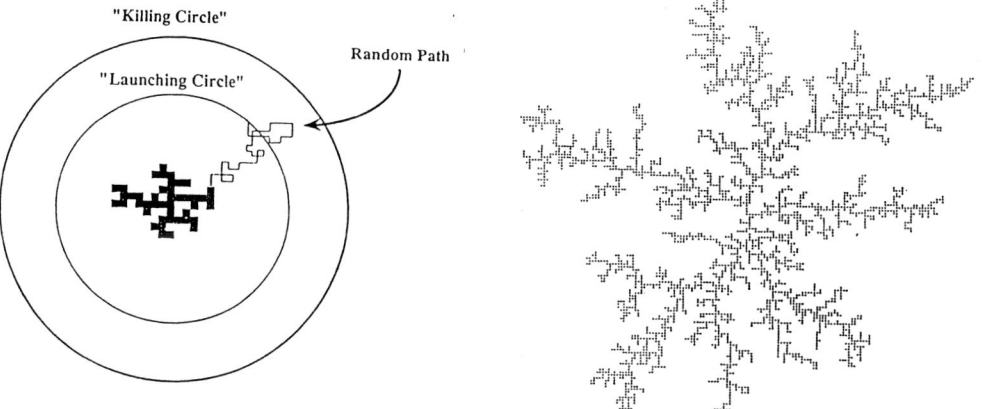

Figure 4.8: *Diffusion-limited aggregation*. The DLA growth of a cluster and its application to drainage river network (figure 8.14, 15 .

In the models for drainage networks, the random walkers (the accreting cells of the DLA model) correspond to units of water flux (rainfall and overland flow) that migrate over a relatively flat surface until they find a gully (network) in which to flow (see Tucotte, 1997 [516]). When the flux enters the gully, it erodes and expands the network (see figure 4.9).

There is a very large body of work (see, for example, Turcotte p198-199 : Van Damme et al, 1986 [518, 519]) concerning DLA random growth. This technique can be used to generate realistic drainage networks (Dubois 1988 unpublished project INSU ATP; Masek and Turcotte, 1993 [345])

Drainage basins and multifractals

Before developping the multifractal characterization of river basins we note that not every spatial distribution of a variable has a multifractal spectrum (Ijjasz-Vasquez *et al.* (1992) [258]). The multifractal spectrum of a random field where $P_r(x)$ increases as r^2 (box size r at point x), as every grid box has the same value of α, $N_r(x)$ is equal to the total number of grid boxes and $f(\alpha) = 2$. So, the multifractal graph is not a curve but a single point ($\alpha = 2$, $f(\alpha) = 2$).

We described the theoretical framework of multifractal analysis in our first book [134]. Here, we use this method to study the distribution and spatial organization of a variable of interest (for example, energy expenditure or mass) over a certain set (a river basin).

In their analysis Ijjasz-Vasquez *et al.* (1992) studyied multifractal spectra of

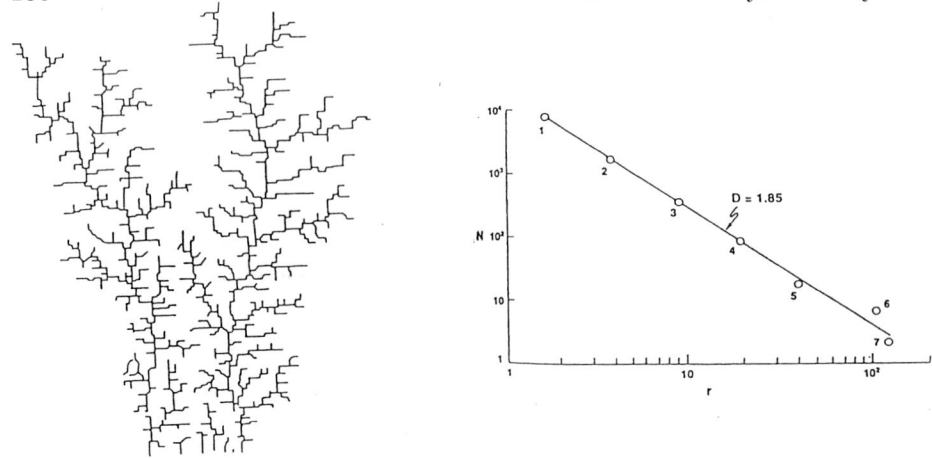

Figure 4.9: *A DLA drainage network.* In the figure the simulation was done on a 256×256 grid of cells (after Turcotte, 1997). The highest-order streams in the simulation are seventh order and the results correspond to actual drainage networks with a fractal dimension $D = 1.85$.

different variables calculated using topographic data from digital elevation maps with 30 m spacing over the basin. Every grid block was considered as a pixel and was assigned four variables [258]:

- The local slope and flow direction following the steepest gradient S_i.

- The accumulated contributing area draining through the point A_i, used as a surrogate for discharge.

- The energy expenditure P_i, which is the product of contributing area multiplied by the slope.

- The channel initiation function $a_i = \beta A_i^m S_i^n$. This function, which models a variety of processes related to channel initiation and river basin evolution, was developed by Willgoose et al. (1991), [540].

The multifractal analysis of these variables was applied to 9 river bassins across the US which were selected for variety in geography and climate.
Let us briefly recall the method. A grid of boxes of size r is superimposed over the set (here drainage basins). Every grid box is assigned the value of the integral over that box of the variable under study (here the four variables previously defined). The result obtained in the box of size r around the point x is denoted by $P_r(x)$. $P_r(x)$ is a probability measure: its value depends on

Geomorphology

the location of the box and also on the size of the box. The behaviour of $P_r(x)$ as a function of r shows the organization of the variable around x; It can be described by:

$$P_r(x) \sim (r/L)^\alpha \tag{4.9}$$

where L is the domain size and α is the scaling exponent (see Halsey et al., 1986 [225])

As different points may have different values of α, to complete the description of the drainage basin and the spatial organisation of the measure $p_r(x)$, we may count the number $N_r(\alpha)$ of grid boxes with common values of α.

$$N_r(\alpha) \sim (r/L)^{-f(\alpha)} \tag{4.10}$$

$f(\alpha)$ is the fractal dimension of the set of boxes with the same α-value and it measures not only the proportion of points with similar characteristics around them, but also the degree of clustering of these points.

The curve $f(\alpha)$ vs α is called the *multifractal spectrum* of the variable under study.

There are different way to construct this multifractal spectrum [134]. In the present study the authors studied the cumulants of order q of $P_r(x)$ defined as:

$$C_q(r) = \sum_i [P_r(x_i)]^q \tag{4.11}$$

where the centers of the boxes have are indexed
Following Halsey et al., (1986) [225], the authors define the function $\tau(q)$ as the exponent at which the cumulant $C_q(r)$ scales with the box size r.

$$C_q(r) \sim (r/L)^{\tau(q)} \tag{4.12}$$

The exponents $\tau(q)$ are related to the generalized dimensions D_q (see Dubois and Gvishiani, 1998 [134]) as:

$$D_q = \tau(q)/(q-1) \tag{4.13}$$

It is possible to relate the exponents $\tau(q)$, α and $f(\alpha)$.
Following [258] we note that

$$C_q(r) \sim \int [P_r(x_i)]^q \sim \int (\frac{r}{L})^{-f(\alpha)}(\frac{r}{L})^{q(\alpha)} d\alpha \tag{4.14}$$

Since r is very small, the value of the integral is dominated by the largest value of the integrand, which occurs when $q\alpha - f(\alpha)$ is minimized, i.e.

$$df(\alpha(q))/d\alpha = q \qquad (4.15)$$

Therefore

$$C_q(r) \sim r^{q\alpha(q)-f(\alpha(q))} \qquad (4.16)$$

Using equations (4.12) and (4.16) we get:

$$\tau(q) = q\alpha(q) - f(\alpha(q)) \qquad (4.17)$$

Taking the derivative with respect to q and using equation (4.15) we have:

$$\frac{d\tau(q)}{dq} = \alpha + q\frac{d\alpha}{dq} - \frac{df}{dq} = \alpha. \qquad (4.18)$$

We summarize here the three following steps used by the authors to calculate the multifractal spectra

1. find the cumulants $C_q(r)$;

2. find their scaling behaviour and calculate $\tau(q)$;

3. use equations (4.16) and (4.18) to calculate α and $f(\alpha)$ with the values of $\tau(q)$.

Applications to river basins

The method was applied to nine very different basins selected for variety in geography and climate. The areas of the basins vary from 2834.0 km^2 (St Joe River, Montana, Idaho) to 98.2 km^2 (Schoharie Creek Headwaters, New York). The multifractal spectrum were calculated for each of the variables. Figure 4.10 shows the mean spectra for each of the three variables P_i, A_i and S_i.

The main conclusions [258] from this study are:

- The multifractal formalism is useful for describing the spatial distribution and scaling properties of river basins that interest hydrologists and geomorphologists.

- The fact that the variables from very different basins present multifractal characteristics and have very similar multifractal spectra shows the existence of a common underlying structure of organization.

Geomorphology

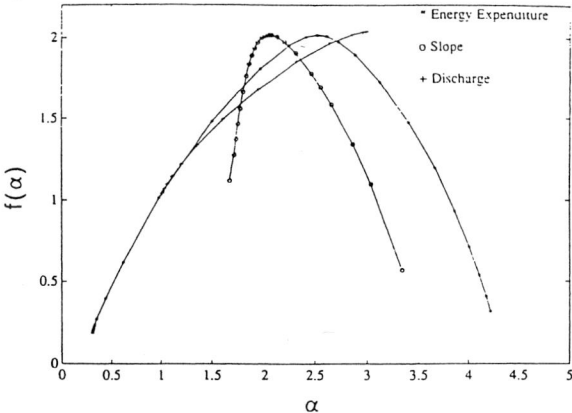

Figure 4.10: *Multifractal spectra of the three main variables in 9 drainage networks.* The spectrum of S_i (local slope) is much narrower than that for P_i (discharge multiplied by the slope). A_i is the accumulated contributing area draining through a point (after Ijjasz-Vasquez *et al.*, 1992).

- The comparison of river basins with models of network structure can show differences which may not be obvious. The multifractal spectrum of Scheidegger's (1967) model [452] may be one of them.

We will come back to multifractal analysis when studying the dynamic system of river discharge in the time domain (chapter 5, section 5.1).

For readers interested in this multifractal approach of fractal river networks and the induced geomorphology (OCN Optimal Channel Network) we suggest to look at three papers by Rinaldo *et al.* (1993, 1995) and Rigon *et al* (1993) [436, 437, 434].

4.3 Gravity Anomalies and Structural Inversion Modeling

In that section we look at scaling properties of field potential used in geophysical prospecting.

Introduction of fractal analysis methods to problems of source detectability

This problem was studied previously (Dubois, 1995, 1998 [129, 131]) to quantify the roughness of the altimetric geoid using the method proposed by Dubuc et al. (1989) [139]. There the fractal analysis used corresponded to a kind of filtering at different wavelength scales. The question we adress here concerns the ability of a field network to resolve a specific problem (e.g., the pattern of anomaly sources). The question could be posed as follows:"What is the best geometry of a measurement network, to solve a certain problem?" An important number of papers have been published about this problem (Lovejoy et al., 1986 ; Thorarinson and Magnusson, 1990 ; Pilkington and Todoeschuk, 1993 ; Pilkington et al., 1994 ; Maus and Dimri, 1994, 1995, 1996, [313, 506, 414, 415, 347, 348, 349]).

Fractal characterization of an inhomogeneous measuring network (the detectability problem) In order to correctly present the problem we follow Lovejoy et al. (1986) [313], who were the first to examine and quantify the consequences of an inhomogeneous geophysical measuring network on the quality of the data obtained. For a given network, we have:

$$\langle n(L) \rangle \propto L^{D_m}$$

where $n(L)$ is the number of stations within a circle of radius L centered at any station, and $\langle n(L) \rangle$ is the average over all stations. In the Lovejoy et al.' study, the network was the worlwide meteorological network (9,563 stations). Avoiding double counting, each station has $9,562/2 = 4,781$ values of L, so there are $9563 \times 4781 = 45,720,703$ independent values that go into the histogram for estimating $\langle n(L) \rangle$. Plotting on a bi-logarithmic graph shows a good linearity implying the relation

$$\langle n(L) \rangle \propto L^{1.75}$$

where $D_m = 1.75$, $L_{max} = 7500 km$, and $L_{min} = 0.3 km$.

If we call E the dimension of the embedding space (the space where the studied object is included) the fact that $D_m < E$ allows us to determine the limits of detectability for the network. A network of dimension D_m allows the detection of a phenomenon of dimension D_p only if the two sets intersect. A theorem of fractal geometry (Falconer, 1992 ; Dubois, 1995) says that this is possible only if $D_p > E - D_m$ allowing us to define the minimum detectable dimension of a phenomenon. In the meteorological network example we obtain $D_p > 2 - 1.75 = 0.25$ which can be interpreted as follows: Any meteorological phenomenon whose dimension is less than 0.25 (i.e. turbulence) is undetectable by the network.

Let us notice that a homogeneous network (regular grid) has a dimension 2. The minimum detectable dimension is then 0, and, all the phenomena larger than the network grid size will be detectable.

4.3.1 Fractal analysis of gravity anomalies

In this subsection we will follow the Thorarinsson and Magnusson study (1990) made in South-West Iceland [506]. Their objective was to determine density values for the Bouguer reduction of two gravity data sets using a new method based on the minimization of the roughness of the Bouguer anomaly surface. A common approach is to estimate this density by minimizing the correlation of the Bouguer anomaly with the topography (Nettelton, 1939) [379], which assumes that the topography is supported by a rigid crust rather than by isostatic compensation.

In their study Thorarinsson and Magnusson operated on two areas, the first a 15×15 km square on Hengil volcano, the second a 150×150 km square on the south-west part of Iceland. In the first area, the gravity stations were spaced about 1 km apart, and in the second one about 8 km. Each zone contained about 300 gravity stations.

The roughness of the gravitational surface (from free air and Bouguer anomaly maps) was defined using the surface variogram from Mark and Aronson (1984). Such surfaces, known as fractional Brownian surfaces (Mandelbrot, 1975), have fractal dimensions between 2 (a plane) and 3 (a solid volume) [336, 326].

For a fractional Brownian surface the expected value of the squared elevation difference is:

$$E[(Z_p - Z_q)^2] = k(d_{pq})^{2H} . \tag{4.19}$$

where Z_p and Z_q are the values of the surface at points p and q, d_{pq} is the horizontal distance between points, and H equals $3 - D$.

A linear relationship between $E[\cdot]$ and d_{pq} on a bilogarithmic graph indicates self-similarity. We can estimate the fractal dimension D of the surface by:

$$D = 3 - (b/2). \qquad (4.20)$$

where b is the slope of the line through the points.

The authors applied this method to free-air and Bouguer anomaly variograms computed on both areas (Hengill volcano area and South-west Iceland area).

The results are shown in figure 4.11

The interpretations of the results are:

1. The free air anomaly variogram in the Hengill area shows a linear relationship for 10 km range, indicating a fractal surface that is interpreted as crustally supported topography. The nonfractal character at larger distances is interpreted as isostatically compensated topography.

2. The free air anomaly variogram in Southwest Iceland shows a linear relationship out to 25 to 30 km, indicating a fractal surface also interpreted as crustally supported topography. The difference between the break distance in the two areas is due to the lower thickness of the crust under Hengill (rift zone) than under Southwest Iceland area.

3. The Bouguer anomaly variogram shows fractal surfaces from 2 to 10 km (Hengill) and 5 to 30 km (Southwest Iceland). Graph 5 shows the minimisation of the fractal dimension as a function of densities used in the topographic correction. The best fits are obtained for $2730 kg.m^{-3}$ (Hengill) and $2490 kg.m^{-3}$ (Southwest Iceland).

To summarize, the fractal analysis enhanced the information which could be obtained from the gravity data.

Gravity Anomalies

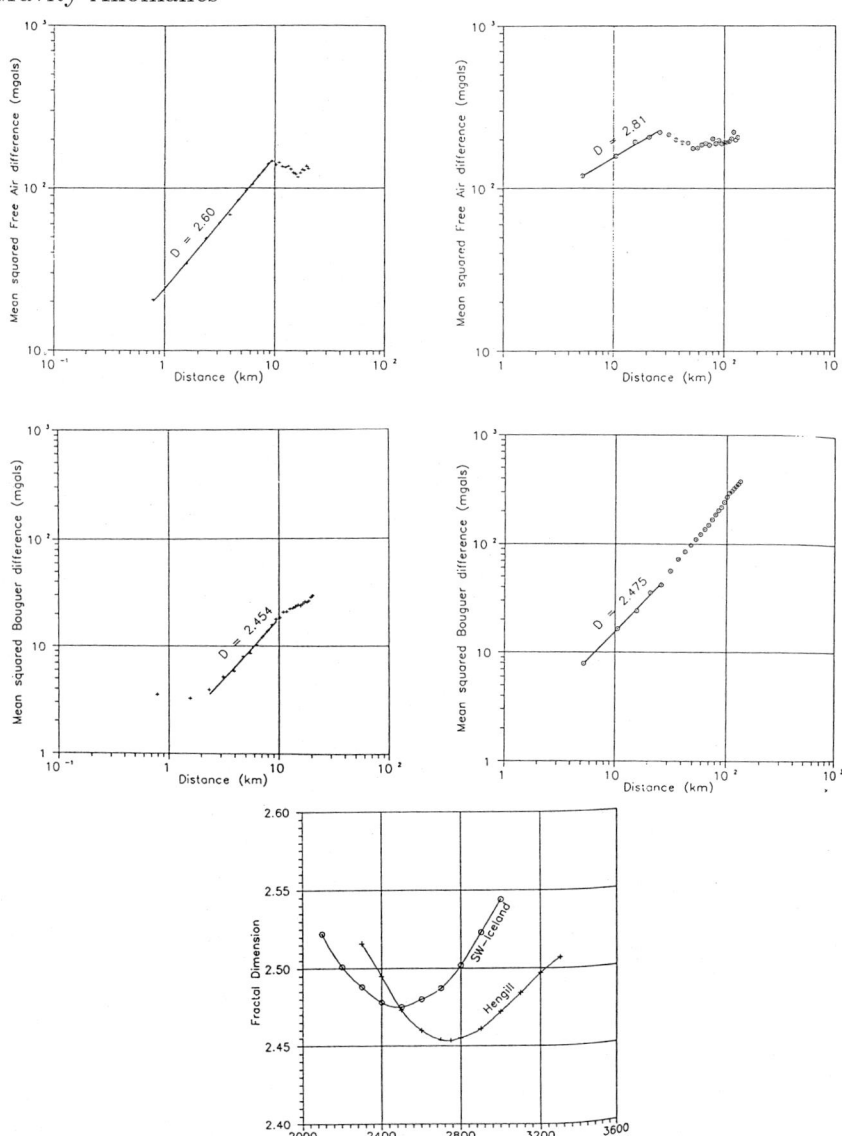

Figure 4.11: *Fractals and gravity anomalies.* Graphs 1 to 4 show, respectively, free air and Bouguer anomaly variograms versus horizontal distance of surface points applied to Hengill volcano and Southwest Iceland area. Graph 5 shows the fractal dimension of Bouguer anomaly surfaces versus crust density. For interpretation see text (after Thorarinsson and Magnusson, 1990).

4.4 Geomagnetism

4.4.1 The Fractal Structure of the Interplanetary Magnetic Field

One of the pioneering study of the fractal properties of the magnetic field was published by Burlaga and Klein (1986) [82] who analysed the magnetic time series recorded during the interplanetary flight of Voyager.
The spacecraft measured the magnetic field near 8.5 AU along its path from July 5 to August 24, 1981.by the space craft from June 5 to August 24, 1981. The field was self-similar over time scales from $\sim 20s$ to $\sim 3 \times 10^5 s$, and the fractal dimension of the time series of the strength and direction of the magnetic field was $D = \frac{5}{3}$ corresponding to a power spectrum $P(f) \sim f^{-5/3}$. These results were obtained using a method very similar to that chosen by Mandelbrot for the fractional Brownian function [333], which is a Gaussian scalar function $B(t)$ whith zero mean and a variance given by

$$(\delta B)^2 = \langle [B(t_2) - B(t_1)]^2 \rangle = (t_2 - t_1)^{2H} \qquad (4.21)$$

where H is a constant between 0 and 1. The function reduces to Brownian motion when $H = 1/2$. For stationary time series with $t_2 = t_1 + \tau$, $B(t)$ has a power law spectrum $P_B(f) \sim f^{-\alpha}$, where

$$\alpha = 2H + 1.$$

The case $H = 1/3$ corresponds to Kolmogorov variance and gives the Kolmogorov spectrum, $f^{-5/3}$, which describes inertial range turbulence in an incompressible fluid.
The function $B(t)$ can be viewed geometrically as a curve which has structure on every scale and which is statistically self-affine (see section 4.2).
Bulargia and Klein presented a method to calculate the fractal dimension D of the time series by computing its length $L(\tau)$, and obtained stable values for the fractal dimension of a large-scale fluctuation of the magnetic field. They defined the length $L(\tau)$ of the curve $B(t)$ as:

$$L(\tau) = \sum_{k=1}^{N} |\overline{B}(t_k - \tau) - \overline{B}(t_k)| / \tau \qquad (4.22)$$

where $\overline{B}(t_k)$ denotes the average value of $B(t)$ between $t = t_k$ and $t = t_k + \tau$.

Geomagnetism

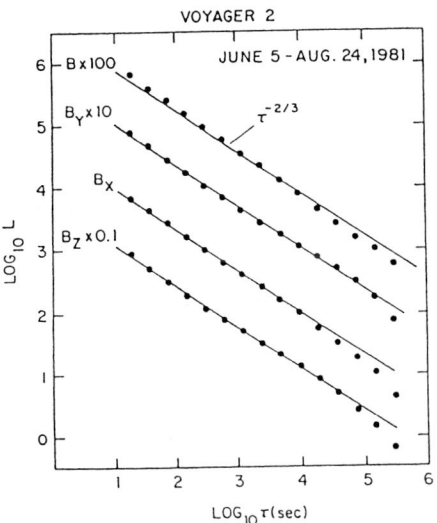

Figure 4.12: *Fractal structure of interplanetary magnetic field* - The fractal lengths of the curves representing the magnetic field magnitude and components as a function of time, computed using various averaging intervals τ. Lines with a slope of $-2/3$ are drawn through the points (after Burlaga and Klein, 1986).

For statistically self-affine curves, the length is expressed as $L(t) \propto \tau^{-D}$. Using this relation, one can estimate the value of D as the slope of the log-log plot of the length $L(\tau)$ vs the time interval τ (see figure 4.12). The fractal dimension D and the power index α of the power spectrum are related by Berry's expression $D = (5 - \alpha)/2$ [82].

Since the Kolmogorov spectrum for homogeneous, isotropic, stationary turbulence is $f^{-5/3}$, the Voyager 2 measurements are consistent with the observation of an inertial range of turbulence extending over approximately four decades in frequency.

4.4.2 Fractal dimension and power law for geomagnetic time series

Let us briefly mention the fractal characteristics of geomagnetic times series as observed by Barraclough and De Santis (1996) and Telesca *et al.* (1998) [41, 499]. We will reexamine geomagnetic time series when we develop the properties of dynamic systems in chapter 5 (especially section 5.4).

Barraclough and De Santis (1996) use a non-linear forecasting approach to investigate time variations in the geomagnetic field at three European locations over the last 130 years. The analysis of data in terms of first-differences (secular variation) of horizontal magnetic components made in phase space seems to exclude the pre-eminence of any stochastic or periodic behaviour. The results give some evidence that the geomagnetic field evolves as a non-linear chaotic system with unpredictable behaviour after times greater than a few years. We will further examine the chaotic aspect in chapter 5, but suffice to say for now that fractal structure is observed in the geomagnetic field [41, 82].

In the paper by Tedesca et al. (1998), the time series are data observed at three different magnetometric stations in Norway. The authors used the Higuchi method (1990) (see [241]) which gives an interesting relationship between the fractal dimension D and the spectral power law scaling index α. The value of D can be obtained much more precisely than the value of α, because of the better correlation between the curve length $L(\tau)$ and the time interval τ than between $P(f)$ and the frequency f. The frequency domain analysis of the series under study reveals a power spectrum following the power law $P(f) \propto f^\alpha$, with the scaling exponent α representing the irregularity of the time series.

We shall investigate in more detail the dynamical aspect of the geomagnetic time series in the next chapter.

4.5 Tectonics, Seismicity, Volcanology

4.5.1 Renormalization group theory

We can use renormalization group theory method to study fracturing and fragmentation of rocks, using the SOFT algorithm to model earthquake occurence. We can use the same approach to model the geomagnetic dynamo as well as fluid percolation in rocks. We will present the principles of the method here by applying them to the percolation problem (see figure 4.13. The foundation of the theory for the geophysical systems studied here lies in the idea that macroscopic phenomena are generated from the evolution of microscopic scale structures (Narteau, 1996). A simple system is considered at the smallest scale, then the problem is rescaled to utilize the same system at the next larger scale. For the rescaling to work we need to assume two rules:

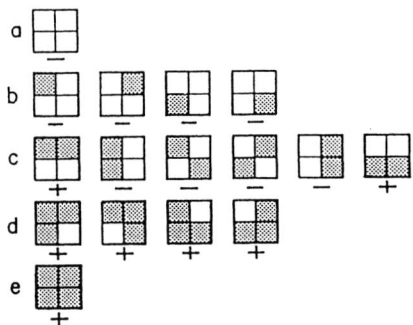

Figure 4.13: *A four elements permeability model in a renormalization approach.* Let us start from a cell (large square) consisting of four elements (smaller squares) . Each elements is permeable (grey) or impermeable (white) to a liquid crossing from the left to the right. According to the four possible combinations of permeable or impermeable elements, we compute the possibility that the cell is permeable or not. The cell is permeable if there is a continuous permeable path from left to right. The configurations that are permeable from left to right are denoted by (+) and the confugurations that are impermeable from left to right are denoted by (−). See the text for details about the 16 possible configurations (after Turcotte, 1992).

- All the elements, for any degree, have the same properties.

- The transfer processes through the different scales are independent of the scale order.

Any observer who describes the system at a certain degree could not distinguish the absolute size of the different structures whitout knowing the scale. One can say that the system is renormalized at each degree.
In order to study the evolution of system properties through different ranks, it is necessary to use probability laws. That is what we shall do now.
This method was proposed in the geophysics field by Madden (1976, 1983), [316], (see also: [229, 408, 431, 487, 543].
We will illustrate this process using a two dimensional example dealing with percolation, developed by Turcotte (1992).
We consider a cell consisting of four elements (figure 4.13). At the lowest first order the probability p_1 that the first-order cell is permeable, is determined in terms of the probability p_0 that the individual first-order element is permeable, the probability that this first-order is impermeable is $1 - p_0$. The cell is defined to be permeable if there is a continuous permeable path from left to right (see figure). The problem is then renormalized and four first order cells become the four second order elements inside a second order cell. We compute the probability p_2 that this second-order cell is permeable as a function of first order probability p_1 and so on.
Let us first obtain the relation between p_1 and p_0 considering all possible configurations (figure 4.13)
A cell (a two-dimensional square) containing four impermeable or permeable elements has 16 possible configurations which are drawn in figure 4.13.
On the first line we have configurations in which all four elements are impermeable. The probability is $(1 - p_0)^4$ and there is only one configuration out of 16 possible.
On the second line we have the possible 4 configurations with only one element permeable.
The third line shows the possible configurations with two permeable elements. The probability of this case is $p_0^2(1 - p_0)^2$, and two of the six configurations result in a permeable cell.
The fourth line shows the four possible configurations with three permeable elements. The probability of this case is $p_0^3(1-p_0)$, and all four configurations result in a permeable cell.
Finally the last line shows the sole configuration for a cell with four permeable elements. The probability of this case is p_0^4.

Tectonics, Seismicity, Volcanology

Now let us examine which cells are permeable out of the 16 possible configurations. There are six permeable and ten impermeable configurations (underlined by signum $+$, and signum $-$ in the figure). Thus, we may give the relation between the probability that a cell is permeable as a function of p_0, which is the sum of the permeable configurations, two on line 3, four on line 4, and one on line 5, :

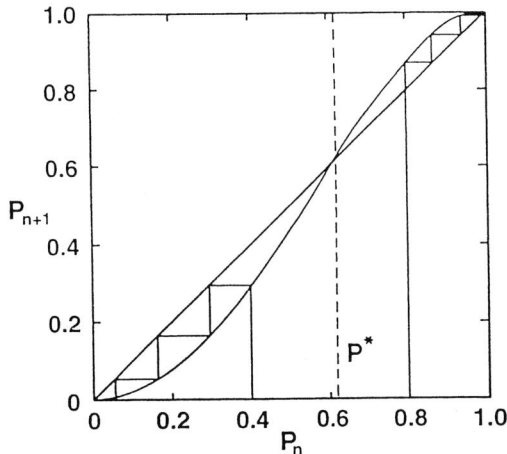

Figure 4.14: *The probability function of a cell being permeable as a function of the permeability probability of its four elements.* The relationship between these probabilities is given by the quadratic map (Feigenbaum logistic mapping, see the text). This function has three fixed points, two are stable, 0 et 1, the third is unstable for a une probability value of 0.618 (after Turcotte, 1992).

$$p_1 = 2p_0^2(1-p_0)^2 + 4p_0^3(1-p_0) + p_0^4 = 2p_0^2 - p_0^4. \qquad (4.23)$$

The same relation may be obtained for the rank 2 cells, whose 4 elements are rank 1 cells.
In terms of probability we get:

$$p_2 = 2p_1^2 - p_1^4 \qquad (4.24)$$

and for ranks n and $n+1$,

$$p_{n+1} = 2p_n^2 - p_n^4. \qquad (4.25)$$

We notice that the probability for a given order as a function of the probability of the previous order is the same as in the Logistic mapping, for $\mu = 1/2$, which is the Feigenbaum attractor [148, 149].
The fixed points of this logistic map are the solutions of the equation

$$x = 2x^2 - x^4$$

Inside the interval $0 < X < 1$ these 3 fixed points are: 0, 0.618, 1. It is easy to see, from the previous section that the fixed point 0 and 1 are stable and that the point at $x = 0.618$ is unstable.
These behaviours are illustrated in figure 4.14

4.5.2 Fragmentation, Fracturation

The observables

Let us fragment a rock with, for example, a hammer as if we were breaking up a lump of sugar. We observe that the fragments sizes vary considerably, from dust grains to fragments nearly as big as the original sample. It appears that the fragment sizes are distributed, with a number of fragments that increases as the fragment sizes decreases. The distribution of fragment sizes is a statistical problem which will be developed here.
We will propose some models for the fragmentation mechanism which are based on the renormalized group theory as developed in the previous subsection. In these models we will define a fragmentation mechanism related to the distribution of joints or the preexistence of weakness planes inside the rock.
A large variety of statistical relations has been established to correlate the distribution of fragment sizes, including log-normal, Pareto, Rosin and Rammler, Weibull and power-law distributions.
Turcotte (1986) [514] has shown that the power-law distribution is equivalent to a fractal distribution. In table 4.1, reproduced from Turcotte (1989), we present a series of values obtained when fragmenting different miscellaneous objects: gabbro, basalt, coal blocks, meteorite fragments, clays, gravels and sands.

An example of fragment distributions, the Weibull distribution

Let $N(m)$ be the number of fragments of mass greater than m, we obtain $N(m) = cm^{-b}$, where c is constant.
Remarking that $m \sim r^3$, we get, $N(m) = cr^{-3b}$.

TABLE 4.1

Objects	D
Gabbro fragmented by a projectile	
lead projectile	1.44
steel projectile (Lange et al., 1984) [296]	1.71
Meteorite fragments (McCrosky, 1968) [350]	1.86
Plane of weakness model (Turcotte, 1986) [514]	1.97
Disaggregated gneiss (Hartmann, 1969) [231]	2.13
Disaggregated granite (Hartmann, 1969) [231]	2.22
Fragmentation by a chemical 0.2kT explosion (Schoutens, 1979) [463]	2.42
Fragmentation by a nuclear 62 kT explosion (Schoutens, 1979) [463]	2.50
Coal fragments (Bennett, 1936) [52]	2.50
Interstellar dust (Mathis, 1979) [346]	2.50
Fragmented basalt (Fujiwara et al., 1977) [163]	2.56
Clay sand (Hartmann, 1969) [231]	2.61
Gravels and sand from alluvial terraces (Hartmann, 1969) [231]	2.82
Pilar of strength model (Allègre et al., 1982) [6]	2.84
Glacial alluvia (Hartmann, 1969) [231]	2.88
Rocky meteorites (Hawkins, 1960) [232]	3.00

Table 4.1: Some examples of fractal dimensions observed when fragmenting miscellaneous objects (after Turcotte, 1986).

We find here the classical definition of a fractal set $N \sim r^{-D}$ (see Dubois and Gvishiani, 1998 [134]), where $D = 3b$. This power-law distribution is therefore equivalent to a fractal distribution.

A Weibull-type distribution is sometimes observed experimentally. It is expressed as:

$$\frac{M(r)}{M_T} = 1 - \exp(-(\frac{r}{\sigma})^\alpha), \qquad (4.26)$$

where $M(r)$ is the cumulated mass of fragments of radius (i.e. (volume)$^{1/3}$) smaller than r, M_T is the total mass and σ, the mean size of fragments. For $r/\sigma \ll 1$, we come back to a power-law,

$$\frac{M(r)}{M_T} = (\frac{r}{\sigma})^\alpha. \qquad (4.27)$$

Differentiating, we get:

$$dM \approx r^{\alpha-1} dr. \tag{4.28}$$

Differentiating the definition $N \sim r^{-D}$ we obtain $dN \sim r^{-D-1} dr$. Between increments dM and dN, we have a relation, $dN \sim r^{-3} dM$, so we can combine these equations to get:

$$r^{-D-1} \sim r^{-3} r^{\alpha-1}, \tag{4.29}$$

That is, $D = 3 - \alpha$.

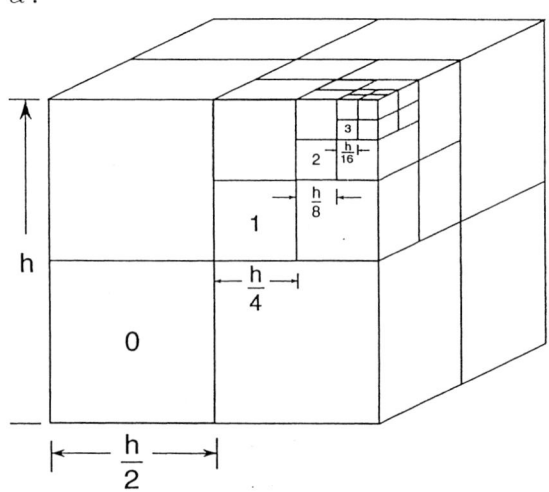

Figure 4.15: *The fragmentation model of a cube by Allègre et al. (1982)*. One starts from a cube with dimension h which may be fragmented in eight cubic elements with dimension $h/2$, which may again each of them be fragmented in eight cubes with dimension $h/4$, etc (after Allègre *et al.*, 1982 and Turcotte *et al.*, 1986).

Thus this Weibull-type distribution is equivalent, for $r/\sigma \ll 1$, to a fractal distribution.

Here let us remark that the exponential distribution is very distinct from the fractal distribution: the former is linear on a semilog graph and the latter is linear on a log-log graph. We will develop this point when studying the probabilities of occurence of natural events: the exponential distribution will be random, while the fractal distribution will be deterministic.

Some models

For a better understanding of the possible mechanisms let us follow an approach proposed by Allègre et al., (1982) and Turcotte, 1986 [6, 514]) that uses the renormalization group iteration method (Madden, 1976, 1983 [316, 317]).

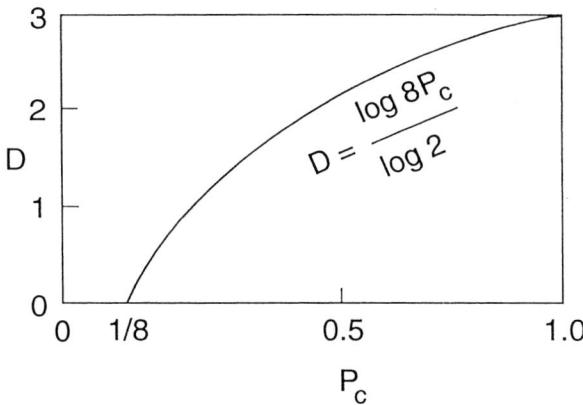

Figure 4.16: *Dimension and probability of fragmentation.* Variations of D as a function of p_c, fragmentation probability varying from $1/8$ to 1 (Turcotte, 1986).

Application to fragmentation of the renormalization group method

The basic hypothesis of the renormalization group approach is the assumption that the probability p_c that a cell will fragment into eight elements is the same at all orders. The original body, a cell (a cube in the simplest case) has a linear dimension of h, the eight largest fragments have a linear dimension of $h/2$ (fig. 4.15), these inturn fragment to give eight fragments with a linear dimension of $h/4$ etc.(see [6, 514]).
Considering now m levels of fragmentation, the total number of particles is

$$N_m = (1 - p_c)[1 + 8p_c + (8p_c)^2 + ... + (8p_c)^m]. \qquad (4.30)$$

From the relation $N \approx r^{-D}$, it follows that

$$\frac{N_{m+1}}{N_m} = \frac{h^{-D}/2^{-D}}{h^{-D}} = 2^D. \tag{4.31}$$

This ratio is a function of p_c:

$$\frac{N_{m+1}}{N_m} = 8p_c, \tag{4.32}$$

Therefore,

$$8p_c = 2^D, \quad \text{and} \quad D = \frac{\log 8p_c}{\log 2}. \tag{4.33}$$

Because $1/8 < p_c < 1$, D varies between 0, (for $p_c = 1/8$) and 3 ($p_c = 1$) (remember that, from one rank to the next, p_c is the probability for a cube to be fragmented into 8 elements and that, once given, p_c remains the same at all ranks).

The variation of D as a function of p_c is outlined in figure 4.16. We can see further into the details of different models when we are able to compute the relation between the probability p_n that a cell of the n-order is fragile and the probability p_{n+1}, that an n-order element (i.e., a cell of the $n+1$-order) is fragile. The characteristic dimension of an n-order cell is $(h/2)^n$ and the dimension of an $n+1$-order element is $(h/2)^{n+1}$.

This relation, which is the basis of the renormalization group modeling method, was established by Allègre et al. (1982) [6] in the "sound pillar" model (also called "pillar of strength" model) and by Turcotte (1986) [514] for the "weakness plane" model. Let us describe the two models.

The "sound pillar" and the "weakness plane" models

The weakness of a particular cell (cube of n-th order) is determined by the weakness of its elements (cubes of (n+1)-th order). We developed the principles of the method and showed how to calculate the relation between p_n and p_{n+1} in section 4.4.1. For more detail see Allègre et al (1982), Smalley et al (1985), Turcotte (1986, 1989), Dubois (1995, 1998) [6, 470, 514, 515, 129, 131].

In each cell, one may have from 0 to 8 fragile elements, which leads to $2^8 = 256$ possible combinations. By excluding multiplicity, these can be reduced into 22 different topological configurations (Figure 4.17). The cells will be considered as "sound" or solid when they contain a "sound" pillar, which is a corner stone made from two solid elements (Figure 4.18). In all other cases, the cell is fragile. The sound pillar occurs in 6 of the 22

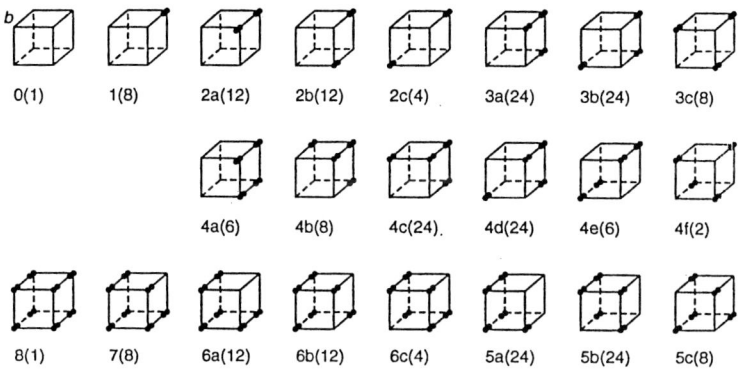

Figure 4.17: *The 22 different topological configurations in the pillar of sound elements model.* There are 256 possible combinations in a cube fragmentation, which may be reduced to 22 topological different configurations (Allègre et al., 1982).

---

configurations (Figure 4.17). The relationship, which can be deduced from this, is (Turcotte, 1986) [514]:

$$p_n = p_{n+1}^8 + 8p_{n+1}^7(1-p_{n+1}) + 16p_{n+1}^6(1-p_{n+1})^2 + 8p_{n+1}^5(1-p_{n+1})^3 + 2p_{n+1}^4(1-p_{n+1})^4. \quad (4.34)$$

The graph of relation 4.34 is presented in figure 4.19, where the bisecting line $p_n = p_{n+1}$ is drawn. This is a typical first-return mapping (see our first book [134]). The points 0 and 1 are stable fixed points; the intersection of the curve with the bisecting line occurs at $p_{n+1} = p_n = p_c = 0.896$, the point separating the two regions $p > p_c$ and $p < p_c$. p_c is the critical probability corresponding to a catastrophic fragmentation of the object. Plotting the next iterations of the renormalization group shows that, for values of $p_{n+1} < p_c$, one tends toward the point 0 with a decreasing fragility. For example when starting from $p_{n+1} = 0.6$, one finds $p_n = 0.2723$, then, $p_{n-1} = 0.0118$ and $p_{n-2} = 3.910^{-8}$, etc. The object remains solid.

Conversely, for the values $p_{n+1} > p_c$, the successive iterations lead to the point 1 : as n decreases, the object gets fragmented. There is, for $p_{n+1} = p_c$, a bifurcation. One may compute the fractal value associated with this catastrophic fragmentation, which is $D = 2.84$.

For the "weakness plane" model, one again considers 22 topological configu-

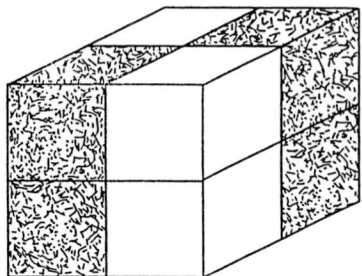 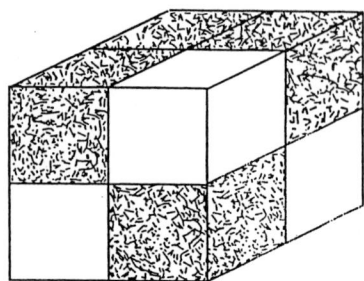

Figure 4.18: *The sound pillar model.* The initial cell is solid if it contains a sound pillar, i.e. a cornerstone made from two sound elements (the cube on the left). On the contrary, the second scheme shows a fragile cube. (after Allègre et al., 1982).

---

rations which are different from figure 4.17, and the cells are now considered fragile only when the fragile elements make up a plane crossing the cell (Turcotte, 1986).

There are 9 configurations out of 22 (Figure 4.20) corresponding to this situation. The relation between p_{n+1} and p_n is now:

$$p_n = p_{n+1}^8 + 8p_{n+1}^7(1 - p_{n+1}) + 28p_{n+1}^6(1 - p_{n+1})^2 + 56p_{n+1}^5(1 - p_{n+1})^3$$

$$+38p_{n+1}^4(1-p_{n+1})^4, \quad p_n = 3p_{n+1}^8 - 32p_{n+1}^7 + 88p_{n+1}^6 - 96p_{n+1}^5 + 38p_{n+1}^4. \quad (4.35)$$

The graph of the first-return mapping (p_n, p_{n+1}), is plotted on figure 4.20 and one finds $p_c = 0.490$ corresponding to a catastrophic fragmentation of dimension $D = 1.971$.

These models correctly account for the fragmentation processes implying power-law fragment distribution, as is observed in the experiments.

Furthermore, these models are internally consistent, as they verify the starting hypothesis that breaks can appear at joints or along weakness planes. The choice of a cube as initial cell was made for an easier understanding, but other types of geometries could have been chosen (e.g. crystal lattices in crystallography).

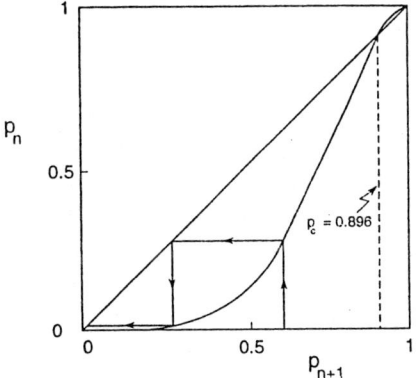

Figure 4.19: *The graph* p_{n+1}, p_n, *in the sound pillar model.* In this first-return mapping, the intersection of the curve with the first bisecting line gives: $p_{n+1} = p_n = p_c = 0.896$, where p_c is the critical probability corresponding to a catastrophic fragmentation of the object.

Other examples

Obviously, it is possible to choose in fragmentation modeling a finer fragmentation size using, for example, a cube of size h which can be divided into 64 cubic elements of dimension $h/4$ at each fragmentation stage.

In another example Smalley et al. (1985) [470] applied the renormalization group approach to the stick-slip behaviour of fault systems which are supposed to be controled by a given distribution of an array of asperities (Aki, 1981) [3]. They claimed that when an asperity fails, the stress on the failed asperity is transferred to one or more adjacent asperities. For a linear array the stress is transferred to a single adjacent asperity and for a two-dimensional array to three adjacent asperities. The authors use a renormalization group method to investigate the properties of a scale invariant hierarchical model for the stochastic growth of fault breaks through induced failure by stress transfer.

At stresses less than the critical stress, virtually no asperities fail on a large scale, and the fault is locked. Above the critical stress, asperity failure cascades away from the nucleus of failure; this catastrophic failure is interpreted as an earthquake and it corresponds to the transition from stick to slip behavior of the fault.

The relation between p_{n+1} and p_n, for cells containing two or four asperities with a Weibull quadratic distribution for their resistance was computed, giving a critical breaking probability $p_c = 0.2063$, for two asperities and $p_c = 0.1707$ for 4 asperities. For $p_1 < p_c$, the successive iterations lead

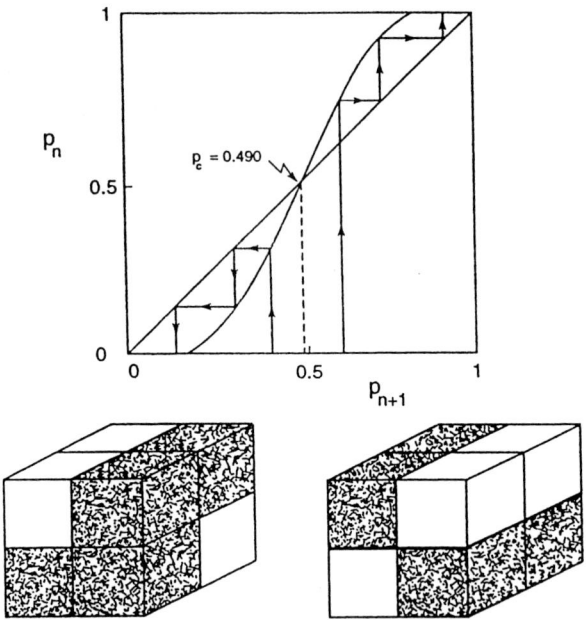

Figure 4.20: *Model of the weakness plane (Turcotte, 1986)*. The first-return mapping p_{n+1}, p_n where $p_c = 0,490$ is plotted above the model.

to $p_1^\infty = 0$ and there is no break. Conversely, for $p_1 > p_c$ the successive iterations lead to $p_1^\infty = 1$ and the system breaks apart; i.e. the fault forms (Figure 4.21).

Blocks dynamics model (Blanter *et al.*, 1997), and scaling laws
The Blanter *et al.* (1997) model [62] deals with scaling laws in block dynamics and dynamic self organized criticality. It is a hierarchical model of blocks which can move in two orthogonal directions (see Figure 4.22). The evolution of the system with time is studied.

The model combines two concepts which were considered separately in previous models: 1) the interaction between fracture probabilities along two orthogonal directions (according to the process described in Allègre and Le Mouel, 1994) [7], and 2) the consideration of energy dissipation in building earthquake models with renormalization techniques (Allègre *et al.*, 1995) [8]. The combined model thus adresses the seismic cycle.

Another difference from the SOFT model (for a description of the SOFT (Self Organized Fracturation and Tectonics) model see our first book [134])

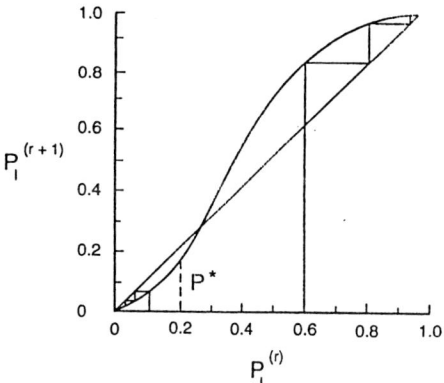

Figure 4.21: *Two asperities model.* The critical probability of breaking for a two asperities model is $p_c = 0,2063$ (Smalley *et al.*, 1985).

is that the authors consider a definite domain with finite dimension whereas the SOFT model consider implicitly supposes an infinite domain.
In this model, blocks which start moving can be interpreted as earthquakes. We observe successive cycles with strong earthquakes followed by clusters of aftershocks, then by quiet periods. The analogy with the SOC process of the "sandpile, slider-blocks model" is very striking (see [233]).
The model behaviour consists in comparing the distribution in time of the bigger seismic events and of the aftershocks with the seismic cycles, as they may be observed in different seismic active zones.

Energetic balance in scaling organization of fracture tectonics

Like the model developed previously, this new model (Allègre *et al.*, 1998) is a development of the SOFT process [9]. It is based on energy splitting combined with a renormalization group approach.
In conformity with a scale hierarchy in a fault zone as described by King (1983) [280], the authors consider that an earthquake appears as a critical phenomenon which occurs when the fracturation becomes self organized at different scales (Ito and Matsuzaki, 1990; Keilis-Borok, 1990) [263, 276].
The SOFT model has shown that the balance of energy can be such that all the received energy has the time to dissipate at intermediate-level scales, giving raise only to moderate magnitude earthquakes. However a small energy increase can lead to the occurence of a strong earthquake.
In the new model the authors introduce a redistribution of energy over the

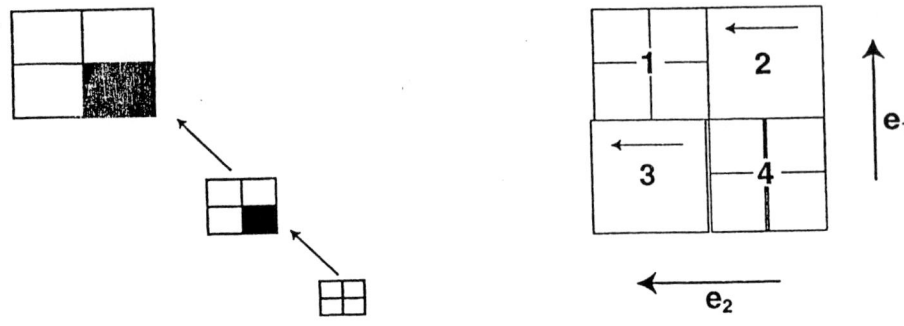

Figure 4.22: *Hierarchical model of blocks and interaction between these moving blocks at the same level.* Four blocks of a given level compose a block of the upper level. On the second scheme is shown the interaction between moving blocks of the same level. (Blanter *et al.*, 1997).

entire domain by a creep mechanism. This allows them to obtain more realistic aftershocks sequences, with a temporal decrease in intensity, according to Omori law, and also to reproduce the entire seismic cycle.

To summarize the previous developments about fracturation using the renormalization group approach and SOFT modeling, we examined four successive stages.

1. An introduction of scaling techniques in brittle fracture of rocks which consider an anisotropic model with normal and tangential probabilities of fractures, and interactions between them (Allègre and Le Mouel, 1994 [7]).

2. The scaling organisation of fracture tectonics was then presented in a possible interpretation of earthquake mechanism. In this model, energy splitting was combined with a renormalization group approach to model the behaviour of a fault zone subject to earthquakes (Allègre et al., 1995, [8]).

3. A hierarchical model of blocks moving in two orthogonal directions implying dynamic self-organized criticality was proposed to exhibit the general properties of seismicity-the seismic cycle, the foreshock and

aftershock activity, the Omori law for temporal decrease of aftershock activity, and the Gutenberg-Richter law (Blanter et al., 1997, [62]).

4. A development of the SOFT model which displays some general features of real seismicity, using numerical experiments in both the single domain case and the case of energy exchange between several domains (Allègre et al., 1998, [9]).

4.5.3 Tectonics, Fractals and Multifractals

This domain is a very large field of theoretical and applied studies. Analysis of the geometry of fracture fields shows that, despite their random appearence, it is possible to reveal an internal statistical invariance, when changing the study scale. This is due, of course, to the fragmentation laws studied earlier and to the empirical laws established from observed geomorphological features. It is logical to expect fracture fields to be described by fractal analysis.

General observation of fault and fracture fields

Before developing methods and techniques for the study of fault and fracture fields, let us observe them at different scales.
As it was noticed by Allègre et al., (1982) [6], geologists can look at fractures at different scales, on thin slices through a microscope, on rock outcrops, on aerial photographs, on geological maps, or on satellite images. The confusion between images is easy, as is shown in figure 4.23, in which the scales of each scene are not given. This is a qualitative evidence for the self-similarity of fractal sets and we will demonstrate this property for a granite batholith in Montana.

Measuring the length of fractures or faults

The method, which was described in different papers (Sornette et al., 1990 ; Davy et al., 1990 [479, 116]), consists of plotting a circle of radius r centered at M, a location inside a fracture/fault network, on a section plane of the rock with a clear fracture print. One may also operate on an aerial photo or on a surface map. Inside the circle we observe some faults or fault segments, the length of which are $l_1, l_2, \ldots, l_n$. One sums these values for the whole radius r circle to obtain $L(r) = l_1 + l_2 + \ldots + l_n$, then one increases r and computes a new $L(r)$ value, and so on (fig. 4.24).

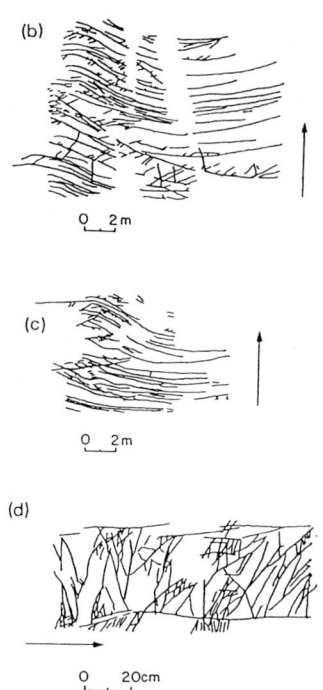

Figure 4.23: *Self similarity in fracturation*. Four images of fracture and fault fields observed at different scales (Velde *et al.*, 1991).

Generally, one observes that $L(r)$ increases as a power law of r, $L(r) \approx r^{Df}$. The exponent Df is the fractal dimension of the network.

Practically, one plots $\log L(r)$ against $\log r$ (fig. 4.24). If the distribution is a power law, the representative points fall on a straight line. One draws the regression line and one measures its slope Df. From what was have already said stated about fractal sets, we know that Df is such that $1 < D < 2$. Davy et al. (1990) [116] looked at an experimental model of the India-Asia collision and found $D = 1.73 \pm 0,05$. One of the main question answered by this approach is how vast undeformed regions are preserved in the middle of heavily fractured areas. Several authors had interpreted these regions as resulting from the stronger local resistance of. But, fractal analysis shows that the existence of these zones may naturally arise from the fractal character of continental tectonics, and they are not necessarily associated with lateral heterogeneity of continental lithosphere.

Tectonics, Seismicity, Volcanology

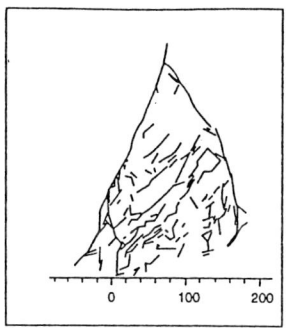

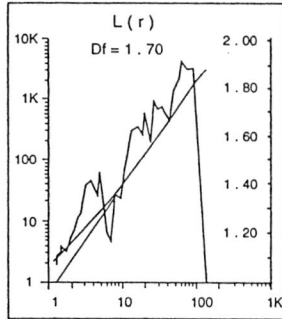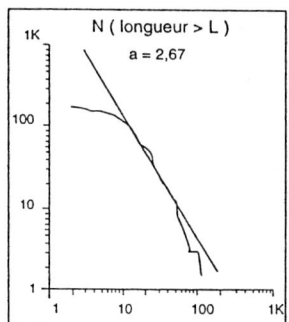

Figure 4.24: *Fractal Analysis of a fault field*. The fractal dimension of a fault network measured on a plane (see the tectonic map above) is obtained by measuring the length l_1 of the fault network in a circle of radius r_1, its length l_2 in a circle of radius r_2, ..., and looking at the slope D of the resulting log-log graph. (from Sornette *et al.*, 1991).

Cantor dust method

The method that we develop here is also found in section 3.3. There, it was applied to time series of events such as volcanic eruptions series or seismic crises in the time domain, while here it is applied to the spatial domain. The method described in detail in a paper by Velde *et al.*, (1990) [523], consists of crossing the two dimensional fracture field with a straight line. The direction of the line relative to the fracture field is recorded and the points where the line intersects the fractures are analysed using a box counting method (see Ledésert *et al.*, 1993 a,b) [299, 300].

The study interval is covered by segments of length u_i, the rank of which is i, which cover one or several events (here, the events are the intersection points). The number of segments is $N_i = 1/u_i^D$ if the point distribution is

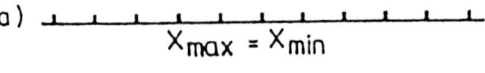

Figure 4.25: *Application of the Cantor dust method to the fractal analysis of a fracture field on a 2D section.* The crossing of the lines produces a points series which is analysed through the Smalley method (see Smalley et al., 1987). The slope is zero when the intervals between the events are constant (a), it is still close to zero when the intervals are slightly different (b), it tends toward 1 (and D tends toward 0) when the events are clustered (c), in intermediates cases, the slope and dimension D have values ranging between 0 and 1 (d) (after Velde et al., 1990).

---

fractal. In the case of a triadic Cantor set it has been shown [134] that for two successive iterations, i and $i+1$, we have $N_{i+1}/N_i = 2$, $u_{i+1}/u_i = 3$ and,

$$D = -\frac{\log(N_{i+1}/N_i)}{\log(u_{i+1}/u_i)} = -\frac{\log 2}{\log 1/3} = 0,6309 \ . \tag{4.36}$$

The parameter generally considered is the fraction x_i of segments (or steps) with length u_i which contains dust. The relation between x_i and u_i becomes,

$$N_i u_i = x_i L, \quad N_{i+1} u_{i+1} = x_{i+1} L \ , \tag{4.37}$$

where L is the length of the study interval. Therefore:

$$\frac{x_{i+1}}{x_i} = \frac{N_{i+1} u_{i+1}}{N_i u_i} = (\frac{u_{i+1}}{u_i})^{1-D} \ . \tag{4.38}$$

In the Triadic Cantor set

$$\frac{x_{i+1}}{x_i} = (\frac{u_{i+1}}{u_i})^{1-D} = \frac{2}{3} \ , \tag{4.39}$$

and,

$$D = \frac{\log 2}{\log 3} = 0,6309. \qquad (4.40)$$

If the fragmentation becomes more general, the varying ratios N_{i+1}/N_i and u_{i+1}/u_i, D varies between 0 and 1.
If there is only one point, we find $D = 0$, which is the Euclidian dimension of a point. When all the segments are filled, $D = 1$, which is the Euclidian dimension of a segment.

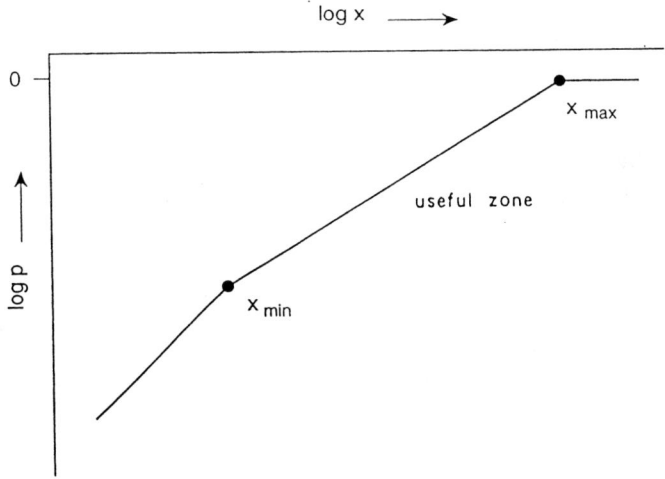

Figure 4.26: *Cantor dust method*. The log-log graph is the result of the analysis performed in the previous figure. The useful zone $X_{\min} < X < X_{\max}$ enables us to measure the slope m and $D = 1 - m$, the fractal dimension of the series.

One of the main problems encountered in this kind of study is related to the number of events, which is often too small to guarantee a significant statistical analysis. To increase this number, one can use a series of parallel lines covering the surface to study (Velde et al., 1990, 1991) [523, 522], with the successive lines read like the lines in a book (fig. 4.25). The intervals between lines must be large enough that the results are not repeated too often from one line to the next.
This condition is satisfied if we choose a spacing between two lines close to the average of u_i. The problem has been empirically studied but it would benefit from a theoretical analysis. To ease the reading of successive series

of points, we developed an automatic analysis system based on a camera looking at photographs of the sections.

The final log-log graph of such an analysis shows up as a line broken into three segments (fig. 4.26). The slope of the first segment is 1 (i.e. a dimension of 0) for u, $0 < u < u_{\min}$. It corresponds to small values of x_i for which x_i/x_{i+1} is close to 1. The slope of $\log x$, as a function of $\log u$, is thus close to 1. This part of the graph is not significant.

For $u_{\min} < u < u_{\max}$, we are in the useful zone where $D = 1 - m$, m being the segment slope.

$u_{\max}$ is the value of u_i for which all u_i cover some dust. Therefore, x equals 1, the segment slope $m = 0$ (because $\log x = 0$), $D = 1$. This means that, for this value of u, the segment is filled.

To recognize a series as fractal, the useful zone should cover at least one order of magnitude (Davy et al., 1990) [116], i.e. $u_{\max}/u_{\min} \geq 10$. This last condition of self-similarity is more than satisfied by a granitic batholith in the Sierra Nevada, where the value of D is constant at scales ranging from the scale of the satellite imagery down to the scale of thin slices, i.e 5 orders of magnitude (Velde et al., 1991) [522]. If the method is applied for different angles of the intersecting line, variations of D may be observed. They look like anisotropy in the fracture field. It is tempting to plot the variations of D versus line angles, as is done when studying the anisotropy of other physical parameters, but the interpretation is not straightforward. Harris et al. (1991) and Velde and Dubois (1991) [230, 521], showed that only two values of D are "seen". This comes from the properties of projections and intersections of fractal sets (see [134]).

Indeed, if the line $\mathcal{E}(\theta)$ is infinite on an infinite plane, and if the perpendicular sets $\mathcal{F}_1$ and $\mathcal{F}_2$ of lines, distributed fractally with the respective dimensions D_1 and D_2, are as large as possible, the points of intersection on $\mathcal{E}(\theta)$ are distributed along two Cantor sets with the respective dimensions D_1 and D_2 (fig. 4.27) (Harris et al., 1991) [230].

Because of the property mentioned earlier, the dimension of the resulting set on $\mathcal{E}(\theta)$ is :

$$\max \left\{ \dim(\mathcal{E}(\theta) \bigcap \mathcal{F}_1), \ \dim(\mathcal{E}(\theta) \bigcap \mathcal{F}_2) \right\} = 0.75, \qquad (4.41)$$

For the example shown in Figure (4.27), the dimension of the resulting set is 0.75. When $\mathcal{E}(\theta)$ is infinite, this is true for all values of θ, except that which gives $\mathcal{E}(\theta)$ parallel to $\mathcal{F}_2$ (i.e. perpendicular to $\mathcal{F}_1$). In this case, the dimension suddenly becomes $0, 25$.

It appears therefore that only two subsets (if any) can be observed. The first

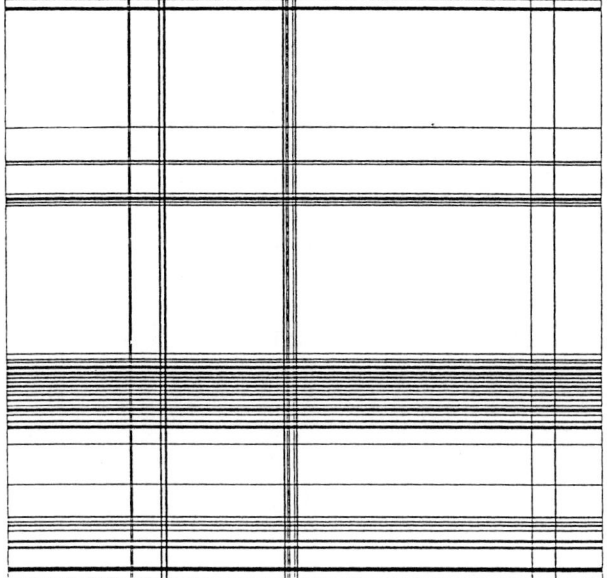

Figure 4.27: *Intersection of two sets* $\mathcal{F}_1, \mathcal{F}_2$. They are made from fracture fields with perpendicular direction, $D_1 = 0.25$ and $D_2 = 0,75$. In general, the dimension of the resulting set $\mathcal{E}(\theta)$, along the angle θ equals max $[\ \dim(\mathcal{E}(\theta) \cap \mathcal{F}_1),\ \dim(\mathcal{E}(\theta) \cap \mathcal{F}_2)\ ] = 0.75$,. The exception is when the intersecting line is parallel to $\mathcal{F}_2$, in which case the dimension jumps to 0.25 (From Harris et al., 1991).

---

subset has the higher dimension and the second subset appears only if $\mathcal{E}(\theta)$ is parallel to the fractures of the first set. The set with dimensions smaller than these two values would be "invisible".

In reality, the fracture field $\mathcal{F}_1$ does not usually consist of parallel fractures, but shows a statistically privileged direction. There will therefore be some noise when $\mathcal{E}(\theta)$ is parallel to this direction.

In summary, two factors can perturb the law established earlier:

1. The fractures of a given set are not completely parallel.

2. The plane of study and the fracture fields are not infinite. This limits the number of intersection points, especially when θ gets close to the critical value (parallel to the direction of the fractures in the set of the highest value of D). This goes back to the situation of short series evoked earlier.

In the examples given above, we could as well have applied the method developped by Grassberger and Procaccia (1983) [195], using the fact that the intervals of "sound" rocks between successive fractures, mapped along the line intersecting the fracture fields, correspond to a series of discrete values. This series can be tested with the correlation function, which gives the dimension of the attractor (in the spatial domain, and not in the time domain as is commonly done for dynamic systems). This method was successfully applied to the position of fractures in drilling data (Dubois et al.,1993).

4.5.4 Tectonics. Study of Surface Faults

Because of self-similarity, it is possible to apply the techniques outlined in the previous section to aerial photograhs, satellite images or tectonic maps. Remember the study by Velde et al., (1990) [523], who applied the Cantor's dust technique to the granitic batholite of Mount Abbot in the Sierra Nevada. Another example is the study by Davy et al. (1990) [116] of fault lengths in the Indian-Eurasian collision zone.

The many box-counting techniques proposed by Falconer (1990) [145] (see also [134]) have been applied to the San Andreas fault system in two articles (Aviles et al., 1987 ; Okubo and Aki, 1987) [22, 388].

The first article [22] uses the technique outlined in Section (4.1), where broken lines (coasts, rivers, contour levels) are measured with segments of variable length between 0.5 km and 1000 km (Figure 4.28). The relatively small values of D, are interpreted to reveal the irregularities of the fault surface expression.

After observing variations along the main fault, the authors identified six different segments, corresponding to very different seismic regimes. The dimensions observed for these segments vary from 1.0008 to 1.0191. The closeness of theses values to 1 shows the generally "smooth" surface expression of the main fault, but significant variations of D appear along the main fault, indicating the presence of heterogeneities. D changes quite a lot between short and long wavelengths, the boundary being around 1 or 2 km. Short wavelengths have a higher value of D than longer wavelengths. The fault's surface may therefore be considered as nearly plane, with roughness appearing only at small scales. The authors conclude that the southern segments, where D is higher, are rougher and thus less susceptible of large earthquakes, because they are too irregular to break during a single event.

The second article, by Okubo et Aki (1987) [388], applies a box-counting technique to the same site — San Andreas fault system —. A minimum number $N(r)$ of circles of radius r is chosen to entirely cover the surface of the San Andreas fault (Figure 4.29). The log-log plot of $N(r)$ as a function of r

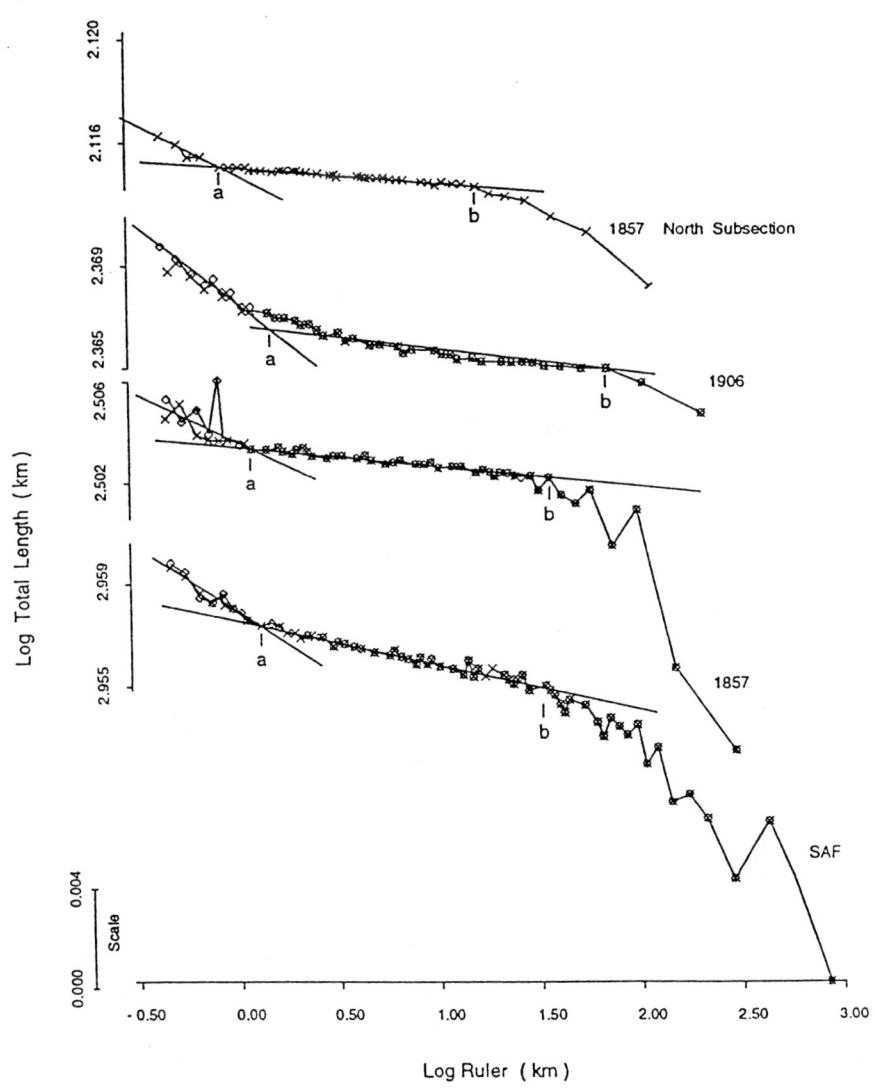

Figure 4.28: *The Aviles et al. method (1987).* In order to quantify the fractal dimension of the San Andreas's fault system, the authors apply the Richardson's method.

222   Fractals and Dynamic Systems

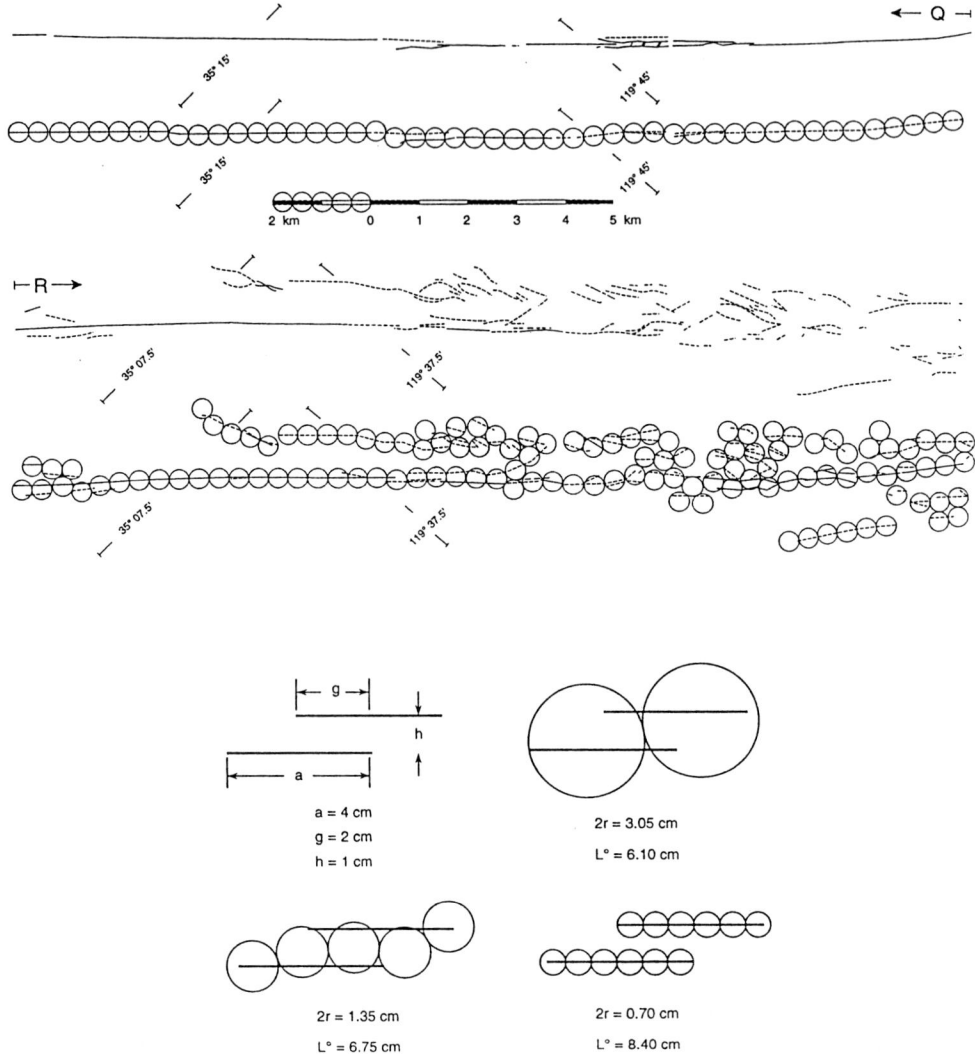

Figure 4.29: *Okubo et Aki's method (1987)*. This method consists of drawing covering circles on the set of fractures and faults associated with the San Andreas fault system.

Tectonics, Seismicity, Volcanology 223

allows the computation of the power-law exponent D. The global dimension for the San Andreas fault system is 1.31 ± 0.02. Dividing the system into six segments, the authors show that local values increase from the northeast, where $D = 1.2$, to the southeast, where $D = 1.43$. The authors link the variations of D to seismicity variations observed by Allen (1968), and make some points about the relationship between seismicity and the fractal geometry of faults. One important point is the existence of a break in the log-log graph for a critical radius r_c. This break implies that the mapped expression of the fault is not self-similar.

Okubo et Aki note that, if the faults' fractal nature persists so that the critical lengths remain fixed in time, then it might be possible to anticipate some of the characteristics related to the degree of complexity of the fractal geometry of the fault trace. Although they limit themselves to the geometry of this surface trace, the authors suggest that the measures of the complexity reflect the complexity of the fault surface itself as:

$$D_{\text{surface}} = D_{\text{trace}} + 1. \qquad (4.42)$$

4.5.5 Multifractals and Wavelets applied to fault fields

Ouillon (1995) [394] notices that there are two practical ways to compute the multifractal spectrum of a fault field. The first consists of digitizing the fault segments from a map. The main difficulty of this method is that at large scale we measure the real exponents of the fault field, which gives us a dimensions equal to one, whereas at small scale we consider each segment individually. At a smaller scale we consider only individual points and the dimensions tend to zero. The second method adopted by the author consists in covering the fault field by a ϵ path grid and measuring the incremental length of every fault segment in each box. In the classical multifractal approach we have to compute the generalized dimensions D_q for a given box size, to estimate the proportion P_i of fractal mass (total segments length) inside in each box i, then to calculate the power q of this proportion, P_i^q. Instead, Ouillon used the method proposed by Roux and Hansen (1990) [440] to directly obtain the values of D_q and α for each value of q. The order q moment of the distribution of boxes is:

$$M_q(\epsilon) = \sum_{i=1}^{N(\epsilon)} P_i^q \quad q \in]-\infty, +\infty[\qquad (4.43)$$

and

$$M_q(\epsilon) \propto \epsilon^{(q-1)D_q}. \tag{4.44}$$

Let $L_q(\epsilon)$ be defined as:

$$L_q(\epsilon) = \frac{\partial}{\partial q} M_q(\epsilon) = \sum_{i=1}^{N(\epsilon)} P_i^q \log(P_i) \tag{4.45}$$

and

$$L_q(\epsilon) \propto M_q(\epsilon) \frac{d}{dq}[(q-1)D_q] \log(\epsilon) \tag{4.46}$$

thus

$$L_q(\epsilon) \propto M_q(\epsilon)\alpha(q)\log(\epsilon) \tag{4.47}$$

from which we get

$$\alpha(q) = \frac{1}{\log(\epsilon)} \frac{L_q(\epsilon)}{M_q(\epsilon)} \tag{4.48}$$

So, it is only useful to draw two graphs for each value of q:

- $(\log(\epsilon), \log(M_q(\epsilon)))$, the slope of which is $(q-1)D_q$ which gives D_q
- $\left(\log(\epsilon), \frac{L_q(\epsilon)}{M_q(\epsilon)}\right)$, the slope of which is $\alpha(q)$.

It is then possible to estimate $f(\alpha(q))$ using the equations of the Legendre transform (see Dubois and Gvishiani, 1998, [134]).

The method was applied to a bidimensionnal geometrical multi-scale characterization of fracturing on data sampled at different scales (from ground up to satellite imagery) on the sedimentary cover of Saudi Arabia. The fractal dimension was found to be always equal to 2. The multifractal method, which allows a more complete characterization, shows that the fracture distribution is uniform in the sedimentary basin, whereas it is unhomogeneous in the basement. The local fractal dimension of the most fractured areas equals 1.75, close to the value commonly observed in Diffusion-Limited-Agregation processes.

Figure 4.30 presents the field data obtained at different scales from scheme I, directly mapped in the field (one to several meters) to scheme V (1/250,000 scale). Figure 4.31 shows the generalized dimensions vs q values. This analysis emphasizes the interest of multifractal formalism to describe and to compare the differents fracture networks. From a careful analysis of the

Tectonics, Seismicity, Volcanology

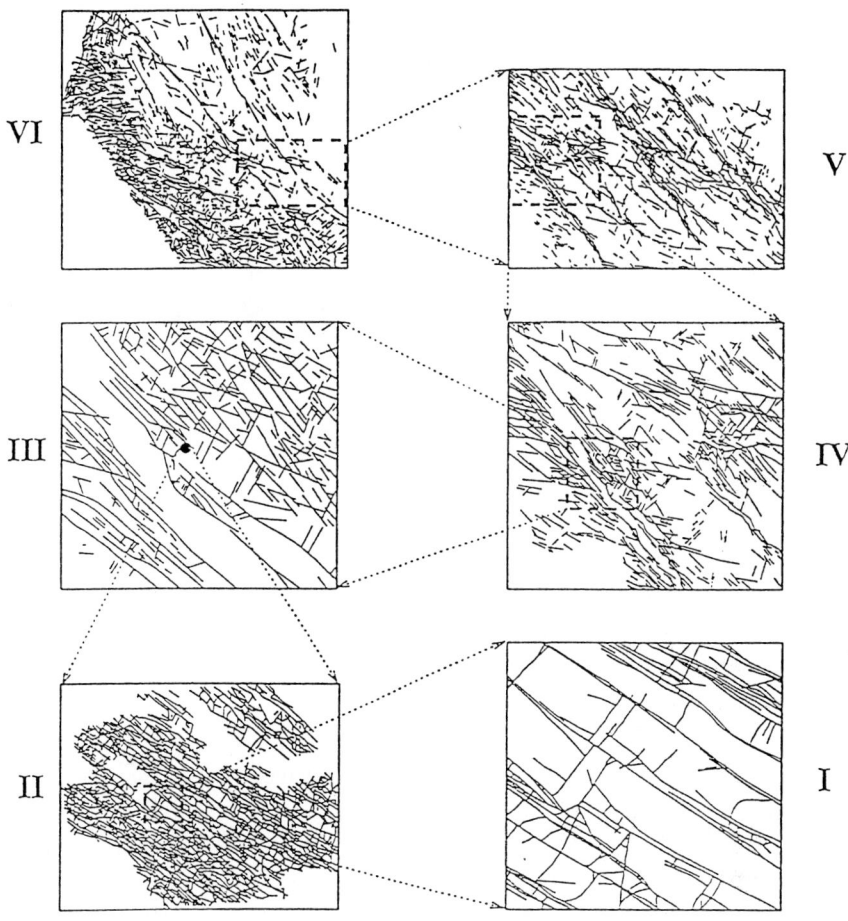

Figure 4.30: *Fracturization Cascade*. The Arabic plate fracturization appears here at different scales, from centimeter (scheme I, scale 1/1) to 100 kilometers (scheme VI, scale 1/1 000 000), (after Ouillon 1995).

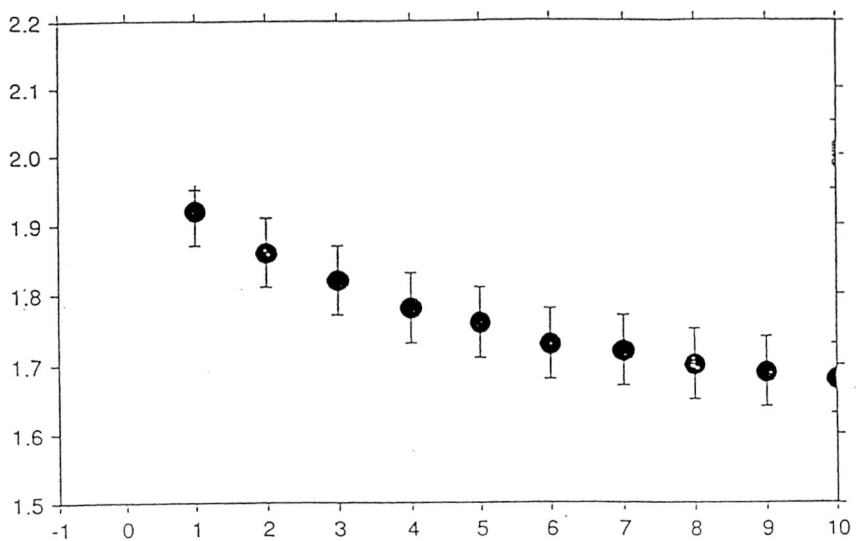

Figure 4.31: *Multifractal analysis of Arabic plate fracturation.* On the first graph, the generalized dimension is plotted versus q values for scheme I (1/1 scale). The second graph shows the multifractal spectrum of fracture fields IV and V of the previous figure (after Ouillon, 1995).

different multifractal spectrum the author deduces that the cutoff scales correlate perfectly with the thickness of the main crustal lithologic units: sedimentary bed, sedimentary formation, sedimentary basin, depth of quartz plasticity, depth of feldspar plasticity, depth of the Moho. This suggests that some features we see on the surface have ductile roots extending down to the base of the crust. Surface fracturing thus depends on the local lithology, coupling between involved layers, and inherited tectonic structures [394].

4.5.6 Seismicity, Gutenberg and Richter Law, Multifractals

To begin, let us recall the well known $\log - \log$ relation between the number N of earthquakes in a given region and their energy (Ishimoto and Iida ; 1939, Gutenberg and Richter, 1949, 1954, [261, 202]).

A quantitative definition of the magnitude parameter allowing the energy of an earthquake to be calculated was given in 1935 by Richter [202]. From that definition and from its relationship with the number of events the "Guten-

Sismicity, Gutenberg-Richter law

berg and Richter" law was formulated:

$$\log N = a - bm, \qquad (4.49)$$

where N is the number of earthquakes of magnitude greater than m and a and b are constants (see Figure 4.35).

Before linking this relation to the notion of self invariance, as was done by Aki (1981) [3], it is interesting to further develop the notion of power-law distributions. In the examples presented here, such distributions are observed for the number of earthquakes of magnitude $\geq m$ (Gutenberg-Richter law), and for the time intervals between two events (eruptions, earthquakes, magnetic inversions, etc.) (see Sections 5.3 and 5.4).

Multifractal analysis of the 1992 Erzincan aftershock sequence

Let us summarize an interesting multifractal analysis of an aftershock sequence published by Legrand *et al.*, 1996 [301]. The main shock of the Erzincan (Turkey) earthquake occured on March 13, 1992 ($M_S = 6.9$). A few days after the main shock, a portable seismic network of 25 stations was installed in the epicentral region, and operate from march 31 to April 22, 1992. The good coverage of the aftershock area good resulted in precise determination (to within few hundred meters) of the 1161 recorded hypocenters, including their depths.

In that study, spatial distances between earthquakes were computed in 3D because of the good depth accuracy. Rather than using a moving window of constant time length, the authors used a moving window containing a constant number of events (see Volant, 1993, De Rubeis *et al.*, 1993 [531, 123]). The window contained 200 points guaranteeing precise estimations of fractal dimensions. Consecutive windows were shifted by 20 data points. For every subset the fractal dimension D_q was calculated, with q varying from 2 to 23.

The fractal dimension considered here is the generalized fractal dimension D_q (see Dubois and Gvishiani, 1998 [134]), defined as [195, 234]:

$$D_q = \frac{1}{q-1} \lim_{\epsilon \to 0} \log[\sum_{i=1}^{N(\epsilon)} p_i^q]/\log(\epsilon) \qquad (4.50)$$

where $p_i = p_i(\epsilon)$ is the probability of occupation of the ith box of size ϵ, and q is a positive or negative real number. The definition of the generalized fractal dimension can be seen as an extension to order q of the correlation dimension (order 2 dimension). The value q can be seen as the degree of correlation since p_i^q is the probability of having q points within the ith box.

A generalized correlation function is then defined as:

$$C_q(\epsilon) = \left\{ \frac{1}{N} \sum_{j=1}^{N} [\frac{1}{N-1} \sum_{\substack{i=1 \\ i \neq j}}^{N} H(\epsilon - \| \underline{X}_i - \underline{X}_j \|)]^{(q-1)} \right\}^{1/(q-1)} \quad (4.51)$$

where H is the Heaviside function, $\| \underline{X}_i - \underline{X}_j \|$ is the distance between the two points $\underline{x}_i$ and $\underline{x}_j$, and N is the number of points.
Then

$$D_q = \lim_{\epsilon \to 0} \log C_q(\epsilon) / \log(\epsilon). \quad (4.52)$$

For more details on correlation functions, and generalized fractal dimension, look at our first book [134] or the specific papers on the matter : Grassberger and Proccacia (1983) [195], Hentschel and Proccacia (1983) [234], Kurthz and Herzel (1987) [291], Hirata and Imoto (1991) [242].

Results and Discussion The spectrum of the multifractal dimension D_q versus q is shown in figure 4.32 for the complete data set, with corresponding errors. D_q varies from $D_2 = 2.1$ to $D_\infty = 1.35$. We can see a saturation effect for q greater than about 18, corresponding to the most compact fractal structure.

Figure 4.32 shows the evolution of the different generalized multifractal dimensions with respect to time. The fractal dimensions varies from every q values before a seismic crisis (April 20). For comparison, in the Kanto region, Hirata and Imoto (1991) [242] showed a variation of D_q from $D_2 = 2.2$ to $D_\infty = 1.7$. The value of D_∞ which was obtained by the authors in the Erzinan region is smaller than in the Kanto area. This difference is interpreted by observing that the Erzincan aftershocks are concentrated in a rather flat spatial region, in contrast to the more diffuse distribution in the Kanto region. A decrease of fractal dimensions before a big earthquake has been noticed by several authors (Ouchi and Uekawa, 1986, De Rubeis et al., 1993 [395, 123]). It has also been observed that before large aftershocks, multifractal dimensions change for all values of q, with a significant range of about 0.6.

The authors conclude by noting that these results emphasize the importance of applying multifractal analysis as a robust index to follow the evolution of the spatial distribution of seismicity in relation to the occurrence of large events.

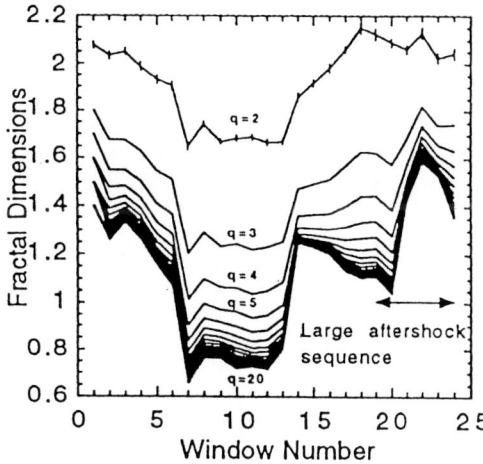

Figure 4.32: *Variation of generalized fractal dimensions D_q versus time.* The time window number corresponds to the 25 windows, each of them containing 200 aftershocks. Their relative shift is 20 data points. For every subset, the fractal dimensions were calculated with q varying from 2 to 23 (after Legrand et al., 1996).

4.5.7 Power Law or Poisson Law ?

A few tests allow us to determine the random or non-random nature of a time series. They are based on recognising the types of time distributions associated with the events.

Histograms and cumulated distributions

We will examine the histograms and cumulated distribution of random number series, a theoretical triadic Cantor series, and, for comparison, a data time series.

Normal Distribution The normal distribution histogram of a random variable x is a bell shaped curve which is called a Gaussian. Figure 4.33 shows histograms built using random number series 100 and 1000 elements long (selected using an algorithm of Khoklov, 1993). One can observe that the bell-shaped curve is better defined for the longer series.

The cumulated distribution gives three graphs of N, total value greater or equal to x, as a function of x then $\log N$ as a function of x and $\log N$ as a

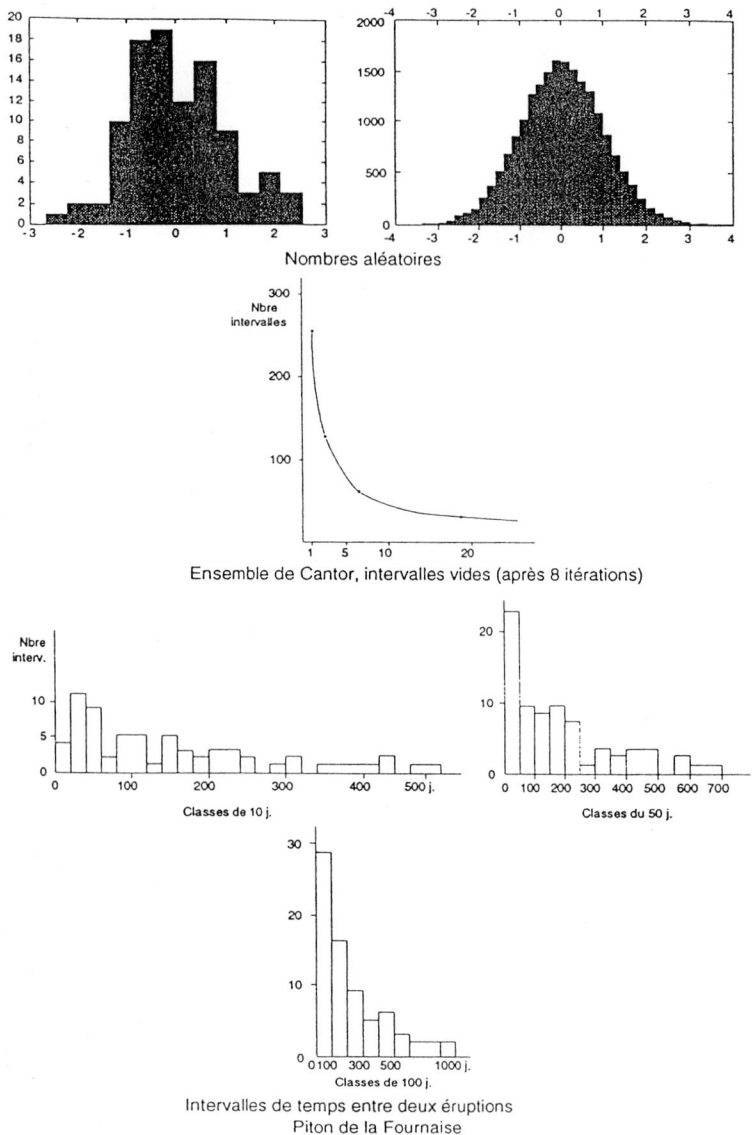

Figure 4.33: *Examples of histograms.* The different examples correspond to series of 100 and 1000 random numbers, a triadic Cantor set, and the series of eruptions at Piton de la Fournaise (Reunion island) from 1930 to 1994 (in 100-days classes).

Sismicity, Gutenberg-Richter law

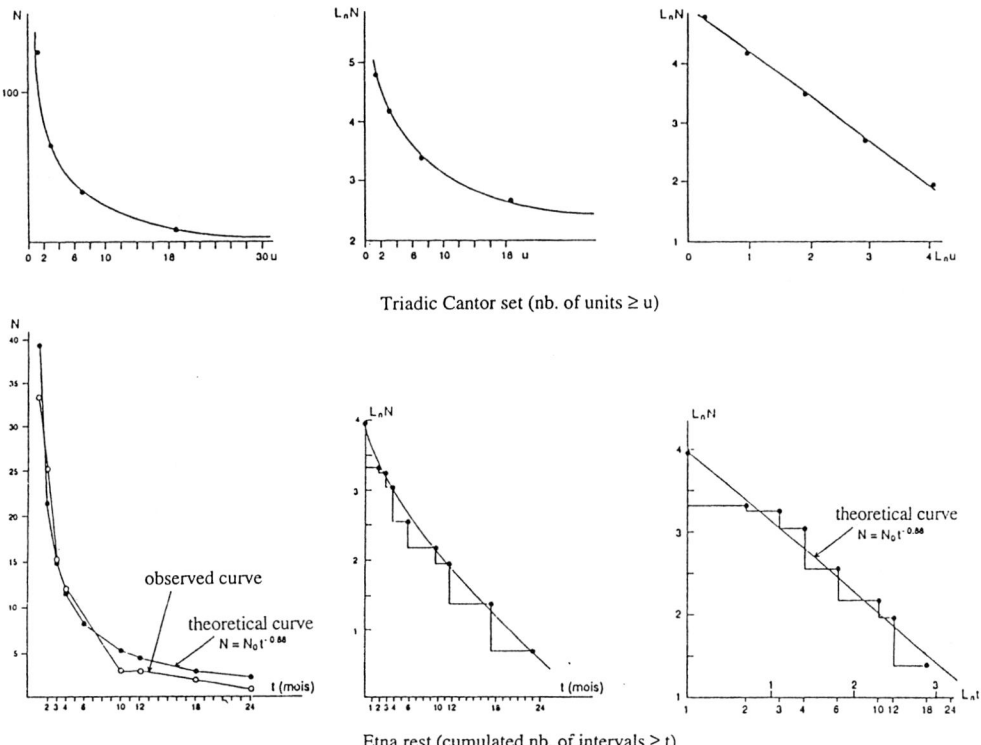

Figure 4.34: *Cumulated distribution.* Normal, semilog and log-log graphs of N as a function of x are shown for the three cases examined in the previous figure: random numbers, triadic Cantor set and series of eruptions at the Piton de la Fournaise.

function of $\log x$. Notice the linear aspect of the semilog curve, characteristic of a normal distribution of exponential or Poisson type. This is the test bringing into evidence the randomness of a numerical series.

Triadic set The computed histogram of "voids" in a 127-point triadic Cantor set after 7 iterations (series of 127 points) is very different from the previous one. Its envelope is exponential and cumulated distributions are very different from those of the normal distribution. The log-log graph is a straight line, which is characteristic of a power-law distribution. The slope of the line is the exponent of the power-law (fig. 4.34).

Observed data Let us consider a series of volcanic eruptions on the Piton de la Fournaise volcano (La Reunion island), setting as our variable the time interval between the beginning of successive eruptions (fig. 4.33). The histogram has a shape similar that of the Cantor set. The cumulated distribution graphs are also linear when plotted on a log-log scale, as for the Cantor set. This similarity was observed during a preliminary analysis (Dubois and Cheminée, 1988 [132]), and led us to propose a power-law distribution rather than a Poisson-type model proposed by Wickman, 1966, 1976 ; Klein, 1982 ; Kono, 1973, [537, 282, 286]. Those authors deduced (often from analyse of too few data points) exponential distributions implying a random process. In Contrary we show that the distribution follows a power-law, implying a determinist dynamic system generating the eruptions. We will prove further into this point when looking at the probabilistic aspect of these natural time series (section 5.3.1).

The Gutenberg-Richter law expresses the same property, i.e. the power-law distribution of earthquake magnitudes. Indeed, magnitude is the logarithmic expression of energy and the log-log graph shows a linear correspondence between the logarithm of the cumulated number of earthquakes of magnitude $\geq m$ and this magnitude.

The paralel with scale invariance concept and fractal object was achieved by Aki (1981) [3].

The relation was improved by introducing the seismic moment. Hanks et Kanamori (1979) established that:

$$\log M_0 = d + cm, \qquad (4.53)$$

where M_0 is the sismic moment, m is the magnitude d and c are constants with values $c = 1, 5$ and $d = 16$ (Thatcher et Hanks, 1973 ; Purcaru et Berckhemer, 1978 ; Hanks et Kanamori, 1979 [504, 421, 227]).

Kanamori and Anderson (1975) [269] had found the proportionality relation between the seismic moment and the fault length (or dimension):

$$M_0 = kL^3. \tag{4.54}$$

Combining the preceding relations, we get:

$$\log N = -\frac{3b}{c} \log L + \log f, \tag{4.55}$$

or:

$$N(L) = fL^{-3b/c}, \tag{4.56}$$

with:

$$f = \frac{bd}{c} + a - \frac{b}{c} \log k. \tag{4.57}$$

When this relation is verified, the slope of the graph $\log N(L)$ as a function of $\log L$ is $-3b/c$, and b and c respectively are the slopes of the semilog graphs frequency-magnitude and seismic moment-magnitude.

If we compare with the definition given by Mandelbrot (1975, 1982) [325] of the number of fractal objects with a size greater than r:

$$N \sim \alpha r^{-D}, \tag{4.58}$$

we can deduce that the fractal dimension of a seismic fault can be written as $D = 3b/c$, where b and c are the slopes of the lines defined as above.

Aki (1981) [3] considers three groups of possible mechanisms for a series of earthquakes in a given region, according to the value of b.

- If $b = 1$, the fault dimension is 2 (as $c \approx 3/2$), i.e.cits topological dimension.

- If $b = 3/2$, the dimension is 3, corresponding to a cluster of earthquakes whose fault planes tend to fill a volume.

- It is also possible to have values of $0.5 < b < 1$, corresponding to fractal dimensions between 1 and 2. In this case, the fault cannot be considered as a plane, but as rupture lines tending to fill the plane. This situation seems to correspond to the Goishi model of Otsuka (1972) and Maruyama (1978) [392, 340]. The branching model of Vere Jones (1976) [525], consisting in the propagation of a rupture, has a geometry corresponding to micro-fissures in the medium and the propagation of the rupture along the branches of the micro-fissures.

In this approach, the study of objects, earthquakes or earthquake clusters is using a limited number of parameters amongst the 5 available (at the first approximation): 3 spatial coordinates, time, and the energy produced (the earthquake's magnitude). Here we only use earthquakes occuring in a given region and exceeding a given energy. The time parameter is not analysed as extensively as it will be in later studies of the probabilistic aspect, on its own for the Cantor dust, or with other parameters in the Russian methods. The computation of D, for several types of earthquakes allows ue to better understand the possible mechanisms which are at the origin of different earthquakes clustering. The study of the variability of b which corresponds to our above classification has given rise to several works (Mogi, 1962 ; Scholz, 1968, [362, 462]) on laboratory samples as well as in the field .Let us notice the field work of Main (1987) [318], on seismicity models before the Mt. St. Helens eruption of 1980. Main (1987) uses the similarity dimension with the method with the method we have described in the previous paragraphs. Immediately before and after a large earthquake, the value of b changes from $b < 1$ to $b > 1$. This is explained by the releasing of stress on the main branch of the self-similar fault system. Before the break, $b < 1$ and $D < 2$. After the release around the main fault, only secondary branches participate to the release of residual stress, yielding higher values of $b > 1$ and $D > 2$.

It has also been remarked that a decrease in b may be the precursor of large earthquakes (Smith, 1986 [472]), or of the breaking of a rock sample in laboratory experiments (Ohnaka et Mogi, 1982 [387]), or of a volcanic eruption (Gresta and Patane, 1987 ; Yuan et al., 1984 [198, 544]).

This technique was applied by Nouaili et al. (1987), and Dubois and Nouaili (1989) [384, 135] to subduction zones by dividing the seismic zones into three regions: 100-300 km depths, 300-500 km, and 500-700 km. The different subduction zones will exhibit different behaviours and so will the different depths inside the same zone. Theses results are based on the investigation of 11 subduction zones around the Pacific Ocean (Figure 4.35. It starts from the realisation that the branching model, characterized $1 < D < 2$, is the most frequent process in subducting lithospheres.

This confirms the mechanisms proposed by some seismologists (Frohlich et Willemann, 1987) [162] who remark that earthquakes do not occur preferentially along the nodal planes of the main shocks, but that the breaks occur in a non-planar zone distributed around the initial focal point. The value of D implies a stress regime associated with the dipping of a dense lithosphere into a viscous and less dense asthenosphere. This is for example the case for the Indonesia-Banda S subduction zones, with lateral stress adding to the main contribution from subduction.

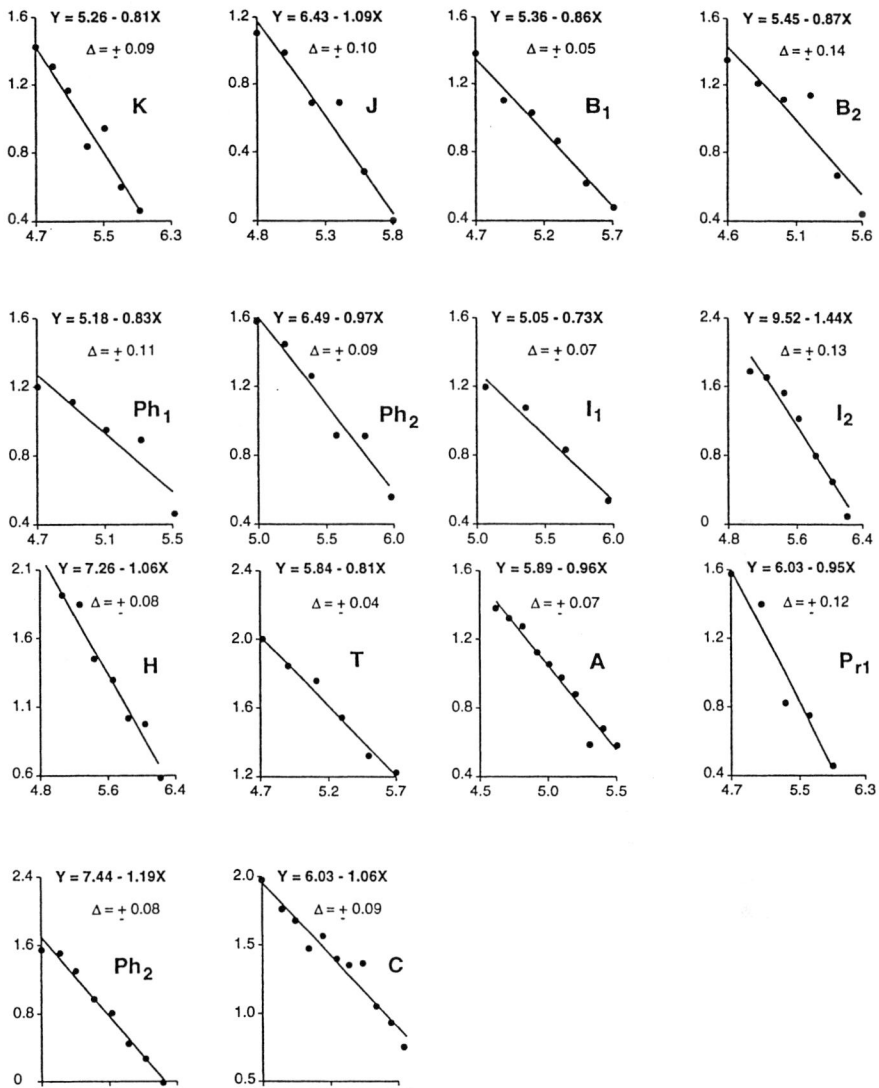

Figure 4.35: *Illustration of the Gutenberg-Richter law.* On earthquakes series in the different subduction zones around the Pacific Ocean, with focal depths between 100 and 300 km. K, J, B, Ph, I, H, T, A, Pr, and C respectively design the zones of Kermadec, Japan, Bonin, Philippines, Java, New Hebrides, Tonga, Aleutians, Peru, Chile. Δ is the standard deviation on b, $Y = \log N$, $X = m$ (Dubois and Nouaili, 1989).

High values of D appear at the edges of the dipping lithosphere, in the Tonga Islands. They express the fact that the rupture planes of all earthquakes tend to occupy the whole space. This supplements of observations of Giardini et Woodhouse (1984) [177] of a north-south deformation of the edge of the lithosphere, where the mesosphere introduces lateral stresses which complicate focal mechanisms between 500 and 700-km depths.

Conversely, in the contact zone between the subducting and subducted plates, the values of D are generally close to 2, implying a simpler geometry of the contact area (Tonga Islands, cf. Nouaili et al., 1987, [384]).

In summary, the variability of D along the subducting lithosphere is a good expression of the fracturation processes at play in the lithosphere, and our interpretations of it are in agreement with the geodynamic setting of the principal circum-Pacific subduction zones.

4.6 Fragmentation, tectonics, seismicity, synthesis trial

The link between the different results in the study of the physics of rock breaking was done in a recent study by Nagahama et Yoshii (1994) [374].

We notice that long before the beginning of the fractal approach, power laws had been established relating the different physical parameters (Gaudin, 1926 ; Schuhmann, 1960 ; Bond, 1952 ; Bergstrom et al., 1963 ; Charles, 1957 ; Tartaron, 1963, etc., [168, 464, 65, 56, 98, 501]). Nagahama et Yoshii (1994) [374] investigate the different scaling laws intervening in fracturation. The first one expresses the power law dependency between the cumulated number, $N(r)$ of fragments the size of which is greater than r,

$$N(r) \sim r^{-D_s}, \qquad (4.59)$$

where, D_s is the fractal dimension of size fragment distributions (cf. section 4.5).

The second scaling law, in the $\Re^3$ space, quantify the amount of roughness of breaking surfaces. This dimension (cf. section 4.2) D_R is the fractal dimension of the surface's shape. Its value ranges from 2, when the surface is smooth, to 3, when the surface is so complex that it tends to fill the whole volume (Mandelbrot, 1977, 1982, [333]).

As it has been already shown the cumulated mass $M(r)$ of fragments with a size smaller than r can be expressed as a power-law:

$$M(r) \sim r^h, \qquad (4.60)$$

where, h is a positive constant (Gaudin, 1926 ; Schuhmann, 1940 ; Fujiwara et al., 1977 ; Turcotte, 1986 [168, 464, 163, 514]).

If all fragments from a given set have a common fractal dimension D_R for their roughness, their cumulated mass becomes (Nagahama et Yoshii, 1994 [374]):

$$M(r) \sim \int_0^r r^{D_R - D_s - 1} dr \sim r^{D_R - D_s}. \qquad (4.61)$$

Comparing equations 5 et 6, we get:

$$h = D_R - D_s. \qquad (4.62)$$

Nagahama et Yoshii [374] then use the energy per mass unit necessary for a fragmentation. According to Avnir et al. (1983) [23],

$$E \sim r^\omega, \quad \omega = D_R - 3. \qquad (4.63)$$

From the empirical law of Walker-Lewis, we have:

$$E = C r^{-n+1} \qquad (4.64)$$

where, C and n are constants (Bergstrom et al., [56] 1963),

As Charles [98] (1957) had establish that $h - n + 1 = 0$, we get, from 4.63 and 4.64: $\omega = D_R - 3 = -n + 1$, as $h = D_R - D_S$,

$$\omega = h + D_S - 3 = D_s + n - 1 + 3. \qquad (4.65)$$

The preceding equations give: $2\omega = D_s + 3$, and $D_R - 3 = \frac{1}{2}(D_S + 3)$, e.i.:

$$D_R = \frac{1}{2}(D_S + 3). \qquad (4.66)$$

That relation (4.66) means that the fractal roughness dimension, D_R, in a 3-D space, equals the mean value of the fractal dimension of the size distribution, D_S, and the Euclidean dimension of the space in which the observation is conducted (here, this dimension is 3). This shows there is a constraint in the fractal geometry of fragmentation.

If we consider the lithosphere is a fragmented object, and that tectonic faults correspond to the limits of these fragments, the roughness of fault surfaces can be easely computed.

Other relations using Gutenberg and Richter law have been proposed by Nagahama and Yoshii (1994) [374].

It has been shown (Aki, 1981, [3]), as we have notice above, (section 4.5), that:

$$D_S = \frac{3b}{\delta}, \qquad (4.67)$$

where, b is a Gutenberg and Richter constant and δ a constant which depends on the relative durations of the seismic source and on the delay time of the recording system.

In most seismological studies, a value of $\delta = 1, 5$ (Kanamori and Anderson, 1975, [269]). This yields: $D_S = 2b$ (see above 4.5.7).

This last equation shows that the values of b are related to the fractal dimension of lithospheric fragmentation.

Going back to the roughness dimension, the equations 4.67 and 4.66 can be combined into:

$$D_R = \frac{3}{2}(\frac{b}{\delta} + 1), \qquad (4.68)$$

which applies to the roughness of a seismic break zone. The authors deduce that this relation will lead to important developments in our knowledge of the breaking process, and hence of earthquakes, inside a tectonic field (Nagahama et Yoshii, 1994, [374]).

Chapter 5

Dynamic System Properties and Long Time Series

5.1 Geomorphology, Hydrology

5.1.1 Correlation Function and Rivers Flows

The Correlation integral function method has often been used to analyse the dynamical systems governing turbulent processes in fluid dynamics (Brandstater et al., 1983; Guckenheimer and Buzyna, 1983; Malraison et al., 1983 [74, 201, 322]), series of volcanic eruptions (Sornette et al., 1991; Dubois and Cheminée, 1993 [481, 137]) and climatological time series (Nicolis and Nicolis, 1984 ; Essex et al., 1987 ; Tsonis and Elsner, 1988 [381, 142, 511]). The method is presented in Dubois and Gvishiani (1998) (section 9.2) [134]. The method was recently applied (Beauvais et al., 1998; Aubert et al., 1999; Aubert, 2000 [47, 19, 18]) to daily time series of the Oubangui river discharge. The geometrical organization of the original non-filtered time series in a 2-dimension pseudo-phase space only shows the annual cycle of rising, high, and falling regimes that cannot really account for the underlying dynamical system. Instead, the results obtained on the filtered time series, i.e. free of the annual cycle, indicate that the system is governed by ~ 10 degrees of freedom which may correspond to climatic variables such as the mean annual rainfall, and the intensity, magnitude and duration of rainstorms. The filtered time series is also described by an attractor exhibiting a relatively clear geometrical organization.

The interactions between the geomorphic and hydroclimatic patterns regulate the river discharge and tend to attenuate the effects of the stochastic dynamics induced by variations of the climatic parameters. Even though the highest frequencies of the power spectrum can be related by a power law that may define a discrete fractional Brownian motion characteristic of a pure stochastic process, the Correlation integral function method indicates that the river discharge time series may be driven by a dynamic in between stochasticity and deterministic chaos. Further research into nonlinear time sequence analysis are needed to really separate stochastic from deterministic processes governing the dynamics of short and noisy natural time series, as well as efforts to relate the resulting generalized dimensions to physically meaningful processes.

Figure 5.1 illustrates the main results obtained by analysis of the Oubangui river discharge time series.

5.1.2 Wavelets Applied to Floods

We have already applied the wavelet transform to syntactically classify seismograms (see section 2.2.2).

The wavelet transform has a fractal basis (Argoul et al., 1989 [12]) and is particularly useful when applied to local, nonperiodic, multiscaled phenomena such as stream flow (Turcotte, 1997 [518]).

The method consists in expanding an arbitrary real function $S(x)$ onto wavelets constructed with a simple function g, using dilations and translations (Dubois, 1998 [131]). The wavelet transform of $S(x)$ by the wavelet g is defined as:

$$\mathcal{T}_g(a, x) = \omega(a) \int \left(\frac{g(x-y)}{a} \right) S(y) dy \quad a > 0,\ x \in \Re, \tag{5.1}$$

where the weight function $\omega(a)$ represents the visual enlargement. The wavelet g is a regular function, localised around $x = 0$.

For a large class of functions $S(y)$, the wavelet transform can be inverted, provided g satisfies a certain number of conditions, including:

$$\int g(y) dy = 0 \tag{5.2}$$

The wavelet transform can be considered to be a *mathematical microscope* (Arneodo et al., 1988 [13]). The location and the enlargement factor correspond to x and a^{-1}, and the optical performance is determined by the choice of the analysing wavelet.

For the stream flow the variable is time t and in the following example the function g (the mother wavelet) is the "Mexican hat" wavelet, which is the second derivative of the Gaussian distribution:

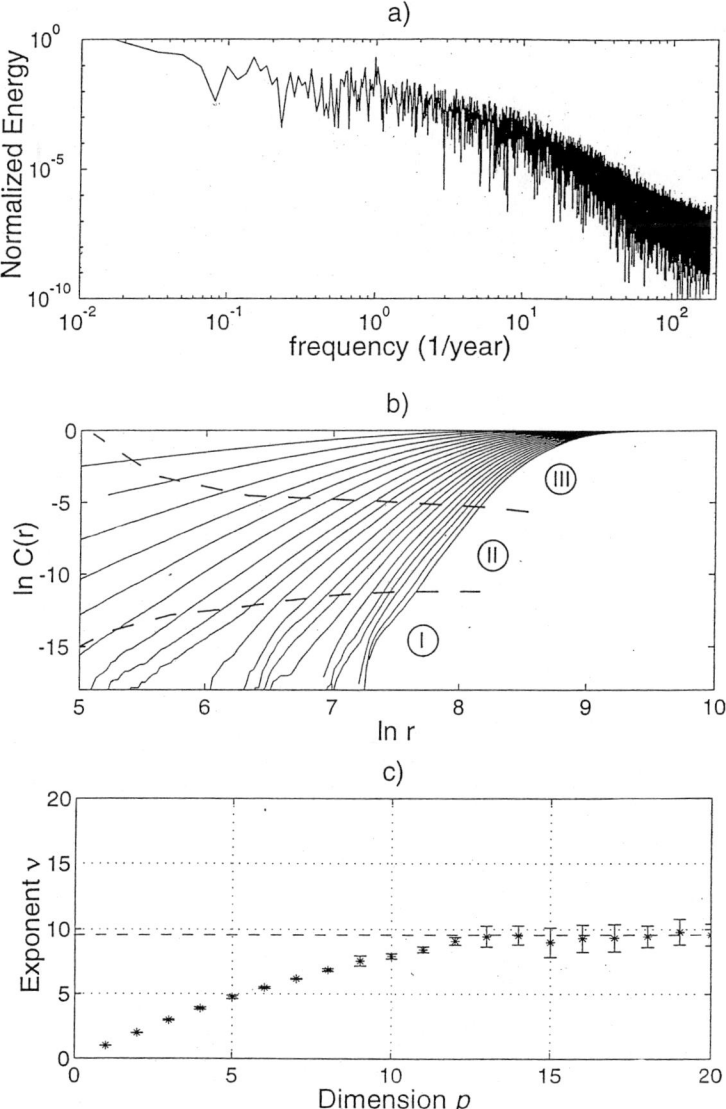

Figure 5.1: *Oubangui river discharge time series.* a) Power spectrum of the time series after the annual cycle is removed. b) Correlation integral function plot of the series. c) Variograms of the correlation exponent ν vs the embedding dimension p for the series. The dotted line represents the saturation level of ν (Beauvais et al., 1998).

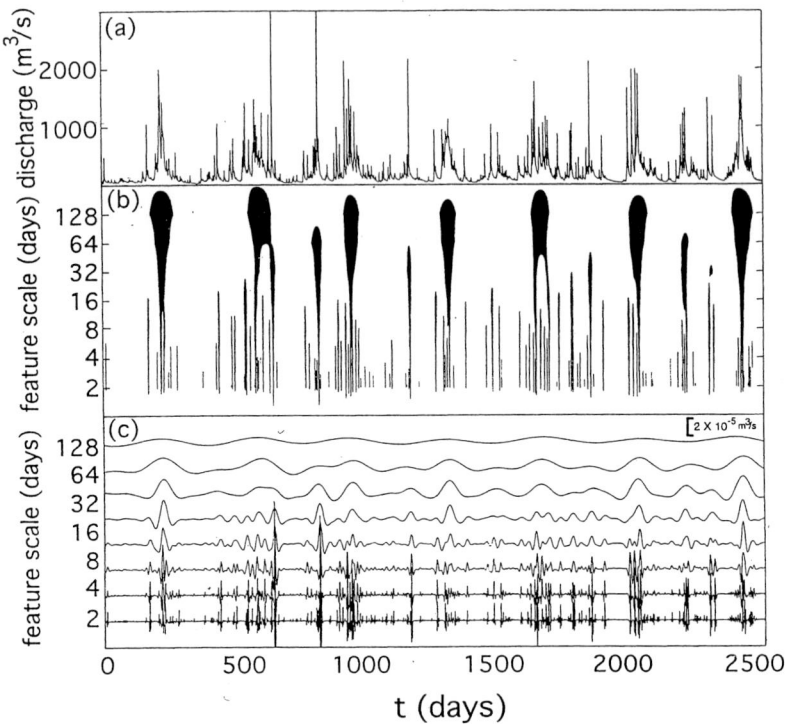

Figure 5.2: *Wavelet transform of a river discharge record.* A seven years time series of the Ammonoosuc River daily discharge. a) The time series. b) The wavelet scalogram; regions of the a vs t plot where the magnitude threshold is exceeded are plotted in black. c) Wavelet transform magnitudes for $a = 1, 2, 4, 8, 16, 32, 64,$ and 128 days (Turcotte, 1997).

$$g(t') = (\frac{1}{2\pi})^{1/2}(1 - t'^2)\exp(-\frac{1}{2}t'^2) \qquad (5.3)$$

To illustrate the application of the Mexican hat wavelet transform we consider a seven-year, daily discharge record for the Ammonoosuc River, Maine (Turcotte, 1997 ; Smith *et al.*, 1997 [518])

Figure 5.2 shows the river discharge record versus time for a time series of 2500 days, the wavelet scalogram and the wavelet transform magnitudes for different values of the enlargement parameter a. The record is characterized by long-period high flows associated with the annual spring snow-melt events and by short-period high flows associated with rainstorms throughout the year.

5.2 Seismology

5.2.1 Cantor Dust application

The probabilistic (Cantor Dust) approach was one of the precursors of fractals and dynamics system studies (Mandelbrot, 1975, 1982; Dubois and Cheminée, 1988, 1991; Smalley et al., 1987 [325, 328, 132, 133, 471]). We will show here its extension to seismicity and volcanic eruption series, using the Cantor dust methods, the correlation function and the first-return map. Let us start from a truncated application of the Cantor dust method, with internal e and external E scales verifying $e > 0$ and $E < \infty$. The intermission order (length of empty intervals between two events) is randomised to make the events statistically independent. If u is the length of an intermission, the length distribution satisfies:

$$Pr(U \geq u) = u^{-D}. \tag{5.4}$$

In other words, the probability of reaching or exceeding u is u^{-D}. In a time series where events (earthquakes or volcanic eruptions) have been identified, and where the power-law exponent D has been found on a log-log graph, the probability that a new event $i+1$ happens at a time t after the event i is t^{-D}, $0 < D < 1$.

We will show that the clustering of events depends on D; the smaller D is, the more the events tend to cluster together.

We choose a clustering threshold, t_0. The intermissions "$> t_0$" and "$< t_0$" are separated and their relative durations are divided by t_0. When D is small, the relative durations of intermissions $> t_0$ have a high probability of being larger than their lower limit 1. Indeed, the conditional probability that $T > 5t_0$ is 5^{-D}, and therefore tends towards 1 when D tends towards 0. However, the relative durations of intermissions $< t_0$ become smaller than 1. Events therefore tend to cluster around t_0, when D is small. Conversely, when D is close to 1, intermissions tend to be more regular.

The probabilistic approach was applied by Smalley et al. (1987) [471] in the course of a study of the time evolution of seismicity in several regions of the New Hebrides. They showed that the intermission distribution law was significantly different from a random Poisson-type distribution, and that it fit a power law, i.e; a fractal law. The observed values of D range from 0.126 to 0.255, according to the study area, showing that earthquakes cluster more in some regions than in others (Figure 5.3).

Smalley et al. (1987) underline the very preliminary nature of this approach, as seismic events possess 5 dimensions (time, the 3 spatial dimensions, and

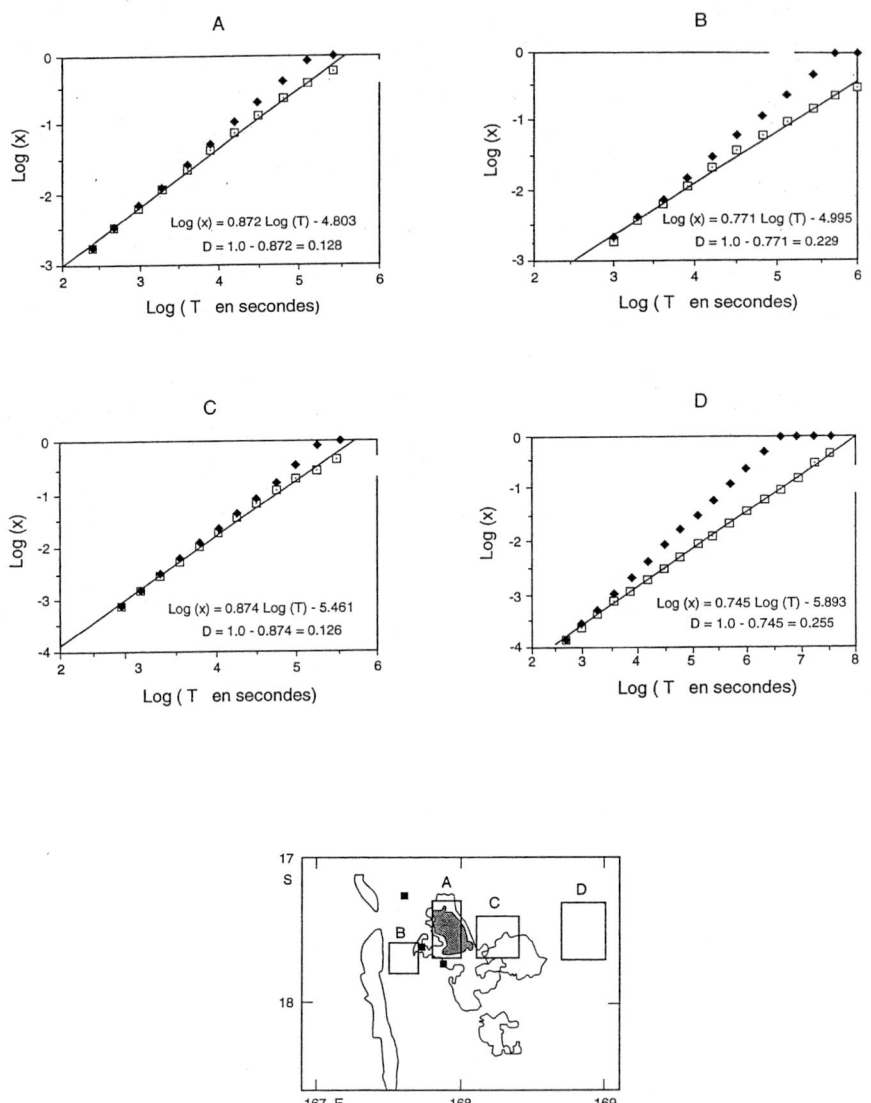

Figure 5.3: *Fractal dimension of seismicity in the time domain.* For earthquake series in Vanuatu, the D values vary with the study area (Smalley *et al.*, 1987).

Seismology   245

magnitude), and as clustering can in principle be analysed in any subset of these. Time was chosen as a parameter to allow an empirical look at the possible correlation between variations in D and the triggering of major earthquakes.

5.2.2 SOC applications to seismology

One chapter of our previous book [134] was devoted to Self-organized criticality (SOC), a concept which may be relevant for understanding temporal and spatial scaling in a wide class of dissipative systems with extended degrees of freedom (see Bak *et al.*, 1987, 1988 [31, 32]). Here we describe in detail the evolving sandpile experiment, a dynamic system which has been studied extensively for self-organized criticality [31, 32, 233, 256, 268]

Following Turcotte (1994) [517] we notice that distributed seismicity is often taken as a classic example of a natural system that exhibits self-organized criticality. There is a continuous input of energy through the relative motion of tectonic plates. This energy is dissipated in a fractal distribution of earthquakes. Scholz (1991) [453] argued that the earth's entire crust is in a state of self-organized criticality. Similar arguments have been developed by Sornette *et al.* (1989, 1990) [477, 479].

Artificial block-spring models of rapidly driven seismicity can be used to illustrate this idea. Before the definition of the SOC concept, such a model was proposed by Burridge and Knopoff (1967) [83]. Huang and Turcotte (1990, 1992), Narkounskaia and Turcotte (1992), and Turcotte (1994) [251, 252, 375, 517] developed the idea for the simplest model for fault interactions: a pair of interacting slider blocks (illustrated in Figure 5.4).

Grasso and Sornette (1998) tested the SOC on induced seismicity. They emphasize the existence of a threshold that enables the system to accumulate and store the slowly increasing stress until the instability is reached and an earthquake is triggered (see [90, 308, 197]). In this way they review the major reported cases of induced seismicity in various parts of the world and find that both pore pressure changes and mass transfers leading to incremental deviatoric stresses of < 1 MPa are sufficient to trigger seismic instabilities in the uppermost crust with magnitude ranging up to 7.0 in otherwise historically aseismic areas.

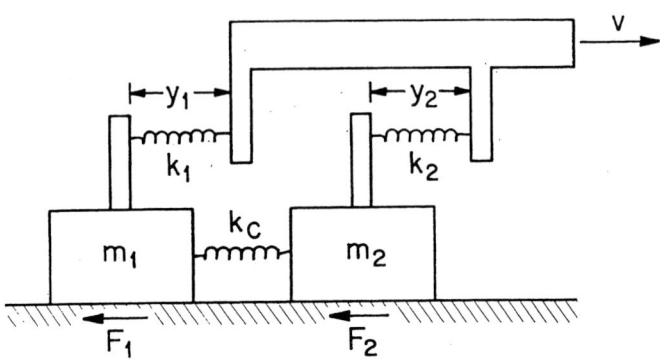

Figure 5.4: *The two-block model.* A constant velocity driver extends the springs until a block starts to slide. In some cases the sliding of one block induces the second one to slide (after Turcotte, 1994).

5.3 Volcanology

5.3.1 Application to Volcanic Eruptions

When observing the time series of eruptions for a volcano, the different events seem strikingly irregular and impossible to quantify. Several authors (Wickman, 1966, 1976 ; Klein, 1982) have noted that the eruptive behaviour of most volcanoes is indistinguishable from a random behaviour, which apparently precludes any possibility of developing efficient predictive methods [537, 538, 282].
Wickman (1966, 1976) proposed some strictly statistical models where eruptions were treated as random events and volcanic activity as a stochastic process. He able to reproduce, at least in a descriptive way, the essential features of the activity of several volcanoes. For basaltic volcanoes, activity is defined here as lava fountains (or explosions) in craters or active lava flows. Repose periods are times without such activity.
The model was applied successfully by Klein (1982) to model historical eruptions of Hawaiian volcanoes. Carta et al. (1981) studied the last 300 years of Vesuvius eruptions as a chronological succession of four phenomenological

states: reposes, persistant activity, intermediate eruption and final eruption [92]. The duration of each state was distributed according to the same empirical law. They observed that the volcano behaved as a quasi-stationary system with four equilibrium states. Mulargia et al. (1987) tried to develop an original procedure based on two sets of Kolmogorov-Smirnov statistics to demonstrate a change point or scan point and thus to identify different regimes in a time series [372].

As was shown by Sornette et al. (1991), several distinct tests are needed to determine if a particular process is random in time [481]. These tests can be divided into two main categories:

- The first category of tests shows that the events follow a well defined probability distribution such as Kolmogorov-Smirnov. The distribution may be flat, corresponding to a uniform probability, or have any other shape.

- The second category of tests mainly detects correlations in the series of events. For example, the sequential correlation test computes the usual correlation function which, when equal to zero, indicates that the events are independent of each other. Another test consists of examining the lengths of monotonous series and comparing the distribution of segment lengths to the ones that would be observed for a purely random variable. The "sub-series" test applies the prior test to a sub-series obtained by, for example, extracting one of every two events from the initial series.

However, a very important fact should be noted: some determinist systems pass all tests for "randomness" (Aneodo and Sornette, 1984). This means that a positive answer to all tests on the randomness of a series does not imply that the system is not deterministic [14].

Two types of systems have been identified:

1. All systems with several degrees of freedom which are intimately coupled and can develop a random dynamic, resulting from the complexity of superposition and coupling of different evolutions of each degree of freedom (e.g. Brownian movement). The evolution of the physical variable studied is undistinguishable from a random process. Any prediction is completely illusory. In [134], this was called chaotic chaos.

2. Systems with a small number of degrees of freedom which may present very complex dynamic behaviours, and have been defined as related to deterministic chaos.

The main conclusion of these discussions is that the observation by Wickman (1976) and Klein (1982) that volcanic eruptions time series pass the randomness tests, does not require that their generating systems are not deterministic. This is why we applied three new tests to this time series; the Cantor dust method, Grassberger and Procaccia's correlation function and the first-return map. These methods were developped in [134]; here we shall focus on their applications and results.

5.3.2 Cantor Dust and Correlation Function applications

Cantor Dust Method

This method has been applied to eruption time series from several volcanoes (Dubois et Cheminée, 1988, 1991) [132, 133]. In the case of basaltic-type volcanoes such as the Piton de la Fournaise or Hawaii, the fractal dimensions show two characteristic values depending on the length of the test periods. For piton de la Fournaise: $D = 0.45$, for rest periods between 1 and 10 months and $D = 0.67$, for rest periods between 12 and 48 months. For Kilauea, $D = 0.58$ for rest periods below 2 years and $D = 0.75$ for more than 2 years. For Mauna Loa, $D = 0.44$ for periods below 6 years and $D = 0, 84$ for periods longer than 6 years (Figure 5.5).

This double periodicity was interpreted as an expression of the filling and emptying of one (or several) magmatic chambers with distinct flow rates.

For both Piton de la Fournaise and Hawaii, the smaller D value corresponds to a tendency for eruptions to cluster which seems to be related to the superficial mechanical respons of the edifice to higher pressures created by magma cooling in a shallow reservoir (Tait et al., 1989) [491]. The second, larger, value of D, corresponds to a more regular rythm and, for the Piton de la Fournaise, may be related to pycritic eruptions from a deeper source.

Conversely, a basaltic volcano like Etna shows only one regime, and a study of the period from 1930-1987, revealed a single value of $D = 0.85$, corresponding to regular eruptions.

When applied to subduction zone volcanism, this method reveals a large variability in the fractal dimension D. Generally, however, andesitic volcanoes of Peleian type (Mt Pelée, Soufrière de la Guadeloupe, Mt St. Helens, Fuji) have low D values, $0, 2 < D < 0, 4$. Indonesian volcanoes, with their frequent and important activity, have values close to 0.7 (except for Krakatau, for which there is insufficient data).

Volcanology

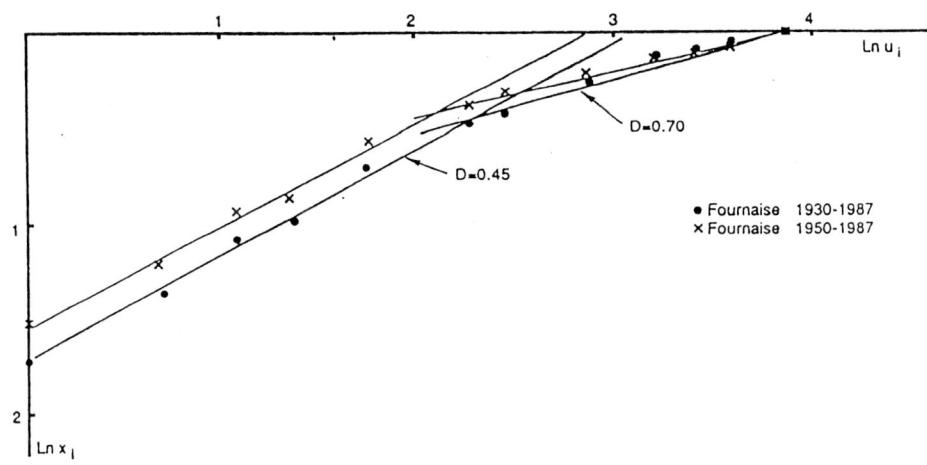

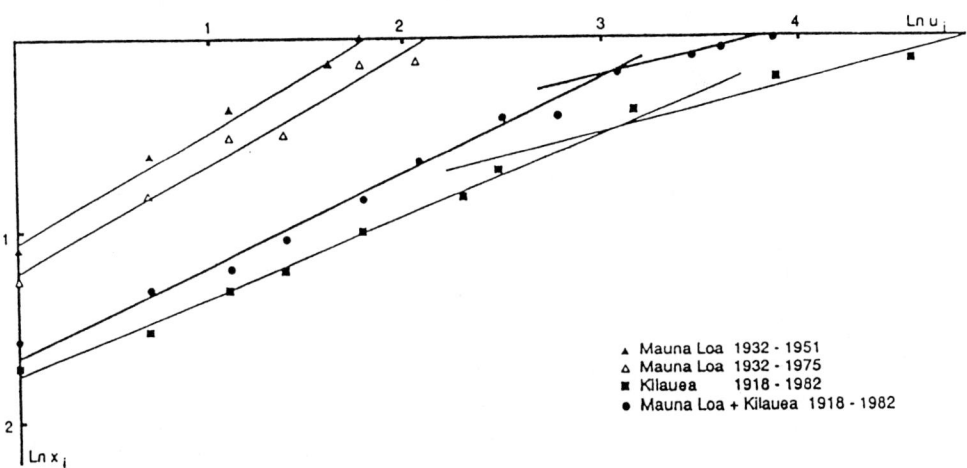

Figure 5.5: *Fractal analysis using the Cantor dust technique.* The data shown come from time series of volcanic eruptions at Piton de la Fournaise (La Réunion) and Hawaii (after Dubois and Cheminée, 1991).

Correlation Function Method

The main difficulty with the correlation function method is the large number of points necessary. This problem was discussed in the first section of this chapter, where we applied this technique to hydrology and river discharges. The number of points necessary increases as the dimension of the attractor increases; it is possible to use as few as 100 points if the dimension of the attractor is less than 3.

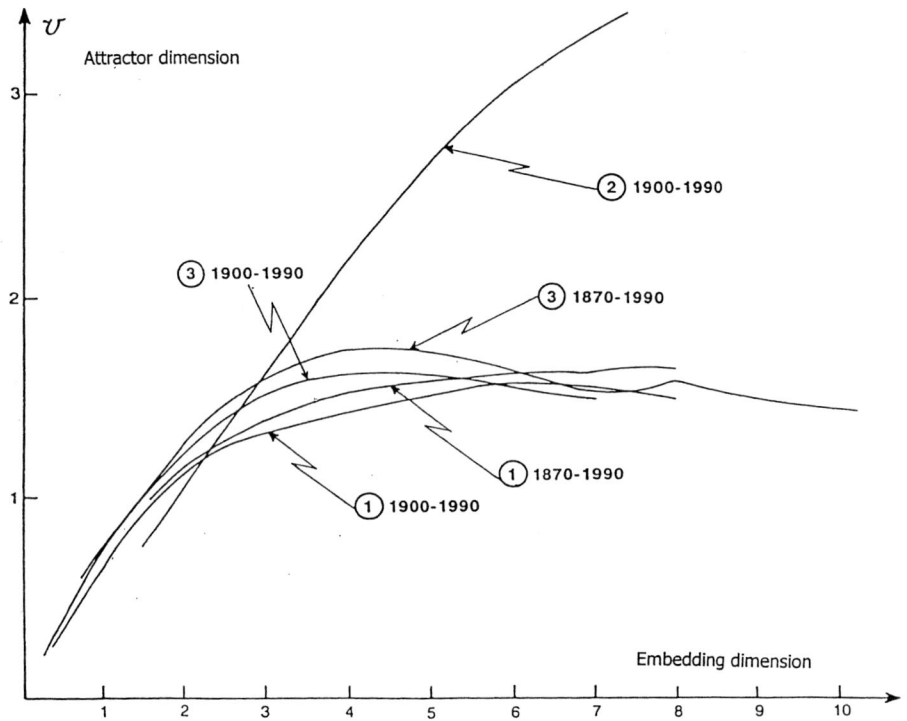

Figure 5.6: *Application of the correlation function method*. The results shown here concern the series $t_1(i)$, $t_2(i)$, $t_3(i)$. $t_2(i)$ appears to be random (ν increases continuously with the embedding dimesion d), $t_1(i)$ is noisy but, for $t_3(i)$, saturation appears for $\nu = 1,7$ (From Dubois et Cheminée, 1991).

For this reason, the technique could only be applied to three volcanoes: Piton de la Fournaise, Hawaii and Vesuvius (Dubois et Cheminée, 1991 ; Sornette et al., 1991 [133, 481]).
This analysis uses three variables: $t_1(i)$, $t_2(i)$ and $t_3(i)$. $t_1(i)$ is defined as the time interval between the end of eruption i and the beginning of eruption $(i+1)$, $t_2(i)$ is the duration of eruption i and $t_3(i)$ is the time interval between

Volcanology

the beginning of eruption i and the beginning of eruption $(i+1)$.
Of course, we get:

$$t_3(i) = t_1(i) + t_2(i). \quad (5.5)$$

Figure 5.6 shows the results of the correlation function method applied to Piton de la Fournaise volcano. The $t_2(i)$ series appears distinctly random (increasing with d, the phase space embedding dimension). The saturation of $t_3(i)$ in the phase space around the value of 1.7 confirms the results obtained with the Cantor dust method. This test shows that the $t_1(i)$ series results from deterministic dynamic system with an attractor of dimension 1.7. The dynamic system therefore has 2 degrees of freedom, which confirms the double periodicity observed previously.

It is more difficult to apply the correlation function test to the two active Hawaiian volcanoes. Each volcano has only had a limited number of eruptions in recent history (38 for the Kilauea, 46 for the Mauna Loa). To apply the test, we merged the two series into a dataset that should tend to give us details about the dynamic behaviour of the whole Hawaiian complex (Sornette et al., 1991). The resulting $D_{\max} \approx 4.6$ which seems to correspond to a dynamic system with more than twice the number of degrees of freedom as the Piton de la Fournaise. This may result from two independent dynamic systems, one controlling the Kilauea eruptions and the other the Mauna Loa eruptions, with each of their respective dimensions close to 2. These two systems should be weakly coupled to avoid the coupling of phases, which would reduce the attractor's total dimension from the sum of the 2 dimensions to a smaller value indicative of the degree of coupling.

A third test of this method was performed on a relatively long series of 160 eruptions of the Vesuvius volcano (Dubois et Cheminée, 1991). For the interval 1694-1872, an acceptable saturation is obtained for $D_{\max} = 2.9$. This implies that the dynamic system generating the eruptions has 3 degrees of freedom, which is consistent with the results from other approaches (e.g. Carta et al., 1981 [92]).

First-Return Map

This method, described in Dubois (1998) and Dubois and Gvishiani (1998) [131, 134], consists of plotting the point couples (x_{i+1}, x_i) of the series:

$$x_i \vert_{i=1}^{\mu}, \quad (5.6)$$

In the present case we use $t_3(i)$ the series of time intervals between the beginning of eruption i and the beginning of eruption $(i+1)$ for Piton de la Fournaise (Sornette et al., 1991 [481]).

Figure 5.7 shows the points t_{i+1} plotted as a function of t_i.

If the points with coordinates (t_{n+1}, t_n) were distributed along a simple curve, this would mean that t_{n+1} is a simple non-linear function of t_n, i.e. that the dynamics behind the series t_i is a simple deterministic law. On the other hand, if the series is completely random, the (t_{n+1}, t_n) plane would be filled densely and uniformly. Figure 5.7 shows that the situation is intermediate for Piton de la Fournaise.

The dimension of the attractor for this series (1.7) was found earlier using the correlation function. It would therefore be illusory to try to represent simply this curve. One way to proceed would be to find a suitable Poincaré section. For an attractor constructed on a series of events, the equivalent of a Poincaré section consists of extracting sub-series from the whole series (cf. Bergé et al., 1984, about the Rössler attractor [55]).

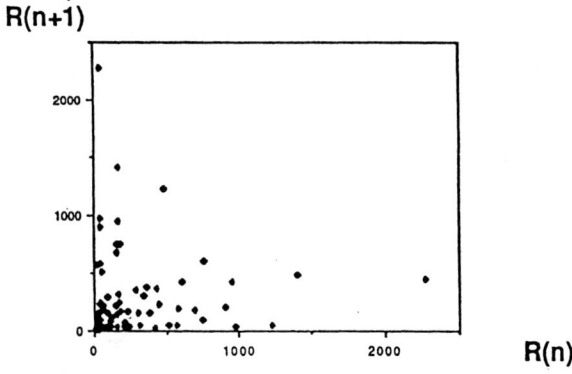

Figure 5.7: *Example of a first-return map.* This test was used on the t_3 series computed for Piton de la Fournaise. The result shows a situation intermediate between a purely random distribution and a determinist distribution (Sornette et al., 1991).

Sornette et al. (1991) [481] observed that most points on figure 5.7) close to the origin resembled noise, suggesting that eruptions occuring too close together cannot be distinguished and that they may correspond to the same higher stage of the volcano's dynamic evolution. In these conditions, sub-series can be extracted after eliminating eruptions for which $t_i - t_{i-1} < t_{\min}$. Several trials for values of $t_{\min}$ between 100 and 200 days, allowed the data

Volcanology

to be "cleaned" to produce clearer graphs (Figure 5.8).

A second way of processing the data consists of selecting the eruptions i for which t_{i-1} and t_{i+1} are smaller than t_i. The graph of the remaining values (24 out of 72) is shown in figure 5.8. A curve fits well the resulting points, characteristic of deterministic chaos (Bergé et al., 1984 [55]).

A similar approach was used on data from the Hawaiian volcanic complex, whose attractor's dimension is larger. The resulting curve is different from the one for Piton de la Fournaise and shows two maxima (Figure 5.9). This may be due to its high attractor dimension, $D \approx 4$, suggesting the interaction of two different non-linear dynamic deterministic systems.

Finally, this method was applied to recent series from Anak Krakatau, a young volcano emplaced in 1929 on the site of the old volcano which exploded in 1883. The simplicity of the resulting curve shows that the attractor's dimension is certainly close to 1, and hence that the current eruptive mechanism is linked to the evolution of a single, shallow, magma chamber, relict from the large 1883 eruption. This confirms the hypothesis generally made by volcanologists (Camus and Vincent, 1982, 1987).

It is also worth noting an important study by Shaw (1988) on Kilauea, using the first-return map on the Kilauea's summit reservoir.

5.3.3 Volcano behaviour and Self Organized Criticality

In our first book (Dubois and Gvishiani, 1998, [134]), we have defined the concept of Self-Organized-Criticality (SOC), which is relevant for understanding temporal and spatial scaling in a wide class of dissipative systems with extended degrees of freedom.

Volcanoes can be described as self organized systems. Grasso and Bachèlery (1994) [196] have shown that volcano behaviours are natural phenomena where power laws and scale invariant geometry are observed.

The authors considered the volcano as an isolated geological object relative to the plate boundary dynamics, and they explored basic statistical distributions of volcanic observables in order to extract constraints on the mechanics of the Piton de la Fournaise volcano.

They studied 6 observables:

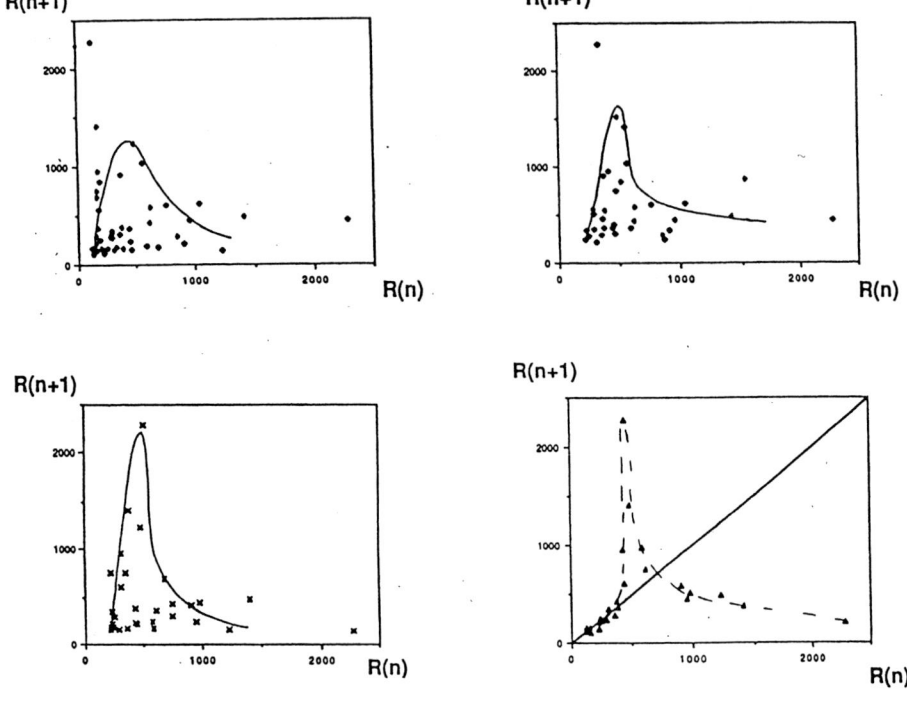

Figure 5.8: *First-return maps obtained with sub-series.* In these sub-series, eruptions too close to the previous one were removed from the dataset, for: a) $R_{min} = 100$ days; b) $R_{min} = 200$ days; c) $R_{min} = 200$ days and $(i-1)$ and i are removed as well; d) only the longest series was kept (Sornette et al., 1991).

1. The piton de la Fournaise induced earthquakes inside a volume $2 \times 3 \times 3\ km^3$ (magnitudes obtained using o duration-magnitude scale).

2. The fissure length on a $2 \times 3 km^2$ surface area, that overlaps the seismic zone, obtained using a photogrammetric cover.

3. The dyke thicknesses estimated *in situ* at the bottom of deep eroded cliffs.

4. Eruption durations for the 1920–1992 period.

5. Eruption volumes for the same period.

6. The interflow periods.

Volcanology

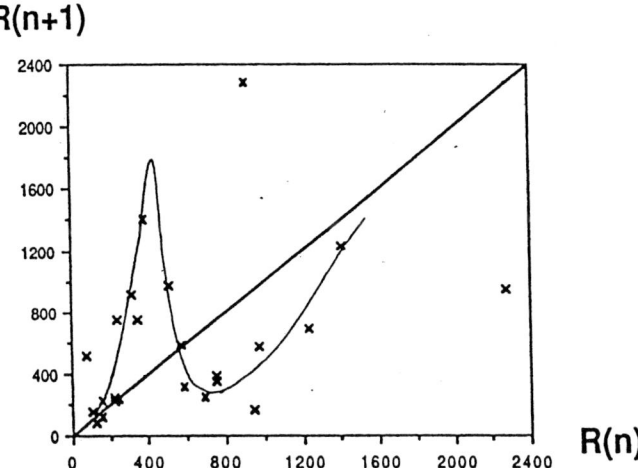

Figure 5.9: *First-return map for the Hawaii eruptive series.* The same procedure was used as in the previous figure. The presence of two maxima may be explained by attractor dimension value, $D \approx 4$.

All these observables have power law distributions, evidence for self-similarity over a finite scale range. The similarity in scaling relations for duration and size argues that the observed breaks in scaling laws, for both duration and volume distributions, are characteristic of the inteplay between the storage capacity and the associated fluid transport processes. The observed breaks of slope for large flows values are typical of finite scalability. The choice of a departure from power-law distribution, not rejected by the data, has the advantage that it is related to an underlying SOC system which allow to rationalize the underlying mechanics of the distribution of volcanic observables.

On the same topic, we note the modeling of eruptions dynamics given by Lahaie et al. (1996) [292].

To analyse the hierarchical organization of Piton de la Fournaise eruptions, the authors proposed a model based on the self-organized criticality concept. The model is a network of multi-magmatic lenses at critical pressure, controlled by weak perturbations which release energy through a power-law distribution of eruptions [132, 133] and evolve into a marginally stationary state. They show that the complex behaviour of the volcano during the 1920–1992 period could emerge from a few basic rules controlling individual parts of the system (lenses), without requiring a specific size for the macro-reservoir of a significant magma supply or change in magma supply.

Acoustic emission at Stromboli volcano Recently Diodati et al. (2000) [125, 124] analysed acoustic emissions recorded over several years at Stromboli volcano. They emphasized the deterministic behaviour of the system generating the acoustic events time series.

5.3.4 Multifractal analysis

A multifractal analysis was applied by Shaw and Chouet (1989) [457] to a series of tremors recorded at Kilauea over a period of 22 years. The method used was the FSA (Fractal Singularity Analysis) proposed by Meakin et al. (1985) [354]. This method begins with a first-return map where the duration, x_{n+1}, of an incoming tremor is plotted as a function of the duration x_n, of the present tremor. This method, described in Dubois (1995) [129], consists of estimating the probabilities from the recurrence times $m_i = p_i^{-1}$; m_i being the number of steps needed to come back to a distance l from a given starting point.

The values of m_i are determined for one value of l at each point, elevated to the power $(1-q)$ and averaged. This is done for all points, for several values of l and q, and allows the computation of the slope τ, for q constant of log-log graphs of the averages (m^{1-q}), as a function of l. We get:

$$\alpha = d\tau/dq, \tag{5.7}$$

and,

$$f(\alpha) = q\alpha - \tau. \tag{5.8}$$

One then plots $f(\alpha)$ as a function of α (Figure 5.10).

For their study of Hawaiian tremors, Shaw and Chouet (1987) used 577 events recorded over 22 years. The uncertainties in the application of FSA show that α is unstable below 0.6 and above 1.5. This is due to relatively small number of data points for this kind of analysis. Other studies benefited from more points (Figure 5.10): 2500 points for the convection movements analysed by Jensen et al. (1985) [266] and 70 000 points for the electronic transport processes analysed by Gwin et Westervelt (1987) [222]. The CGMN (Critical Golden Mean Nonlinearity) curve by Halsey et al. (1986) is drawn in the figure.

Volcanology

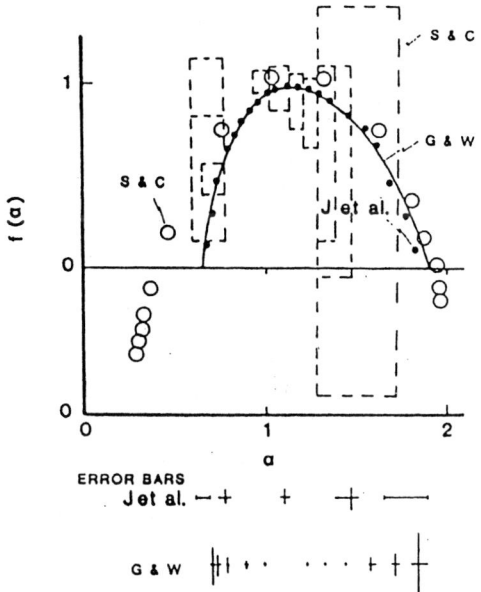

Figure 5.10: *Application of FSA (Fractal Singularities Analysis)*. A series of 577 tremors recorded in Hawaii over 22 years is analysed (Shaw et Chouet, 1987). The results are compared to the conclusions from several models including convection (Jensen *et al.*, 1985) and CGMN (Golden Mean Number; Halsey *et al.*, 1986).

It is worth noting that multifractal analysis can be applied to many domains. Seismicity is one example, with the construction of singularity spectra in the time domain at a given location (interval between earthquakes), or in the spatial domain at a given time (Geilikman *et al.*, 1990) [169].

5.4 Geomagnetism Study at Different Time Scales

5.4.1 Geomagnetic Reversals

Study of inversions : Cox scale

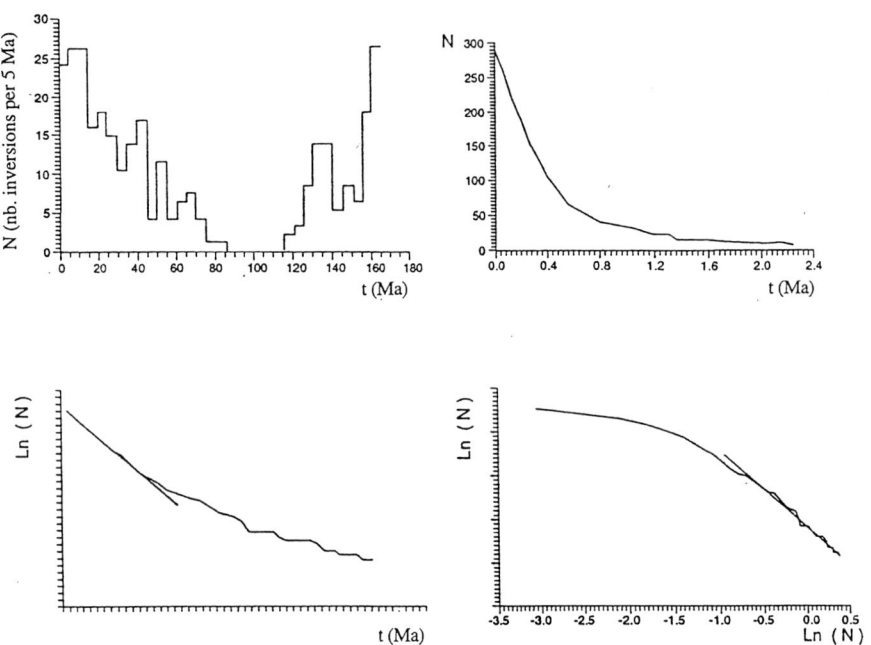

Figure 5.11: *Inversions of the Earth's magnetic field.* These four graphs represent, from left to right and top to bottom: the frequency of inversions every 5 Ma from -165 Ma to present time, and the distribution of cumulated values of duration $\geq t$, with N as a function of t, $\log N$ as a function of t and $\log N$ as a function of $\log t$, (from Dubois et Pambrun, 1990).

This study is a summary of the results obtained by Dubois et Pambrun (1990) [138] for a series of inversions of the Earth's magnetic field (Cox scale) [107].

The study first looked at the distribution of time intervals between two successive inversions. 296 periods were investigated (148 direct and 148

inverse orientations of the magnetic field), with durations between 0.01 and 35.25 Ma.
Figure 5.11 shows the distribution of these periods every 5 Ma, from -165 Ma to the present period. It also shows the cumulated values of inverse and direct periods durations $\geq t$, on three graphs:

1. N as a function of t,

2. $\log N$ as a function of t,

3. $\log N$ as a function of $\log t$.

The semi-log graph shows a quasi-linear slope of -3 between 0 and 0.55 Ma. It is used to deduce that the distribution is exponential ($N = N_0 e^{-3t}$) in the interval [0, 0.55]. This result is in agreement with the conclusions of McFadden (1984) [351], who likens it to a Γ distribution. This distribution applies for 200 of the 296 events. For $t > 0.55$ Ma, the distribution moves away from the regression line, but the log-log graph shows a quasi-linear slope of -1.6 between 0.55 and 2 Ma. A law of the type $N = N_0 t^{-1.6}$ can be deduced for the remaining 90 events.

A random Poissonian character is compatible with the first distribution, but maybe not with the second one. Investigations performed on dynamic systems (Arnéodo et Sornette, 1984) [14] demonstrate that these two types of distribution may both be the expression of chaotic deterministic systems with more or fewer degrees of freedom.

This led us to test the determinism of the system generating the series, using the correlation function of Grassberger and Procaccia (1983) [195]. This correlation function is applied to a pseudo-phase space constructed using the duration between inversions, X, and their occurence number. In a three-dimensional space, for example, the point representing occurrence i will be plotted with the values x_i, x_{i+1} and x_{i+2}.

The results are shown in Figure 5.12. The value of the slope of $\log C(r)/\log(r)$ (see [134]) does not saturate when the phase space dimension (ie embedding dimension) increases (periods [0 to -168 Ma] and [0 to -23 Ma]). However, between -23 and -168 Ma, the saturation indicates that the system has an attractor of dimension slightly greater than 2.

This reminds of the value obtained for the simple Rikitake model.

Chaos therefore appears to be deterministic in this period, which confirms the observation on the distribution law for periods greater than 0.55 Ma. When introducing the inversions separated by short periods, in recent times, it seems that the system is perturbed by white noise.

260   *Fractals and Dynamic Systems*

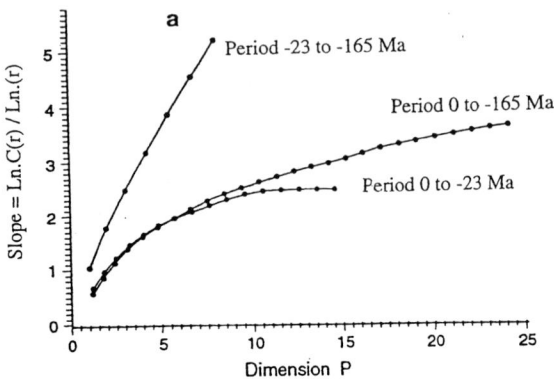

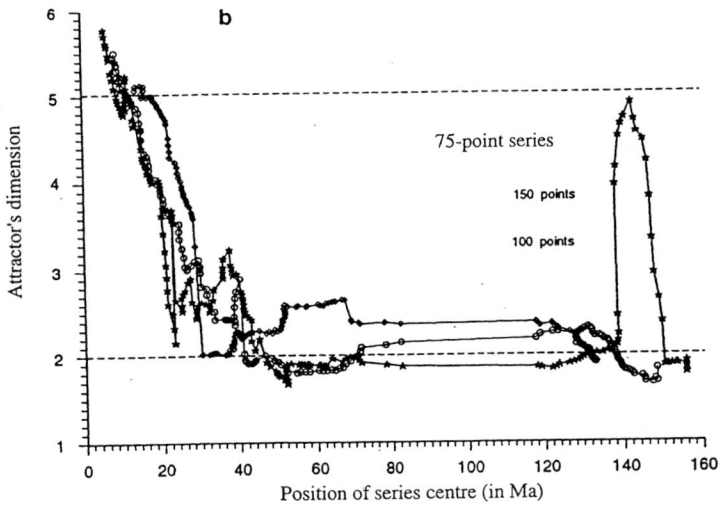

Figure 5.12: *Application of the method of Grassberger and Procaccia (1993) to the series of Earth magnetic field inversions (Cox scale).* The slope of the correlation function is represented as a function of the phase space dimension for the whole dataset (top) and for a moving window (bottom) (from Dubois and Pambrun, 1990).

One of the main problem encountered in this kind of study is the small number of data points (around 300), and the error bars on the dates of the oldest inversions. Developments in paleomagnetic analysis continually improve the dating of rocks and sediments, and we may soon be able to go further than the 168 Ma of the Cox scale. We hope for improved studies in the near future, using continuous series up to 300 or 350 Ma (cf. research on the Carnian and Norian series by Besse and Courtillot (1991) and Gallet et al. (1992) [60, 167]).

Other studies
A debate on this problem arose when Marzocchi and Mulargia (1992), Marzocchi et al. (1995), Marzocchi (1997) [342, 344, 341] emphasized, in studying the periodicity of geomagnetic reversals, that the results indicate that geomagnetic reversals since 85 Ma BP occur according to a generalized Poisson process with an exponentially increasing mean that may have a superimposed periodic modulation. They performed a statistical comparison of synthetic geomagnetic reversal series with real reversal series and concluded that the model is inadequate to represent reality.
However, one may emphasize that the results obtained by Marzocchi et al., (1997) do not contradict the Dubois et Pambrun (1990) results, because they studied only the Cande and Kent's (1995) series, extending between $85Ma$ and the present [343, 138, 85]. The series contains mainly the Poisson distribution part observed by Dubois et Pambrun $(0, -23Ma)$. The determinist is observed inside the total period [- 23, - 140 Ma] only when using a moving window technique, as was observed by Cortini and Barton (1994) [105].
In another analysis of a more complete reversal time series, Cortini and Barton (1994) concluded that, because of the small size of the geomagnetic reversal data set (only 282 points), and the poor definition of the scaling region, the correlation dimension for the magnetic period sequence is quantitatively not meaningful. The clear difference in correlation integrals between measured and randomized reversal sequences suggests that the geomagnetic reversal dynamics is not random and that low-dimensional chaos, if not uneguivocably detected, can be suspected. They compared Earth's magnetic field data (geomagnetic reversal records) and the prediction of the disk dynamo model.

5.4.2 Temporal Variations of the Magnetic Field Vector

Continuous recording of the components of the magnetic field vector only started at the beginning of the nineteenth century, and good quality data goes back at most a century. We have investigated the records from the Observatoire National de Chambon-la-Forêt (France) between 1883 and 1992, which are among the most reliable.

We shall still use the correlation function, which is well suited here for regularly sampled time series. The time series cover a domain from one minute to 110 years, i.e. a bit more than 7 orders of magnitude. We limited ourselves to the search for a possible attractor on hourly, daily, monthly and yearly series of the mean values of the field's component, of its module F and of its declination D. The records of Chambon-la Forêt show that:

1. These series are noisy, except for the series of averaged yearly values, which saturates at $\nu = 1,3$ (ν is the slope of the line $\log C(r)/\log(r)$, i.e. the attractor's dimension).

2. The values of ν decrease from 3.5 (for an embedding dimension of 10 in the phase space) for the hourly values, down to 1.3 for the yearly series.

3. Computation with a moving window shows that the dynamic systems generating these series are not stable. This means that the dimension of their attractor varies with time (which may cause the observed noise).

4. These preliminary observations should be extended to other series recorded at other places They show that the number of degrees of freedom of the dynamic systems ruling the magnetic field variations are higher when the study interval is shorter. This can be explained by the fact that these variations are the sum of various internal and external processes: storms, daily and seasonal variations, etc. These variations are filtered better when the values are averaged on longer times scales. The number of degrees of freedom for the system corresponding to hourly variations is necessarily higher than the one for daily variations, itself higher than the one for monthly variations, etc.

Hongre et al. (1999) [247] analysed the Eskdalemuir observatory time series using nonlinear and multifractal approaches. They used the hourly mean values of the magnetic field recorded over 79 years (692,520 data points for each component). They estimated a 5-dimensional pseudo-phase space and a positive Lyapunov exponent, confirming the possibility of low-dimensional deterministic chaos in the magnetic field observations at ESK observatory. The correlation between the solar activity (the Wolf number), the unstable nature of the magnetic field, and the singularity spectrum points out the forcing of the solar cycles on the dynamics of the magnetic field were studied. The signal analysis of the X, Y, Z components of Eskdalemuir Observatory using the Fourier power spectrum reveals main periods and harmonics between 1 h and half a century in addition to a non-periodic part. The multifractal spectra of a 2-year series of daily mean values, in terms of regularities, show that the most probable behaviour of the geomagnetic field is that of a fractal Brownian motion with regularity of 0.6 (a little more regular than Brownian motion). Additionally, the comparison of these multifractal spectra with the fluctuations of solar energetic activity, in terms of the Wolf number, shows evidence of a linear correlation between them for X and Z. The width of the X and Z spectra, which gives the set of possible regularities locally encountered in the series, is determined by the solar activity. That of the Y spectra is more constant confirming that it is less dependent on external influences.

Going further into the study of the Z component, the phase space orbits and the attractor were constructed using a typical time increment of 8 h and an embedding dimension of 5. The secular variation is seen as a transition along the main bisector of the peudo-phase space, while the short time scale behaviour associated with chaotic dynamics, of typical time 8 h, control the distribution of the orbits around the bissector (see figure 5.13).

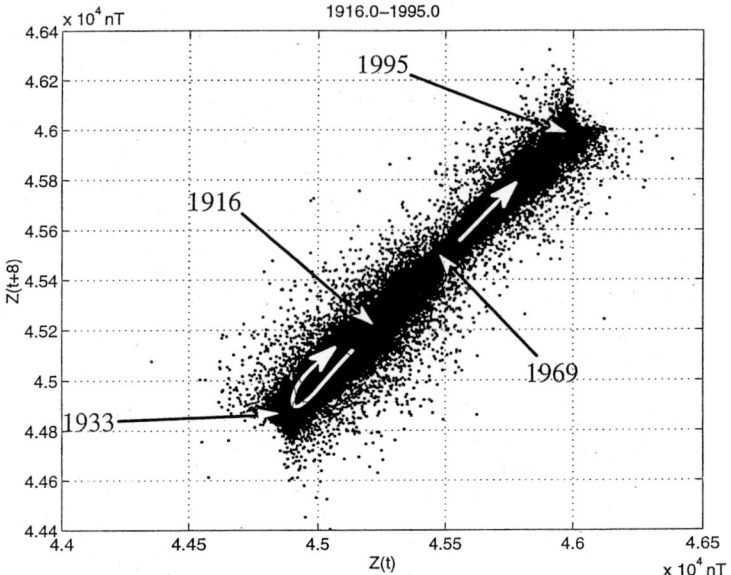

Figure 5.13: *The attractor of the magnetic field at ESK observatory. Z component, the reconstructed attractor.* $T = [1916, 1995]$, $dt = 1h$. (From Hongre et al., 1999).

Secular variations

Researchers interested solely in the internal causes of variations of the Earth's magnetic field can now also use paleomagnetic data, as for the study of inversion series. The studies of cores from lacustrine sediments (Thouveny, 1991) or marine sediments (Tauxe and Wu, 1990), [507, 502], give access to time series dating back as far as 120 000 years.

For lacustrine sediments, the sampling interval is around 200 years. The measures made in observatories and these measure are nearly continuous in time. The lower time limit is similar to the maximum time of observatory data, and the upper limit of 120 000 years is similar to the the lower limit of reversal data.

The time domain of field variations is therefore covered from one second (continuously recorded rapid fluctuations) to 500 Ma, more than 16 orders of magnitude. Paleomagnetic variations recorded in the sediments from Lake du Bouchet (Thouveny, 1991) are shown in Figure 5.14. The variations of declination, inclination and paleo-intensity appear to be chaotic. Analysis using the method of Grassberger and Procaccia leads to the graphs shown in Figure 5.15, plotted for 498 values in each of the three series.

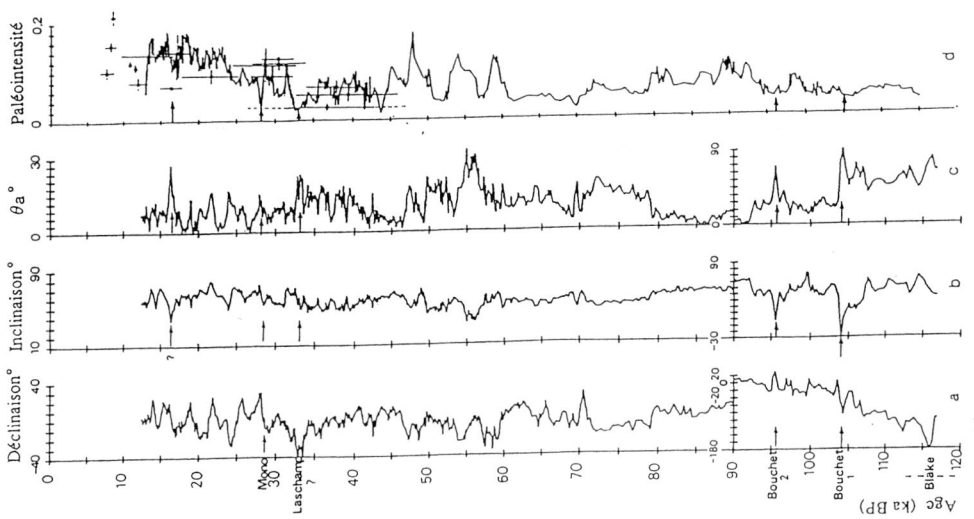

Figure 5.14: *Variations of the Earth's magnetic field in the last 120 000 years. The four graphs correspond to: a) the vector declination; b) the vector inclination; c) the vector amplitude; d) the relative paleo-intensity. (From Thouveny, 1991).*

The main observations are:

1. The dimension of the phase space close tends to saturate to 7 - 9. The values of ν are: 3 for D, 2.4 for I, 2.6 for F.

2. The declination seems the noisiest of the three variables, which seems logical because of the techniques used to place the cores in space (techniques of declination measurement of the core is less precise than inclination measurement).

3. The values of ν range between 2 and 3. This implies the existence of an attractor ruling the behaviour of a determinist chaotic system with three degrees of freedom.

The evolution of the system with time was studied using a moving window of 200 points (Figure 5.16), with a step of 20 points. Even though the system

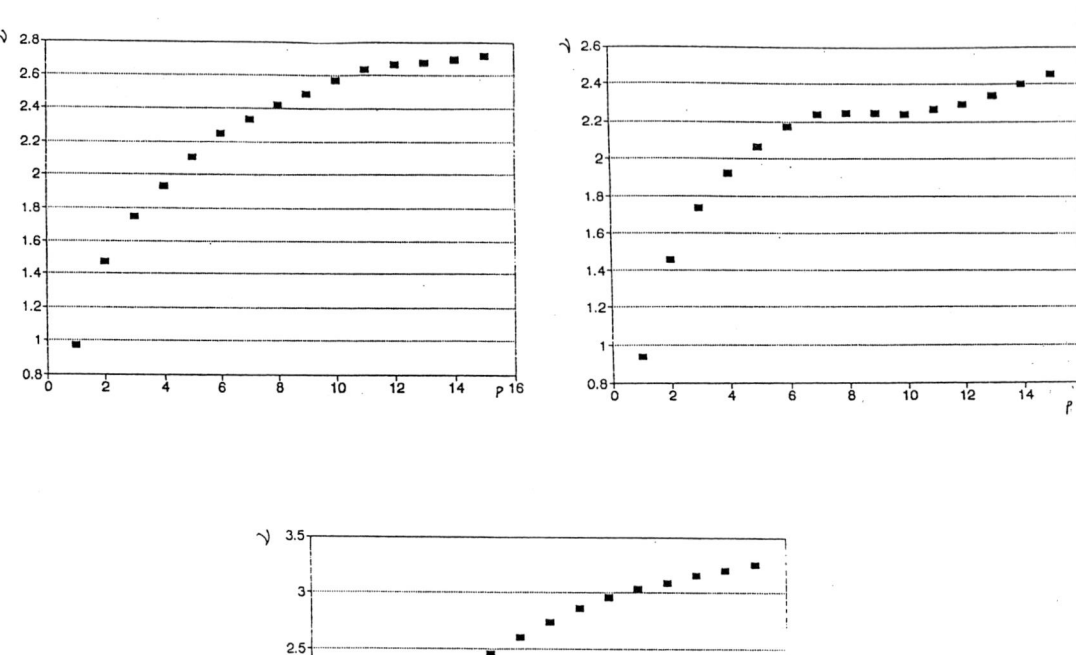

Figure 5.15: *Dimensions of the attractors for the variations of the Earth's magnetic field.* The variations of the slope are measured as a function of the embedding dimension of the phase space (method of Grassberger and Procaccia). Each series contains 498 values of D, I, F, measured on sediments from Lake du Bouchet. There is a good saturation of the inclination for a value of 2.3.

is not perfectly stationary, the attractor's dimension is constrained to be between 2 and 3.

Because of its small dimension (smaller than 3), we have represented the attractor in its phase space with the axes $X(t)$, $X(t+\tau)$ and $X(t+2\tau)$ where X is the value of I and $\tau = 208$ years (Figure 5.17).

We notice that:

1. All phase trajectories fit inside a finite volume (cocoon-shaped).

2. They do not completely fill this volume, meaning that the attractor's dimension is smaller than 3.

The third stage consists of using a theoretical model of the magnetic field and its coefficients g_n^m and h_n^m to compute theoretical variations sampled every 200 years, and to determine the slope of the correlation function on this series.

The computation is:

$$I(t) = \arctan \frac{Z(t)}{\sqrt{X^2(t) + Y^2(t)}}, \qquad (5.9)$$

where:

$$X(t) = \sum_n \sum_{m=0}^n (g_n^m(t)\alpha_m + \beta_m h_n^m(t))\delta_n^m, \qquad (5.10)$$

$$Y(t) = \sum_n \sum_{m=0}^n (g_n^m(t)\beta_m - \alpha_m h_n^m(t))\frac{\gamma_n^m}{\sin\theta}, \qquad (5.11)$$

$$Z(t) = \sum_n \sum_{m=0}^n (g_n^m(t)\alpha_m + \beta_m h_n^m(t))(n+1)\gamma_n^m, \qquad (5.12)$$

and where $n = 1, ..., \infty$ and $m = 0, ...n$; θ and ϕ are the coordinates of the measure point (here, Lake du Bouchet), and α_m and β_m are the Legendre polynomials:

$$\alpha_m = \cos m\phi, P_n^m(\cos\theta) = \gamma_n^m, \qquad (5.13)$$

$$\beta_m = \sin m\phi, \partial P_n^m/\partial\theta = \delta_n^m. \qquad (5.14)$$

In later studies, a generating function $g(t)$ was introduced to calculate the 24 coefficients g_n^m and h_n^m (we stopped at $n = 4$), that accounts for random variations and the exponential decrease in amplitude for increasing n (Hulot

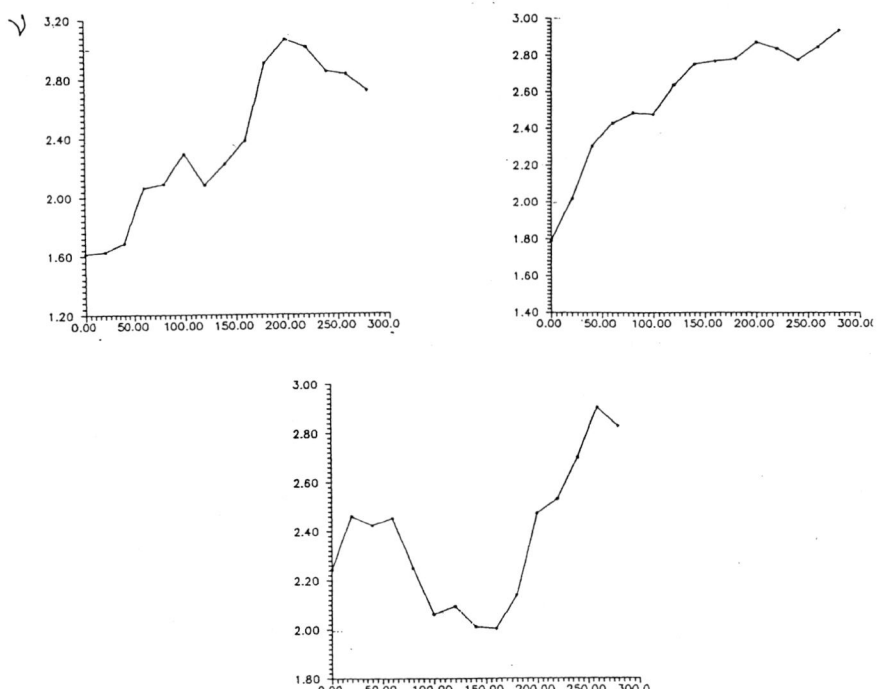

Figure 5.16: *Moving-window analysis of the attractor's dimension along a core from Lake du Bouchet.* The dimension of the phase space has been fixed at 7 for the three components (Atten and Caputo, 1987).

(1992) [253]). The generating function used three sinusoidal terms of periods 220, 550, and 1 200 years, as well as random phases. The resulting variations of the theoretical field are very interesting, and the resulting system has an attractor dimension of $2 < D < 3$.

Other studies

Barraclough and De Santis (1997) studied geomagnetic data based on time series of annual mean values from the Coimbra magnetic observatory, and from reconstructed combinations of three Italian and three UK observatories. The Coimbra dataset covers the time span 1868–1989 [41]. The data were analysed using of a nonlinear forecasting approach. An analysis in terms of first differences (secular variation) of the horizontal magnetic components made in phase space with the simplex technique seems to exclude the pre-eminence of any stochastic or periodic behaviour. The dimensionality of the underlying nonlinear process and the corresponding largest positive Lyapunov exponent give some evidence that the geomagnetic field evolves as

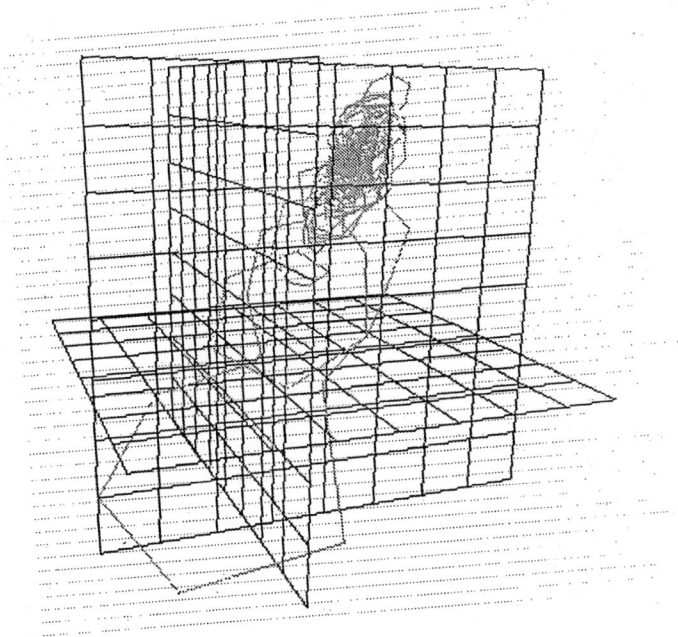

Figure 5.17: *Perspective representation of the attractor of the magnetic field variations.* We have chosen the inclination as variable in a three-dimensional space (the attractor's dimension being 2.3). The algorithm was applied on the CM-2 supercomputer by P. Stoclet.

---

a nonlinear chaotic system with unpredictable behaviour for times greater than a few years, confirming the common practice of updating global models of the geomagnetic field every 5 years.

5.4.3 Theoretical Modeling

Rikitake's model (1958) [435] for geodynamo reversals has long been recognized as an instable dynamic system. Its chaotic behaviour was studied by Cook et Roberts (1970) then by Ito (1980) [103, 262] who showed that a chaotic attractor is observed when the μ parameter (see below) which characterizes the system is ohmic dissipation was confined to a specific interval. Ito shows that this chaotic behaviour is similar to the Lorenz model, which is characterized by the irregular travel of an orbit between two unstable fixed points. The travelling in the Lorenz model corresponds to the polarity reversals in the Rikitake's model. The frequency of the polarity reversals depends strongly on the parameter μ.

In the model that we describe below, from Ershov et al. (1989) [141], we give the simplest generalization of Rikitake's model that removes the degenerate and infinite solutions found in the Ito's version. This is achieved by adding a viscous friction that acts on both disks.

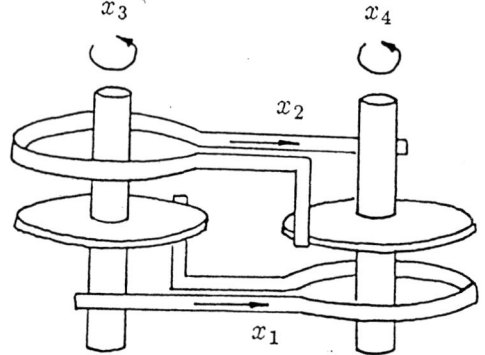

Figure 5.18: *Model of the two-disk dynamo (Ershov et al. (1989)*. This model is the same as Rikitake's (1958). x_1 and x_2 are the dimensionless values of the electric currents in the disks, x_3 and x_4 are proportional to the disks' angular speeds.

Generalised model of the two-disk dynamo (Ershov *et al.***, (1989)**

Ershov *et al.*, (1989) use the model of a dynamo made of two disks with friction, shown in Figure 5.18 [141]. Rikitake's dynamo is a system where the two disks are connected. Their respective axes are submitted to identical torques. Adding viscous friction, which reduce the angular moment of the disks, the system can be described by the following system of equations:

$$\begin{cases} \dot{x}_1 = -\mu x_1 + x_2 x_3, \\ \dot{x}_2 = -\mu x_2 + x_1 x_4, \\ \dot{x}_3 = 1 - x_1 x_2 - \nu_1 x_3, \\ \dot{x}_4 = 1 - x_1 x_2 - \nu_2 x_4 \end{cases} \quad (5.15)$$

where overdotted variables are the time derivates, x_1 and x_2 are electric currents in the two disks, x_3 and x_4 the angular speeds, μ is the ohmic dissipation coefficient (identical for each circuit) and ν_1 and ν_2 are the viscous friction coefficients which can be different for each disk.

Several models of this type have been built and tested (Allan, 1962) [5] . Cook (1972) [102] studied an identical model with different torques on each disk, but with $\nu_1 = \nu_2$.

When $\nu_1 = \nu_2 = 0$, the last two equations can be reduced, and the system becomes Rikitake's model again with $x_3 - x_4 = A$ (constant) obtained by subtracting off the two last equations in the system. Depending on the values of A and μ, Rikitake's system may go toward a limit cycle in which the $x_i(t)$ are periodic, or toward a strange attractor in which $x_i(t)$ oscillate irregularly. For example, for values of $\mu = 1$ and $A = 3.75$, the system exhibits a chaotic behaviour (Cook and Roberts (1979). Using the method of Wolf et al. (1985) (cf. Dubois and Gvishiani, 1998 p. 170 [134, 541]), the Lyapunov exponents are determined to be:

$$\lambda_1 \approx 0,12, \quad \lambda_2 = 0, \quad \lambda_3 = -2,19. \quad (5.16)$$

The existence of positive exponents implies an exponential divergence of the phase trajectories on the attractor. The dimension of the strange attractor is given by the Kaplan-Yorke formula:

$$d_1 = (2 + \lambda_1)/\mid \lambda_3 \mid \approx 2,09. \quad (5.17)$$

When the viscosities ν_1 and ν_2 are not 0, the system can be written as:

$$\begin{cases} \dot{x}_1 = -\mu x_1 + x_2 x_3, \\ \dot{x}_2 = -\mu x_2 + x_1 (x_3 - A), \\ \dot{x}_3 = 1 - x_1 x_2 - \nu_1 x_3. \end{cases} \quad (5.18)$$

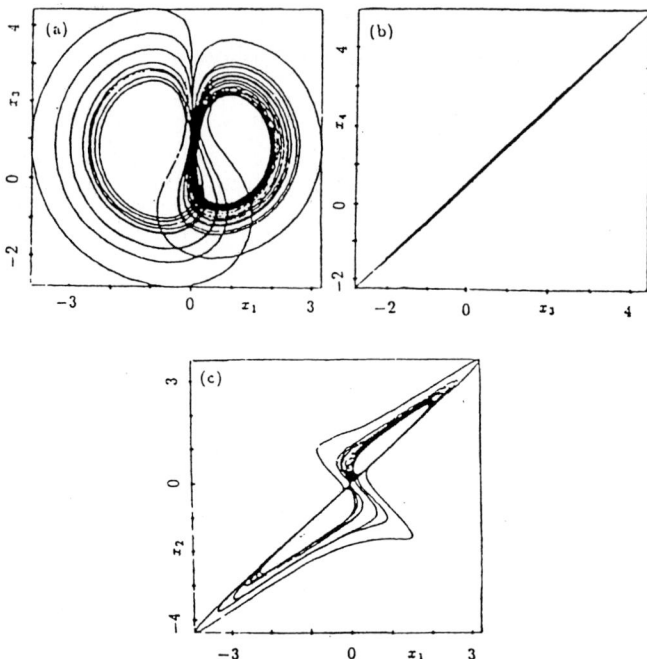

Figure 5.19: *Attractor of the dynamic system associated with the Erschov et al. (1989) model.* For $\mu = 1$, $\nu_1 = 0,004$ and $\nu_2 = 0,002$, the three graphs represent the respective projections on x_1x_3, x_3x_4 and x_1x_2.

---

and
$$\dot{A} = -\nu_2 A - \Delta\nu x_3, \qquad (5.19)$$

where $x_3 = x_4 + A$ and $\Delta\nu = \nu_1 - \nu_2$. The parameter A is constant in Rikitake's system, but it varies with time in this system and does not tend toward a constant as $t \to \infty$.

Ershov et al. (1989) study the system's attractor on the projections x_1x_2, x_1x_3 and x_3x_4 for several values of the parameters μ, ν_1, ν_2, (fig. 5.19). For example, in the case $\mu = 1$, $\nu_1 = 0.004$, $\nu_2 = 0.002$, the attractor is very thin (c.f. the projection of x_3x_4 in Figure 5.19).

The different projections of points representing the dynamic system in phase space vary chaotically from positive to negative values of x_1 and x_2. These two variabes are at the origin of the magnetic fields created by the disks. Fields parallel to the axes add up and produce a resulting field with can be positive and negative, depending on the phase trajectory on the attractor.

Each time the sign changes, there is an inversion. This means the series of inversions are chaotic with time (deterministic).

Many numerical tests performed by the authors (cf. Figure 5.19) allowed them to conclude that:

1. Contrary to Rikitake's dynamo, the friction dynamo "forgets" the initial conditions.

2. For all the parameter values there exists a stable equilibrium.

3. The attractors' dimension is $3 + \epsilon$ with $0 < \epsilon < 0.2$, while it is $2 + \epsilon$ for Rikitake's model.

4. Asymmetric cycles appear, but the attractors are quasi-symmetric.

5. When ohmic resistivity decreases, the chaotic part shows bifurcations with period doubling. This also happens when the friction decreases.

Hide's models (1995, 1997) and Hide et al.'s model (1996)

These models investigate some possibilities for systems such as two identical, coupled, Bullard-type single Faraday-disk dynamos (Hide, 1995) [237], two self-exciting single-disk homopolar dynamos (Hide et al., 1996) [240], a system governing a hierarchy of self-exciting coupled Faraday-disk homopolar dynamos (Hide, 1997) [238] and a system with nonlinear quenching of current fluctuation in a self-exciting homopolar dynamo (Hide, 1997) [239].

- In his study on the structural instability of the Rikitake disk dynamo (two identical single Faraday-disk dynamos), Hide (1995) shows that mechanical friction can render the Rikitake dynamo "structurally instable" and consequently incapable of producing chaotic oscillations [237].

- Two interesting models (see Figure 5.20) controled by a set of nonlinear ordinary differential equations are described as two novel self-exciting single-disk homopolar dynamos (Hide et al., 1996).

    The equation system is

$$\begin{aligned}\dot{x} &= x(y-1) - \beta z, \\ \dot{y} &= \alpha(1-x^2) - \kappa y, \\ \dot{z} &= x - \lambda z,\end{aligned} \quad (5.20)$$

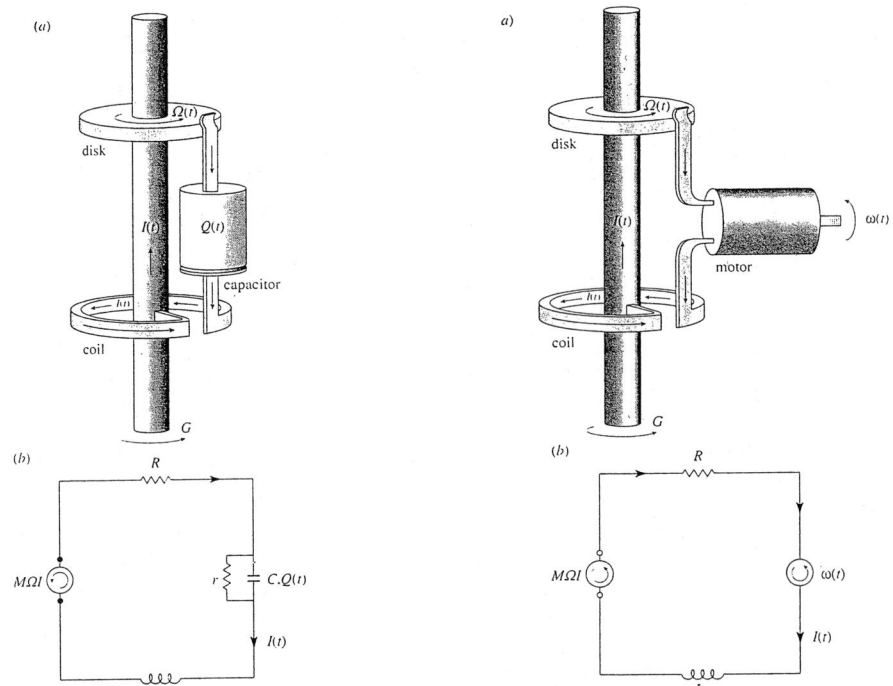

Figure 5.20: *Two self-exciting single-disk homopolar dynamos (Hide et al., 1996)*. a) Single disk dynamo with capacitor in series with the coil and equivalent circuit. b) Single disk dynamo with a motor in series with the coil.

where x, y and z are the phase space coordinates and α, β, κ and λ, are 4 parameters describing the electric circuit characteristics.

Numerical integrations show a vastly more complicated behaviour which is impossible to classify in simple terms.

As for the Rikitake system attractor a bilobed attractor shows orbit jumps from one lobe to another indicating magnetic field inversions.

- Going deeper into model complexities, Hide (1997) studied the nonlinear differential equations governing a hierarchy of self-exciting coupled Faraday-disk homopolar dynamos [238]. The general model designated "system $S(N; J(N))$" consists of n separate units arranged in a ring and numbered consecutively $n = 1, 2, 3, \cdots, N$, where the matrix $J(N) = (J_1, J_2, \cdots, J_n, \cdots, J_N)$, J_n being the number of series or parallel motors present in the n-th unit. Each unit is a self-exciting homopolar dynamo driven by a single electrically-conducting Faraday disk.

Figure 5.21 shows three different types systems comprising single units (a and b) or coupled units (c). Dissipation in the systems is due not only to ohmic heating but also to mechanical friction in the disk and the motors, with the latter process, no matter how weak, playing an unexpectedly crucial rôle in the production of chaotic behaviour regimes.

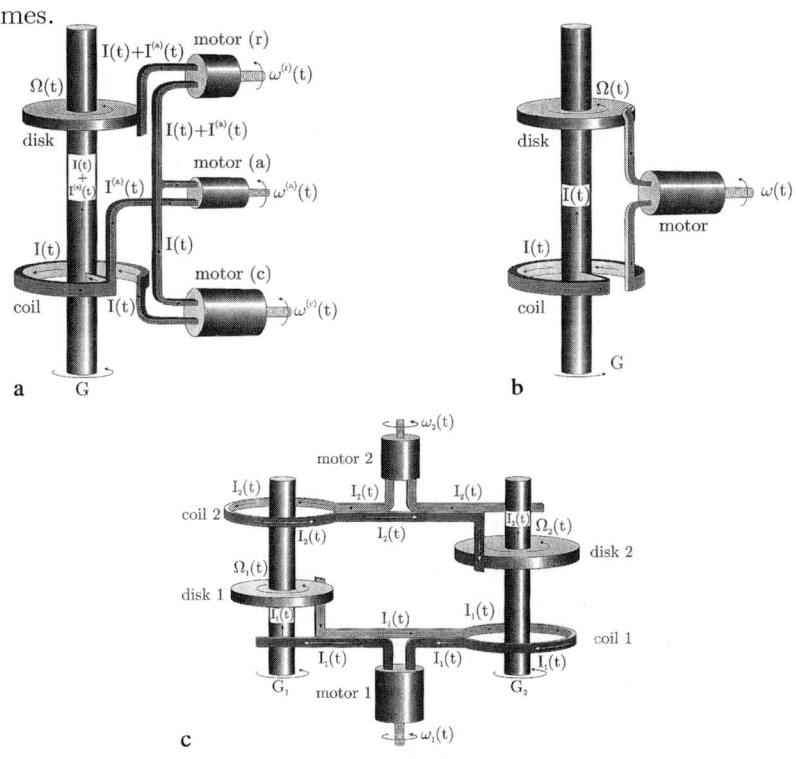

Figure 5.21: *Three systems of self-exciting coupled Faraday-disk homopolar dynamos (Hide et al., 1996).* a) an $S(1;3)$-type system (one single unit an three non-identical motors). b) an $S(1;1)$-type system (one single unit and one single electric motor c) an $S(2;1,1)$-type system (two non-identical coupled units, each with a single electric motor).

- The last example proposed by the author (Hide, 1998) deals with the nonlinear quenching of current fluctuations in a self-exciting homopolar dynamo [239]. It allows long periods of no geomagnetic field inversion, similar to the Cretaceous Long Normal superchron.

It was later shown (Hide et al., 1996) that the electric current I generated by a self-exciting homopolar dynamo with a motor in series with the coil can exhibit multiple-periodicity as well as chaotic persistent temporal fluctuation, even when the torque of the rotating Faraday disc is permanent, as long as the torque T is proportional to I. An unexpected situation was also found: persistent fluctuations are quenched when T is proportional to I^2.

Partial quenching occurs in the intermediate quadratic case when t is proportional to

$$(1 - \epsilon)I + \epsilon S I^2$$

where SA^{-1} is a constant and ϵ ranges from 0 to 1.

In this case the set of nonlinear ordinary differential equations which determines the system behaviour becomes

$$\begin{aligned} \dot{x} &= x(y - 1) - \beta z f(x), \\ \dot{y} &= \alpha(1 - x^2) - \kappa y, \\ \dot{z} &= x f(x) - \lambda z, \end{aligned} \qquad (5.21)$$

where $f(x) = 1 - \epsilon + \epsilon \sigma x$, with σ is the dimensionless measure of S.

As for the previous models one can numerically solve the system. One observes that the representative diagram of the regime is divided into three domains where:

- there is no dynamo action,
- there is steady dynamo action,
- there is fluctuating, and in some case chaotic dynamo action.

Multiple scale dynamo model Le Mouël et al. (1997)

This model was proposed by the authors of the SOFT fracturing model. It is based on the renormalization group model (Madden, 1976 ; Allègre et al., 1982, 1994, 1995, 1998; [316, 6, 7, 8, 9]).
A scaling law approach is used to simulate the dynamo process of the Earth's core. The model is made of embedded turbulent domains of increasing dimensions, the largest of which is comparable in size to the core, which are

pervaded by large-scale magnetic fields. Left or right-handed cyclones appear at the lowest scale, the scale of the elementary domains of the hierarchical model and disappear. These elementary domains behave like electromotor generators with opposite polarities depending on whether they contain a left-handed or a right-handed cyclone.

As for fracturating, the problem consists of computing a relationship inside convenient physical conditions between the probability of a cyclone occuring inside a l rank element and inside a $l+1$ rank element. The geomagnetic field $\mathcal{B}$ is decomposed into a poloidal field $\mathcal{B}_S$ and a toroidal field $\mathcal{B}_T$:

$$\mathcal{B}(\mathbf{r},t) = \mathcal{B}_S(\mathbf{r},t) + \mathcal{B}_T(\mathbf{r},t) \tag{5.22}$$

where $\mathbf{r}$ is the position vector and t is time.

We assume that each term can be expressed as the product of a space function and a time function:

$$\begin{cases} \mathcal{B}_T(\mathbf{r},t) = \mathcal{J}(t)\mathcal{B}_T(\mathbf{r}), \\ \mathcal{B}_S(\mathbf{r},t) = \mathcal{S}(t)\mathcal{B}_S(\mathbf{r}). \end{cases} \tag{5.23}$$

We assume, as shown by Braginsky (1964), Braginsky and Roberts (1987) and Jault (1995) [72, 73, 264] that the toroidal field is generated from the poloidal field through differential rotation while the poloidal field is generated from the toroidal field through cyclonic turbulence.

The Le Mouël et al. model (1997) [303] develops as follows:

A given volume $\mathcal{V}$ of a conducting fluid is represented by the square $\mathcal{D}$ of side $2^{\mathcal{L}-1}$. Figure 5.22 shows out the tendency of the flow to be two-dimensional. This volume is pervaded by a uniform magnetic field $\mathcal{B}_T$. The large square represents the $\mathcal{L}$th degree of the hierarchical model.

The first-degree domains are the squares of side d, the second-degree domains are made of four squares of degree 1, the lth domains are made of four squares of degree $(l-1)$, and so on until l equals $\mathcal{L}$.

Let us now consider a first-degree square, or 1-domain, at time t.

- This 1-square can contain a left handed cyclone (of order 1), which is denoted $\mathcal{C}_+^1$ with probability $p_+(1,t)$.

- This 1-square can contain a right handed cyclone $\mathcal{C}_-^1$ with probability $p_-(1,t)$.

- This 1-square can be void with probability $[1 - p_+(1,t) - p_-(1,t)]$.

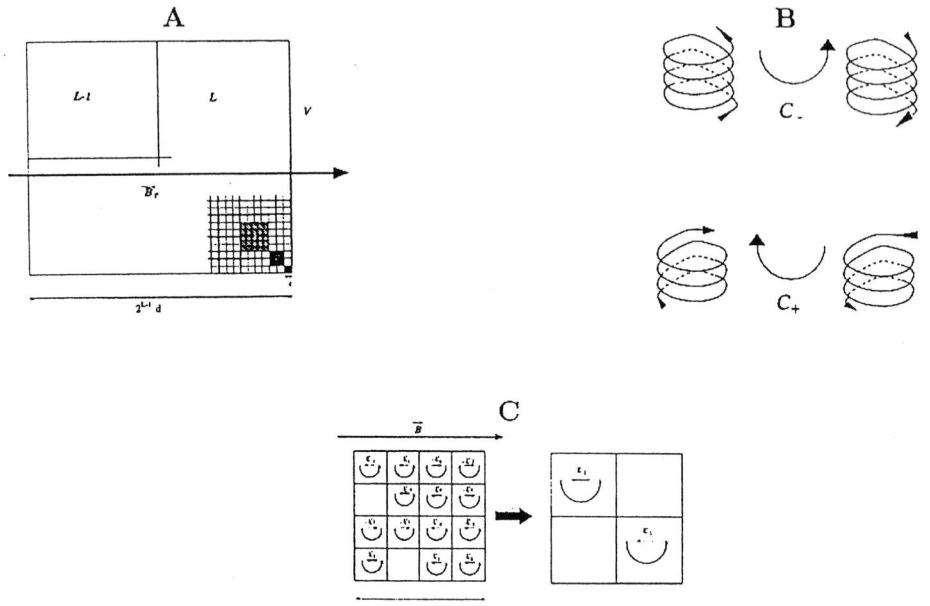

Figure 5.22: *The Le Mouël et al. hierarchical model.* A) The big square ($\mathcal{V}$) contains $2^{2(\mathcal{L}-1)}$ elementary squares of side d. B) Conventional representation of right-handed and left-handed helical motion (cyclones) used in A and C. C) Illustration of the scale transfer technique (after Le Mouël *et al.*, 1997).

This 1-square can be viewed as an elementary generator generating the electromotor force $\mathcal{E}_1$. This generator produce a geomagnetic field the orientation of which depends on which direction the cyclone turns.

The principle of the transfer mechanism leads to an inverse cascade in organization, with step-by-step construction from small to large eddies. Neighboring cyclones of order 1 $\mathcal{C}_1$ suffer both hydrodynamic and electromagnetic interactions. Basically, cyclones attract each other if they generate parallel electric currents, i.e., have the same helicity, but repel each other if they generate antiparallel current (opposite helicity). Transport by viscosity then will succeed in merging two $\mathcal{C}^1$ cyclones, building a helical motion with the same polarity as the two $\mathcal{C}^1$ motions but with a larger scale, if the two $\mathcal{C}^1$ cyclones rotate in the same sense. The same is true when considering $\mathcal{C}^1$ cyclones located in the l-domains or l-cells, which can or cannot form a $\mathcal{C}^{l+1}$ cyclone.

The 2-square made of four 1-squares will be considered to contain a $\mathcal{C}_2$ behaving like a $\mathcal{E}_2$ generator if three of the four 1-squares composing this 2-square contain a $\mathcal{C}_1$ cyclone. The same rule holds for going from the $(l-1)$ scale to the l scale, up to $l = \mathcal{L}$. Figure 5.22 represents this scale transfer.

The scale transfer mechanism depends on the rules imposed in the choice of appearance and disappearance coefficients, which leads to probability relations between the probabilility of cyclones appearing or disappearing between two successive levels. We find again the classical approach of the renormalization groups method. The quality of the model depends precisely on the choice of physical laws applied in the relations that define the appearance and disappearance coefficients.

The authors recognize that the results obtained in their first trials are only a tentative model. The most interesting feature of the model is that very small variations in the helicity generation at the smallest scale yield major changes at the highest scale, leading to polarity intervals, excursions and reversals of the geomagnetic field.

Up and down cascade in a dynamo model: Spontaneous symmetry breaking (Blanter *et al.*, 1999)

In this work a multiscale turbulent model of dynamo is proposed [63]. A secondary magnetic field is generated from a primary field by a flow made of turbulent helical vortices (cyclones) of different ranges, and amplified by an up and down cascade. The model displays symmetry breakings of different ranges although the system construction is completely symmetric.

The general features of the model are:

- the helical vortices (cyclones) can be right-handed or left-handed. They can have two orientations.

- the interaction of an helical motion with the existing magnetic field produces an electric current parallel or antiparallel to the applied primary magnetic field, depending on its orientation ; a secondary magnetic field results. We may say that the vortices of the first orientation give a positive contribution to the magnetic field, the vortices of the second orientation a negative one.

- globally, the intensity and sign of the secondary magnetic field depend on the number, scales and orientations of the vortices involved in the turbulent motion.

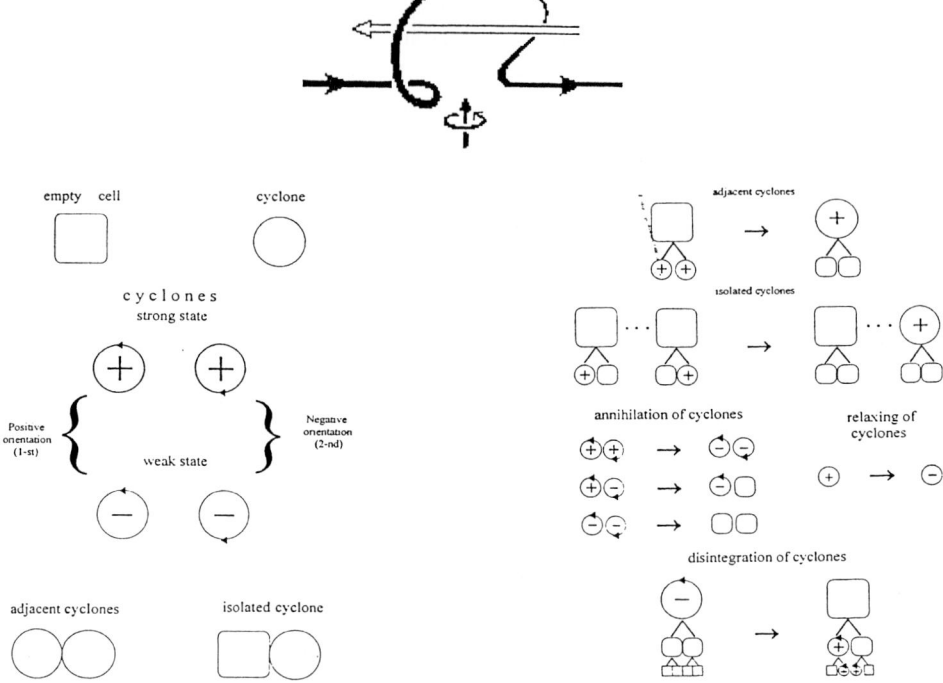

Figure 5.23: *Up and down cascade in a dynamo model.* (after Blanter et al., 1999).

The figure 5.23 shows the properties and basic transformations of the cyclones. The coupling process transforms two coupling votices into one vortex of higher range and generates an inverse cascade. The annihilation process amplifies the instability of interacting vortices: it changes strong living states into weak living states, or destroys weak vortices.

The evolution of the vortices governs the evolution of the secondary magnetic field which they generate. The sign of the secondary magnetic field is determined mainly by the orientation of the vortices of high ranges. When there is a balance between the two opposite orientations of vortices of high ranges, the intensity of the secondary magnetic field is close to zero. A sufficient symmetry breaking in the vortices orientations is needed to generate a magnetic field with a significantly nonzero intensity. This symmetry breaking must last long enough, and quick transitions from one polarity to the opposite one must occur, if we wish to mimic the behaviour of the geomagnetic field.

The behaviour of the model is thus depending on the scaling properties of the turbulence which govern the hierarchical construction of the model. The interactions of vortices at level l lead to appearence of new vortices or to disappearing of vortices at level $l+1$. The coupling of strong vortices is analysed in term of probabilities, the coupling probabilities which are invariant in time, do not depend on the orientation of the coupling vortices, and increase with the level l. The disappearence of coupling vortices at a given level has a physical meaning similar to that of the defects healing mechanism described in Blanter et al. (1997) [62]: a transition of perturbation from lower to higher levels of the system.

The result obtained shows that the interaction of both inverse and direct cascades produces strong and long symmetry breakings for relatively small values of the coupling and disintegration parameters. The maximum possible value of the intensity of the secondary field is reached. The polarity reversals are completed in a very short time compared with the duration of the corresponding constant polarity intervals (Figure 5.24).

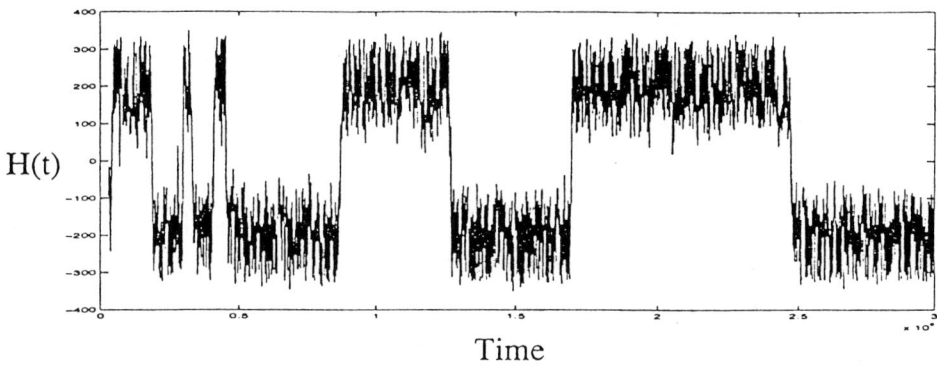

Figure 5.24: *Temporal evolution of the generated magnetic field intensity for a two-side cascade effect.* The secondary magnetic field $H(t)$ generated at time t is the total contribution of vortices of the first orientation considered as positive and the contribution of vortices of the second orientation negative (after Blanter et al., 1999).

5.5 Others

5.5.1 Heat and Water Transport in an Underground cave

The oscillatory convection observed in an underground quarry through self-potential record is an exemple of bifurcation cascade in a natural dynamic system. The result of an experiment was presented by Morat et al. (1998) [367].

The experiment was carried out in the Mériel quarry, 46 km north-west of Paris, dug according to the "rooms and pillars" technique [365, 366, 367]. Temperature, relative humidity and air pressure were recorded in the gallery and the self-potential was recorded in the walls of the galery. Figure 5.25 shows the measurement devices with the elctrodes, the humidity sensors and the thermometers. On the same figure an example of oscillations recorded by thermometers $T_1 - T_6$ (May 19 1994) is shown. The same oscillations are recorded on humidity and self-potential recordings (Adler et al., 1997, [1]). The sinusoidal oscillations are organized in intermittent trains separated by time intervals during which the periodic character is less clear, or even disappears. Phase shifts exist between the different curves. In particular, in the case of I minute period oscillation, the T_5 and T_6 curves are approximately in phase opposition, whereas T_3 leads T_4 by a few tens of degrees; the distance between these electrodes couples are respectively 1 m and 15 cm. From the observation of these phase shifts, one can estimate the vertical velocity u of the flow, assuming it consists of ascending plumes:

$$u = \frac{2\pi \Delta z}{T \Delta \phi} \quad (5.24)$$

where Δz is the vertical distance between the sensors, $\Delta \phi$ the phase shift and T the period. The vertical velocity u is tipically of the order of $2 cm \cdot s^{-1}$. Evidence thatwater is involved in the oscillatory convective process is provided by the recordings of relative humidity made at RH_1, RH_2, RH_3 and RH_4. Again, the phase shifts are consistent with a vertical velocity of the order of $10^{-2} m \cdot s^{-1}$.

Two more observations are worth emphasizing. The periods of the different trains of oscillations tend to be multiples of given periods ($t_1 \approx 30s$, $t_2 \approx 45s$) (see figures ...). A preliminary time-frequency analysis of the signal shows that the periods tend to cluster around 0.5, 0.75, 1, 1.5, 2.5 and 3 min. From one train to the following, there is generally (but not always) a jump of period, e.g. from t_1 to $2t_1$. On the Rayleigh number axis, the system is

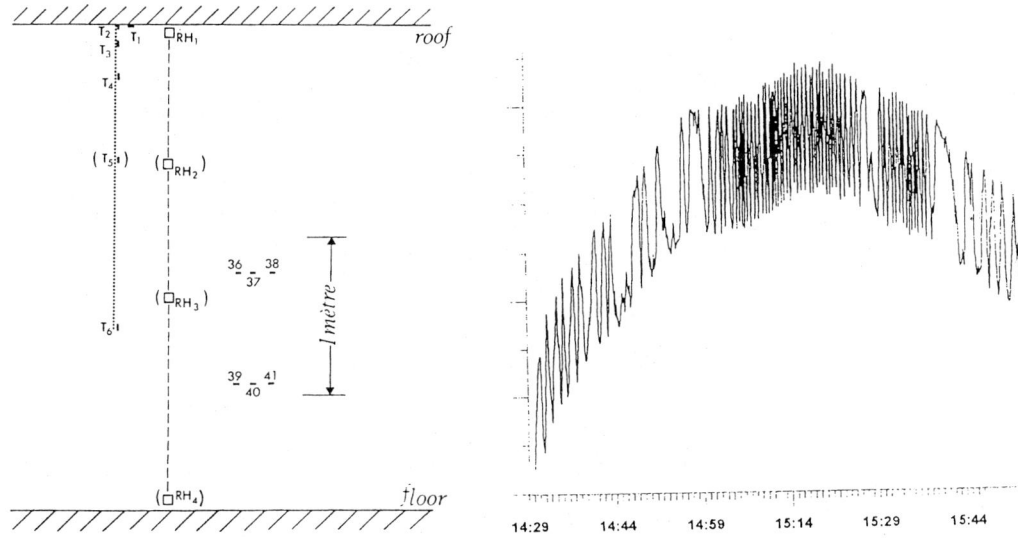

Figure 5.25: *Heat and water transport by oscillatory convection in an underground cavity.* On the left the measurement devices: 36-41: electrodes; $RH_1 - RH_4$: humidity sensors; $T_1 - T_6$: thermometers. On the right an example of oscillations recorded on May 19, 1994 by thermometers $T_1 - T_6$. The same oscillations are observed on humidity and self-potential recordings (after Morat et al., 1999).

presumably close to bifurcation points corresponding to doublings of period (Dubois and Gvishiani, 1998 [134]), and goes back and forth between these points. This suggestion is reinforced by the observation that the larger the period, the less organized the oscillations, as if the system were on the route to chaos by period doubling.

This example is one of the more exciting that can be observed on a natural dynamic system, because the previous identical observations were done on laboratory oscillator experiments.

284   Fractals and Dynamic Systems

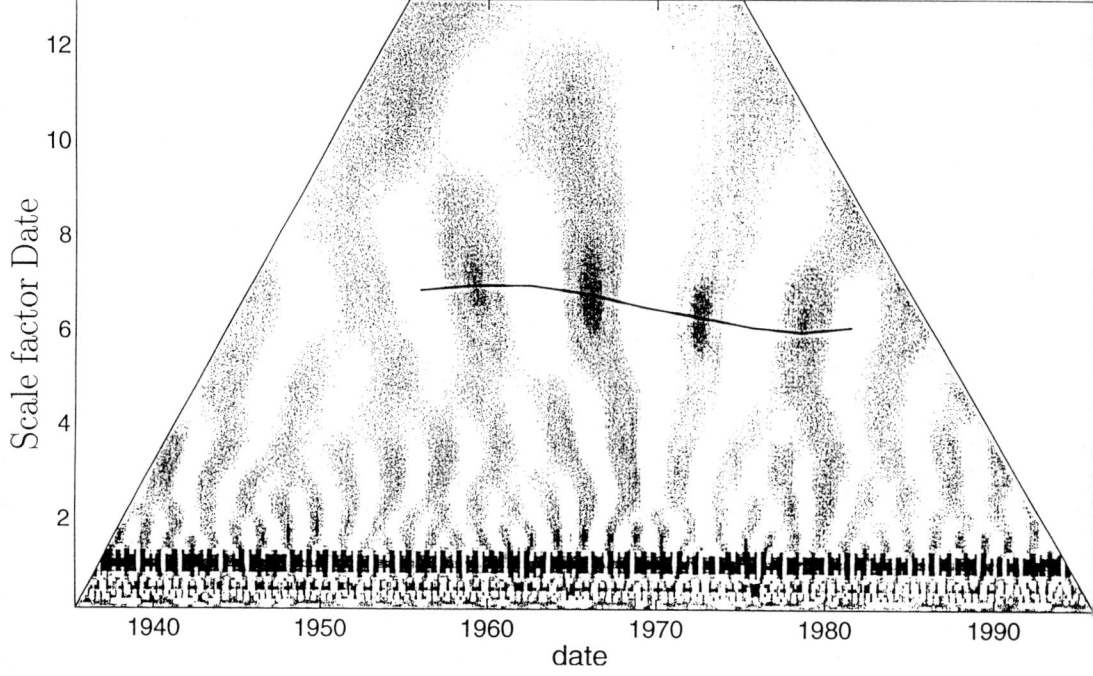

Wavelet transform - Scalogram of Oubangui river discharge, the wavelet is a Gaussian thirtieth derivative (after David Aubert, 2001).

Chapter 6

Conclusions and Perspectives

6.1 About Part I

This book, along with the first volume of this two book series (Dubois and Gvishiani, 1998, [134], Gvishiani and Dubois, 2001), illustrates the increasing importance of artificial intelligence (AI) in geophysical data acquisition management and studies. In the past few years, the number of geophysical observation systems and the corresponding volumes of registered data flow have increased drastically. Because of this, AI techniques have become a major instrument in geophysical research and database management. The results described in Part I of this book show that this trend will continue and probably accelerate in the future. The AI approach will integrate data acquisition, database construction, experimental and theoretical modeling, and other directions of geophysical studies.

One of the important missions of any science and especially of geophysics is to discover, study and carefully describe the important laws and features which are hidden in large and often seemingly random data sets. The recognition of such laws and features is the major goal of the AI techniques introduced in chapter 1.

The emphasis placed on fuzzy sets and fuzzy logic based clustering and dynamic pattern recognition reflects the nature of geophysical observations. In fact, practically all punctual geophysical data records such as earthquake epicenters, magnetic and gravity anomaly interpretations, geodynamic modelling schemes, etc are elements of fuzzy sets by definition.

Their intrinsic relevance to geophysical applications makes fuzzy logic and fuzzy sets-based mathematics important to constructing a new generation of models describing geophysical phenomena. The theoretical and algorithmic results described in chapter 1 contribute to this new approach. The algorithms introduced in sections 1.1, [66, 67, 219, 220, 221], 1.3 and 1.4 [57, 140] help to determine laws, features and data flows which are hidden under the huge random coverage of the initial data. Further advances in "fuzzy modeling" for geophysics will require further advances in the fuzzy mathematics itself. Once new and necessary fuzzy mathematical constructions are available, further advances in fuzzy logic geophysical interpretation can and should be done.

This book is mostly devoted to local and regional applications of the described recognition and classification algorithms and tools. Its path is guided by the availability of the data under consideration. However, preliminary attempts to apply the techniques to global databases, such as the SMDB (Strong ground Motion Data Base; see [548, 550] and 2.1.1) give encouraging results (see [549, 550], 2.1.2). It appears that strong motions records from California, Italy, Japan and China can be well distinguished from each other and from the rest of the SMDB content using the SPARS algorithm.

By definition, the AI technique in general, and particularly the one introduced in this book, is designed to explore large data sets. Therefore, the authors plan next, in collaboration with their French and Russian colleagues, to test global applications for the developed algorithms.

Classical and fuzzy clustering algorithms of the RODIN type, were successfully applied (as described in this book) to classify magnetic anomalies in the gulf of Saint Malo. The next step will be to test this technique over the whole Indian ocean. Linear and circular structure recognition algorithms have been successfully applied in the Wharton bassin near Indonesia (see 1.4). The latter technique will be extended onto bigger and bigger parts of Indian ocean. The results obtained by both methods will be compared and analysed together.

In the same vein, the algorithm described in chapter 1 will be redesigned as a joint system to be applied to global geophysical data sets. For example, we are developing a database to apply the RODIN algorithm (see 1.3) to the worldwide ocean. The linear and circular structure recognition algorithms (see 1.4), will be applied on a worldwide scale to analyze satellite images. A corresponding proposal is under preparation by a French-Russian working group. This next round of applications will aid in future developing the AI systems themselves. The general development path will be the construction of sequencial algorithms that are then merged into one integrated AI automated expert system.

An important question is the relation between the AI and modeling approaches. This book gives new insight into how to more efficiently combine these two basic approaches. The AI approach efficiently tackles large amounts of initial data and reveals hidden, but sometimes rather rough laws and features. On the other hand, the modeling approach can provide a deep and sophisticated analysis of the phenomena revelated by the AI tools and algorithms. A modeling block will therefore be an indispensable part of the above AI based automated expert system.

In part I of this book we describe numerous original applications of the AI techniques introduced in chapter 1. The syntactic pattern recognition scheme (1.1, [221, 549]), the clustering algorithms of RODIN series (1.3), the fuzzy logic based clasification algorithms (1.2), the GIS based derivatives techniques (1.4), and the dynamic pattern recognition algorithms [211, 134], are original innovations developed by A Gvishiani and his collaborators J. Dubois, J. Bonnin, J. Le Mouel, M. Diament, A. Galdeano (France) and M. Zhizhin, S. Agayan, A. Beriozko, Sh. Bogoutdinov (Russia).

The applications to seismology, engineering seismology, natural hazard assessment, marine geophysics, magnetic anomalies studies and geodynamics described in chapters 2 and 3 of part I were also developed by the authors of the book in collaboration with the above-listed co-authors. These applications were originally published in [100, 101, 547, 552].

This book is the second of a two volume series. The AI part of the first volume [134] gives a detailed description of theory and algorithms of dynamic pattern recognition and stability theory for limit classification developed by Gvishiani in late 80s and early 90s [203]. Chapter 3 of the second volume presents an important application of this theory to seismic hazard assessment in three regions of moderate seismicity: the Western Alps, the Pyrénées and the Great Caucasus. This applications were developed by A. Gvishiani, A. Soloviev, V. Kossobokov, E. Rantzman, A. Gorskkov, A. Troussov (Russia) and A. Cisternas, J. Sallantin, C. Weber (France) [100, 101].

This book is an independent monograph, with its own interrelated theoretical and application sections. At the same time, as the second volume of the series, it presents numerous applications of the dynamic pattern recognition techniques and stability theory described in [134] to concrete geophysical data problems. Furthermore, it opens the gates for applications beyond geophysics and Earth sciences. For example, applications to applied psychology are being explored.

In July 2001, speaking at the international conference on gravimetry in Sèvres (France), one of the most famous Russian geophysicists of the 20th century, Vladimir Strakhov, concluded that "... in general, geophysics is too complicated for geophysicists". The authors hope that the AI approach presented in this book will help simply the task of geophysical analysis. With AI techniques in hand geophysics may become less complicated for geophysicists.

6.2 About Part II

The study of dynamic system in part II of this book reveals that some natural processes result from the interaction of dynamic systems with relatively high number of degrees of freedom, usually more than 2, which give them a disorderly appearance. These systems are ideally suited for a non-linear approach.

In this book we examined some of the major works dealing with geomorphology, hydrology and matters related to rock fragmentation and fracturation, rock mechanics, tectonics, seismology. We also investigated applications to volcanology, geomagnetism, electric current circulations.

We observe that all large dynamic systems of the Universe, of the Life and of Human Societies evolve toward complexity [297, 94, 383]. For an example we mention the chaotic obliquity of the planets. Laskar and Robutel showed (1993) [297, 298] that all of the terrestrial planets may have experienced large, chaotic variations in obliquity at some time in the past, based on a numerical comparison of the global stability of the spin-axis orientation (obliquity) of planets against secular orbital perturbations. The obliquity of Mars is still in a large chaotic region, ranging from 0° to 60°. Mercury and Venus have stabilized through tidal dissipation, and the Earth may have been stabitized by capture of the Moon. None of the obliquities of the terrestrial planets can therefore be considered as primordial.

Another important idea is the unifying character of which we have previously mentioned (Dubois, 1998 [131]). This broadening of horizons is a direct consequence of the study of the dynamic systems which interact on our planet. For example the study of fluid circulation links previously separated disciplines, such as hydrology, hydrogeology, the physic and mechanics of fissured environments, geomagnetism, electromagnetic induction and electric currents, electrofiltration, etc. This shows the unifying character that results from the introduction of new paradigms, in the meaning of Thomas Kuhn. This richness can also be found inside individual disciplines. Seismology is one example among many. Smalley *et al.* (1987) [471] conclude their one-dimensional temporal seismicity seismicity with a fractal analysis and conclude:*"the initial goal of our study was to present a fractal analysis of seismic clustering. But earthquakes constitute events with five dimensions: time, three spatial (Euclidean) dimensions, and magnitude. Clustering could have been as well studied in each of these dimensions. One of the aims*

Conclusions and Perspectives

of this type of investigation would be to determine if the change in fractal dimension is associated to the clustering of events, and would then be a precursor signal."

6.2.1 Intermittency and turbulence

This important topic was only briefly developed in this book, so, the case of geomagnetism and geodynamo modelling, but it is becoming an increasingly important tool in dynamic system studies in many Earth science fields.

The statistical description of intermittency in fully developped turbulence is implicitly linked to a particular geometry for the energy dissipation support (Queiros-Condé, 1999, 2000 [422, 423]). Kolmogorov's theory (Kolmogorov, 1941 [283]) assumes a homogeneous field (space filling) for dissipation, Frisch et al. (1978) [159] proposed the β-model using the fractal description then the multifractal approach (Parisi and Frisch, 1985; Frisch (1995) [158]). Queiros-Condé (1999) [422] presented a new geometrical description for the energy dissipation field. Statistics of velocity or energy fluctuations are usually described using the structure function $\langle \delta V_r^p \rangle$ or $\langle \epsilon_r^p \rangle$, where δV_r is the velocity increment between two points separated by a distance r, and ϵ_r the rate of energy dissipation averaged over a ball of size r. If universal scaling exists over an inertial range then it is expected that

$$\langle \delta V_r^p \rangle \sim r^{\zeta_p} \text{ and } \langle \langle \epsilon_r^p \rangle \rangle \sim r^{\tau_p}.$$

Kolmogorov's theory predicts $\zeta = p/3$ but there is a significant departure from this linear behaviour in both experimental (Anselme et al. (1984) [11]) and numerical studies (Vincent and Meneguzzi (1991) [527]).

In a paper by Queiros-Condé, the intermittency of energy dissipation is interpreted through a set of fractal structures Ω_p of dimension Δ_p linked by an inclusion relation: $\Omega_{p+1} \subset \Omega_p (i.e. \Delta_{p+1} < \Delta_p)$.

The author then introduces the entropy jump $\Delta S_p(r) = (\Delta_{p+1} - \Delta_p) ln(r/r_0)$, which characterizes the level of order of the structure Ω_{p+1} developed over the structure Ω_p. He then introduces a deterministic relation between successive entropy jumps using the linear relation $\Delta S_p(r) = \gamma \Delta S_{p-1}(r)$ with $0 < \gamma < 1$. A recursive relation is deduced:

$$\gamma = (\Delta_{p+1} - \Delta_p)/(\Delta_p - \Delta_{p-1}) \text{ or } \gamma = (\Delta_{p+1} - \Delta_\infty)/(\Delta_p - \Delta_\infty).$$

The parameter γ has been determined from other experimental results as: $\gamma = ((1 + 3/\sqrt{8})^{1/3} + (1 - 3/\sqrt{8})^{1/3} \approx 0.68$.

In conclusion, a geometrical description of the energy dissipation in terms of what could be called *fractal skins* of turbulence emerges from this study. Fully developped turbulence would be described by a hierarchy of fractal stuctures linked to each other by a recursive relationship. We find again here the main processes used in modelling fracturation using SOFT model or a multiple scale dynamo model.

Another property was recently emphasized (Queiros-Condé, 2000, [425]) in the context of multifractal theory and She-Lévêque model (1994) [465] describing the intermittency of fully developped turbulence. It was shown that multifractal dimensions can be simply written as:

$$F(\alpha) = 1 + \alpha^* - \alpha^* ln(\alpha^*/2)$$

where $\alpha^* = (2\beta - 1 - \alpha)/ln\beta = 2\beta^p$ (with $p \geq 0$) and $\beta = ((1 + 3/\sqrt{8})^{1/3} + (1 - 3/\sqrt{8})^{1/3} \approx 0.68$. Introducing the fractal dimensions $\Delta p = F(\alpha) + \alpha^* ln(\alpha^*/2)$, leads to the recursive relation $\beta = (\Delta_{p+1} - \Delta_\infty)/(\Delta_p - \Delta_\infty)$ with $\Delta_\infty = 1$. This suggests the existence of an internal symmetry in the multifractal spectrum of fully developped turbulence, which reduces considerably the number of parameters necessary to characterize intermittency statistics.

6.2.2 The problem of short time series, slow and fast dynamics in coupled systems

Short series very often cause practical problems in the processing of nonlinear dynamics datasets, mainly when dealing with dynamic systems. The problem was examined in Dubois (1998) [131]. When computing the similarity dimension D_0, using techniques such as Cantor dust or box-counting, the number of values in the series can go down to around 20. This is the limit that Wickman (1966, 1976) [537, 538] empirically selected for his statistical analysis of volcanic eruption series. Apart from the strict definition of D_0 and its error bar, there is the problem of validity of the linearity test, which confirms or disproves the fractal character of the dataset. The range of the interval between points should also be considered, and should span at least one order of magnitude (Davy at al., 1990, [116]).

Difficulties increase when computing higher order dimensions such as the information dimension D_1 and the correlation dimension D_2. In dynamic systems, the correlation dimension is generally computed by applying a correlation function to points in the phase trajectory (Grasberger and Procaccia, 1983, [195]). The smallest number of points necessary was estimated by

Conclusions and Perspectives

Pisarenko and Pisarenko (1991, 1995) [417, 418] and by Aubert (2000) [19]. They remak that high values of D_2 can be unreliable when the attractor's dimension is estimated for time series of intermediate lengths.
If, for example, $D_2 \geq 5$, the values become suspicious when N is of the order of a few hundred. Limiting rules proposed by Ruelle (1987) [443] and Smith and Jordan (1988) [473], can be written as:

$$N >> 10^{D_2/2} \text{ and } N >> 42^{D_2}.$$

Pisarenko and Pisarenko (1991, 1995) present a statistical method for computing D_2, which frees itself from the increase of N when D_2 increases, and is based on the maximum-likelihood criteria. In the case where there are N points (m-dimensional vectors) $x_1, \cdots, x_N$, if r_0 is the maximum distance between two points and $r_{ij} = | x_i - x_j |$ is the distance between any pair of points, the correlation dimension is estimated as:

$$\overline{D_2} = \frac{1}{1/L \sum_{i>j}^{N} w_{ij} \log r_0/r_{ij}} \tag{6.1}$$

with $w_{ij} = H(r_0 - r_{ij})$ and $L = \sum_{i>j}^{N} w_{ij}$.
This method allows the error bar associated with D_2 to be calculated for series of 200 to 300 points.

More recently Boffetta et al. (1998) [64] developped a time series analysis approach. As we mentioned in section 4.5 of this book, several attempts have been devoted to distinguishing between deterministic and stochastic behavior, where "deterministic" is interpreted as "dominated by a small number of excited modes" and "stochastic" as "dominated by a large number of excited degrees of freedom".
We follow Boffetta et al. (1998) when they mention that two main problems affect the time series analysis:

1. The length of the time series is a crucial point to obtaining reliable estimates of the phase-space properties of the system. Furthermore, there are simple stochastic processes that create a "false positive" answer in the search for low-dimensional chaotic dynamics, providing a finite value of the dimension under time-embedding in most practical cases. Analogously, simple systems characterized by on/off intermittency require additional care in the phase-space reconstruction and analysis.

2. Another problem is encountered in systems with many different timescales. In this case, it has been shown (Aurell *et al.*, 1996, [20]) that the Lyapunov exponents may have a marginal role. The growth of a non-infinitesimal perturbation is indeed ruled by a nonlinear mechanism which depends on the detail of the system. For this reason, it is possible to have a long predictability time for some specific degrees of freedom, even if the largest Lyapunov exponent is positive.

To overcome the last problem the concept of maximum Lyapunov exponent was recently generalized by Aurell *et al.*, 1996, [20] to the case of non-infinitesimal perturbations, introducing the notion of a finite size Lyapunov exponent (FSLE). The method was tested on coupled Lorenz models having time scales that differ by a given factor and showed [64, 20] that the computation of the FSLE allows one to extract information on the characteristic time and on the predictability of the large-scale, slow-time dynamics even with moderate statistics and unresolved small scales.

6.2.3 Self-Organised criticality, SOC

We developed the theoretical aspect of Self-organized criticality (SOC) in Dubois and Gvishiani (1998) [134] (see section 11.2). In the present book we developed some SOC applications in the domain of fracturation and fragmentation modelling (SOFT models, see subsection 4.5.2), of seismology (5.2.2) or volcanology (5.3.3).

All the examples show that there is a close link between scale invariants in the time and space domains. In seismology the SOC concept implies that earthquakes result from the reorganisation of the lithosphere at spatial and temporal levels. This concept casts doubt on the validity of seismic models which only consider a single fault with "bumps" coupled only to their closest neighbours. The SOC theory implies that stress and deformation fields are coupled on large scales, and therefore that movements along a fault should be compatible with the other deformations around the fault.

Most authors recognise that the conditions of SOC are not yet clearly understood, and that they should be the object of further studies using statistical physics models as well as observations of the geophysical processes at play.

6.2.4 Mastering and controlling Chaos

The concept of mastering and controlling chaos was examined in Dubois and Gvishiani (1998) [134] (see section 10.4). There we followed the method of Shinbrot *et al.* (1993) [466] of stabilizing chaotic trajectories.

Conclusions and Perspectives

Methods similar to the OGY technique (Ott *et al.*, 1990, [393]) were shown to allow the maintenance of the phase trajectory on a fixed orbit by acting on one of the system's control parameters. This possibility of control relies on a good knowledge of the system's attractor and on the control parameters. Several researchers demonstrated that this was possible using physics experiments where a system was brought to a stage highly sensitive to the initial conditions. Mastering chaos is relatively easy for coupled, synchronized, chaotic circuits (Pecora *et al.*, 1990, [406]), for a vibrating strip (Ditto *et al.*, 1990, [126]), or for experiments in which the control parameters can be easily modified.

However, the control of chaos is not yet feasible for natural processes, for which the generating mechanism is very often unknown. It is, however, possible to adress the questions mentioned above by building a Poincaré section in the phase space of the system, looking for stable orbits which sometimes mix with the orbits of the attractor, and passively identifying (without influencing the system) which observable parameters can play the role of the control parameter.

At least in the first stage, this field of research will enhance the predictability of the future behaviour of a system, while we wait for the day where it will be possible to modify the key parameter controlling the system,.to, for example push the system from generic instability to a stable regime.

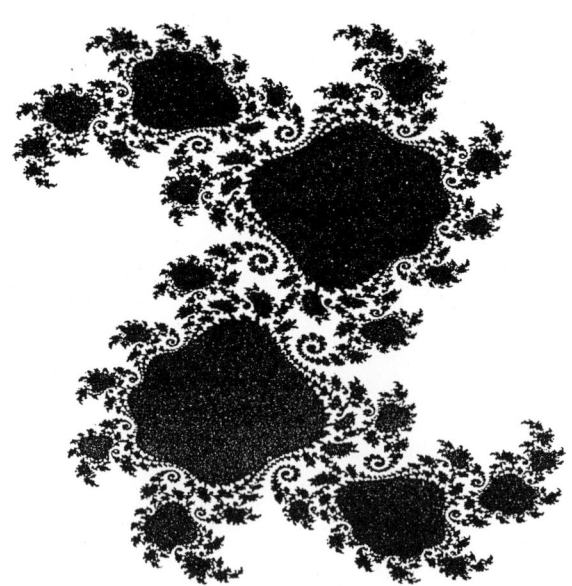

Mandelbrot set $z^2 - \lambda$ and Julia set - The upper figure shows a zoom on a piece of the boundary of the Mandelbrot set; below, the filled Julia set corresponding to the piece of the coastline of the Mandelbrot set (after Barnsley, 1988).

Part III

References

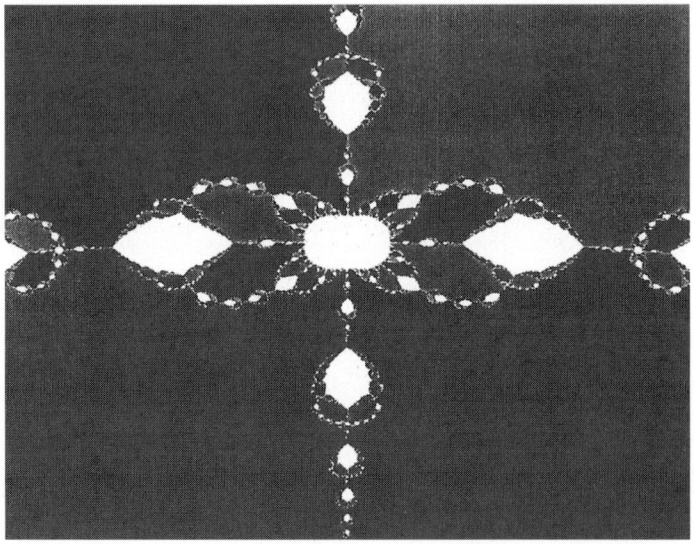

Julia sets associated with one-parameter family of dynamical systems (after Barnsley, 1988).

References

[1] Adler, P.M., Thovert, J.F., Morat, P. et Le Mouel, J.L., 1997. Electrical Signals Induced by the Atmospheric Pressure Variations in Unsaturated Media, *C. R. Acad. Sci. Paris*, **324**, IIa, 711–718.

[2] Aki, K., and Richards, P.G., 1980, *Quantitative Seismology*, (2 volumes), Freeman and Co.

[3] Aki, K., 1981. A Probabilistic Synthesis of Precursory Phenomena, in: *Earthquake Prediction, an International Review*, Simpson, D.W. and Richards, P.G. eds, Am. Geophys. Union, Maurice Ewing Ser., **4**, 566–574.

[4] Alekseevskaya, M., Gabrielov, A., Gelfand, I., Gvishiani, A. and Ranzman, E., 1977. Formal Morphostructural Zoning of Mountains Territories, *J. Geophysics*, **43**, 227–233.

[5] Allan, D.W., 1962. On the Behaviour of Systems of Coupled Dynamos, *Proc. Cambridge Phil. Soc.*, **58**, 671–693.

[6] Allègre, C.J., Le Mouel, J.L. and Provost, A., 1982. Scaling Rules in Rock Fracture and Possible Implications for Earthquake Prediction, *Nature*, **297**, 47–49.

[7] Allègre, C.J., and Le Mouel, J.L., 1994. Introduction of Scaling Techniques in Brittle Fracture of Rocks, *Phys. Earth Plan. Int.* **87**, 85–93

[8] Allègre, C.J., Le Mouel, J.L., Chau, H.D. and Narteau, C., 1995. Scaling Organization of Fracture Tectonics (SOFT) and Earthquake Mechanism, *Phys. Earth Plan. Int.* **92**, 215–233

[9] Allègre, C.J., Shebalin, P., Le Mouel, J.L. and Narteau, C., 1998. Energetic Balance in Scaling Organization of Fracture Tectonics, *Phys. Earth Plan. Int.* **106**, 139–153

[10] Ananin, N., 1966. Seismicity of Western Caucasus, Eastern Part of Black Sea and its Connection with Interior Structure of the Earth Crust, in *Straeni Chernomoiskoi Vpadini*, M. Nauka, 31–48.

[11] Anselme, F., Gagné, Y., Hopfinger, E.J. and Antonia, R.A., 1984. *J. Fluid Mech.*, **140**, 63–89.

[12] Argoul, F., Arnéodo, A., Grasseau, G., Cagne, Y., Hopfinger, E.J. and Frisch, U., 1989. Wavelet Analysis of Turbulence Reveals the Multifractal Nature of the Richardson cascade, *Nature*, **338**, 51–53.

[13] Arnéodo, A., Grasseau, G. and Holschneider, M., 1988. Wavelet Transform of Multifractals, *Phys. Rev. Letters*, **64**, 20, 2281–2284.

[14] Arnéodo, A. and Sornette, D., 1984. Monte Carlo Random Walk Experiments as a Test of Chaotic Orbits of Maps of the Intervals, *Physics Review Letters*, **52**, 1857–1869.

[15] Arnol'd, V.I., 1992, *Catastrophe Theory*, Springer-Verlag. 150 p.

[16] Arthaud, F., Ogier, M. et Seguert, M., 1980-1981. Géologie et Géophysique du Golfe du Lyon et de sa Bordure Nord, *Bull. BRGM*, **13**.

[17] Atten, P. et Caputo, J.G., 1987. Estimation Expérimentale de Dimension d'Attracteurs et d'Entropie, in, *Traitement Théorique des Attracteurs étranges, éd. Cosnard, M., Éditions du CNRS, Paris*, 177–191.

[18] Aubert, D., Beauvais, A. and Dubois, J., 1999. Application of the Correlation Integral Function Tool to Natural Time Series, *EGS Abstract for The Hague Meeting 1999*

[19] Aubert, D., 2000. *Traitement des Séries Temporelles de Débit de Fleuves en Climat Tropical Humide par les Outils de la Dynamique Non-linéaire*. Thèse Doctorat Géophysique Interne, Univesité Paris VII, 191 pp

[20] Aurell, E., Boffetta, G., Crisanti, A., Paladin, G. and Vulpiani, A., 1996. Growth of Noninfinitesimal Perturbation in Turbulence, *Phys. Rev. Lett.*, **77**, **7**, 1262–1265.

[21] Autran, A., Breton, J.P., Chantraine, J., Chiron, J.C., Gros, Y., Roger, P., 1980. Introduction à la Carte Tectonique de la France à 1/1 000 000. *Mémoire du BRGM*, **110**.

[22] Aviles, C.A., Scholz, C.H. and Boatwright, J., 1987. Fractal Analysis Applied to Characteristic Segments of the San Andreas Fault, *J. Geophys. Res.* **92**, 331–344.

[23] Avnir, D., Farin, D., and Pfeifer, P., 1983. Chemistry in Non-integer Dimension between Two and Three, II: Fractals Surface and Adsorbents, *J. Cem. Phys.*, **79**, 3566–3571.

[24] Backus, G., 1970. Inference from Inadequate and Inaccurate Data: I, Proceedings of the National Academy of Sciences, 65, 1, 1-105.

[25] Backus, G., 1970. Inference from Inadequate and Inaccurate data: II, Proceedings of the National Academy of Sciences, 65, 2, 281-287.

[26] Backus, G., 1970. Inference from Inadequate and Inaccurate Data: III, Proceedings of the National Academy of Sciences, 67, 1, 282-289.

[27] Backus, G., 1971. Inference from Inadequate and Inaccurate Data, Mathematical Problems in the Geophysical Sciences: Lecture in Applied Mathematics, 14, American Mathematical Society, Providence, Rhode Island.

References

[28] Backus, G., and Gilbert, F., 1967. Numerical Applications of a Formalism for Geophysical Inverse Problems, Geophys. J. R. astron. Soc., 13, 247-276.

[29] Backus, G., and Gilbert, F., 1968. The Resolving Power of Gross Earth Data, Geophys. J. R. astron. Soc., 16, 169-205.

[30] Backus, G., and Gilbert, F., 1970. Uniqueness in the Inversion of Inaccurate Gross Earth Data, Philos. Trans. R. Soc. London, 266, 123-192.

[31] Bak, P., Tang, C. and Weisenfeld, K., 1987. Self-Organized Criticality: an Explanation of the 1/f Noise, *Phys. Rev. Lett.*, **59**, 381–384.

[32] Bak, P., Tang, C. and Weisenfeld, K., 1988. Self-Organized Criticality, *Phys. Rev. A Gen. Phys.*, **38**, 364.

[33] Bak, P. and Chen, K., 1991. Self-Organized Criticality, Large Interactive Systems Naturally Evolve towards a Critical State ... *Scientfic American*, **264, 1** 26–33.

[34] Bak, P., Chen, K. and Creutz, M., 1989. Self-Organized Criticality in the "Game of Life", *Nature*, **342**, 780–781.

[35] Balakrishnan, A.V., 1976. Applied Functional Analysis, Springer-Verlag.

[36] Balé, P. and Brun, J.P., 1989. Late Precambrian Thrust and Wrench Zones in the Northern Brittany (France), *J. Struct. Geol.*, **11**, 391–405.

[37] Ballu, V., 1992. Analyse Fractale du Fonctionnement de la Dorsale Médio-Atlantique; *Rap. Stage DEA Geophys. Int. Univ. Paris 7, IPGP, 1991-1992*, 167–188.

[38] Ballu, V., Dubois, J. and Needham, H.D., 1993. Does the Sea-floor Bathymetry Reflect the Dynamical System of the Ridge Accretion? *Rap. IPG-IFREMER, Camp. Sigma* 16 pp.

[39] Balmform, N.J., Pasquero, C. and Provenzale, A., 2000. The lorenz-Fermi-Pasta-Ulam Experiment, *Physica D*, **138**, 1–43.

[40] Barfety, J.C., Gidon, M. et Kerchkove, C., 1968. Sur l'Importance des Failles Longitudinales dans le Secteur Durancien des Alpes Internes Françaises, *C. R. Acad. Sci. Paris*, **267**, D, 394–397.

[41] Barraclough, D.R. and De Santis, A., 1997. Some Possible Evidence for a Chaotic Geomagnetic Field from Observational Data. *Phys. Earth Plan. Int.*, **99**, 207–220.

[42] Barenblatt, G.I., Zhivago, A.V., Yu, P., Neprochnov, P. and Ostrovskyl, A., 1984. The Fractal Dimension : a Quantitative Characteristic of Ocean Bottom Relief. *Oceanology*, **24**, 695–697.

[43] Barnsley, M.F., 1988. *Fractals Everywhere*, Academic Press Inc.. 394 p.

[44] Beach, A., 1981. Thrust Tectonics and Crustal Shortening in the External French Alps based on a Seismic Cross-section, *Tectonophysics*, **79**, t. 1.

[45] Beauvais, A., Dubois, J. and Badri, A., 1994. Application d'une Analyse Fractale à l'étude Morphométrique du Tracé des Cours d'Eau : Méthode de Richardson, *C. R. Acad. Sci. Paris*, **318**, **II**, 219–225.

[46] Beauvais, A. and Montgomery, D., 1996. Influence of Valley Type on the Scaling Properties of River Platforms, *Water Ressources Res.*, **32**, 5, 1441–1448.

[47] Beauvais, A., Aubert, D., Dubois, J. and Orange, D., 1998. Are River Discharge Time Series deterministically Chaotic ? *submitted Water Ressources Res.*

[48] Bell, T.H., 1975. Statistical Features of Sea-floor Topography. *Deep-Sea Research*, **22**, 883–892.

[49] Bell, T.H., 1978. Mesoscale Sea Floor Roughness. *Deep Sea Research*, Part **A 26**, 65–76.

[50] Bellman, R.E. and Dreyfus, S.E., 1962. *Applied Dynamic Programming*, Princeton University Press, Princeton, NJ, 267 p.

[51] Bemis, K.G. and Smith, D.K., 1993, Production of Small Volcanoes in the Superswell Region of the South Pacific, *Earth Plan. Sci. Lett.*, **118**, 251–262.

[52] Bennett, J.G., 1936. Broken Coal. *J. Inst. Fuel*, **10**, 22–39.

[53] Benoît, E., (Ed.) 1991. *Dynamic Bifurcations*, Lectures Notes in Mathematics, Springer-Verlag, Berlin, **1493**, 219 p.

[54] Benvenuti, M. and Caputo, M., 1982. Pattern Recognition of the Relation between Seismicity and Gravity Anomalies in Italy, *Accad. Naz. Lincei, Rend. 1 Sc. Fis. Mat.* VIII, LXXII, **6**, 361–372.

[55] Bergé, P., Pomeau, Y. et Vidal, C., 1984, *L'ordre dans le Chaos. Vers une Approche Déterministe de la Turbulence.* Hermann, ed. des sciences et des arts, Paris. 353 p.

[56] Bergstrom, B.H., Crabtree, D.D. and Sollenberger, C.L., 1963, Feed Size Effects in Single Particle Crushing, *Trans. Amer. Inst. Mining Metall. Petrol. Engrs.*, **226**, 433–441.

[57] Beriozko, A.E. and Eliutin, A.V., 1995, Modular Geographical Information System in Seismic Zoning: Joseph Beffort, Luxembourg, *Cah. Europ. Géodyn. et Séismol.*, **9**, 161–174.

[58] Beriozko, A.E. and Eliutin, A.V., 1996, Interactive Earthquakes Catalogs Processing in the Multi-functional Geo-information "ATTILA": How much of a Catalog is Needed for Seismic Zoning: Joseph Beffort, Luxembourg, *Cah. Europ. Géodyn. et Séismol.*, **12**, 73–82.

[59] Berry, M.V. and Hannay, J.H., 1978. Topography of Random Surfaces. *Nature*, **273**, 573.

[60] Besse, J. and Courtillot, V., 1991. Revised and Synthetic Apparent Polar Wander Paths of the African, Eurasian, North American and Indian Plates and true Polar Wander since 200 Ma, *J. Geophys. Res.*, **96**, 4029–4050.

[61] Bitri, A., Brun, J.P., Chantraine, J., Guennoc, P., Marquis, G., Marthelot, J.M., Pivot, F. and Truffert, C., 1997. Structure Crustale du Bloc Cadomien de Bretagne Nord (France): Sismique Reflexion Verticale et Sondages Magnetotelluriques (Projet Geofrance 3D-Armor), *C.R. Acad. Sci. Paris*, **325**, 171–177.

[62] Blanter, E.M., Schnirman, M.G., Le Mouel, J.L. and Allègre, C.J., 1997, Scaling Laws in Blocks Dynamic Self-Organized Criticality. *Phys. Earth Plan. Int.*, **99**, 295–307.

[63] Blanter, E.M., Narteau, C., Schnirman, M.G. and Le Mouël, J.L., 1999, Up and Down Cascade in a Dynamo Model: Spontaneous Symmetry Breaking. *Physical Review E*, **59**, 5-A, 5112–5123.

[64] Boffetta, G., Crisanti, A., Paparella, F., Provenzale, A. and Vulpiani, A., 1998. Slow and Fast Dynamics in Coupled Systems: A Time Series Analysis View, *Physica D*, **116**, 301–312.

[65] Bond, F.C., 1952. The Third Theory of Comminution, *Trans. Amer. Inst. Mining Metall. Petrol. Engrs.*, **193**, 484–494.

[66] Bonnin, J., Gvishiani, A., Zhizhin, M., Mohammadioun, B. and Sallantin, J., 1991. Strong Motion Data Classification Using Syntactic Pattern Recognition Scheme, *Proc. Fourth Int. Conf. Seismic Zonation*, Stanford, vol II, 549–555.

[67] Bonnin, J., Gvishiani, A., Madariaga, R., Mohammadioun, B., Rouland, D. and Zhizhin, M., 1993. Syntactic Pattern Recognition Applied to Seismic Signal Analysis, *Report Contract CEA*, **BC-5918/LB**.

[68] Borissov, B., Reisner, G., Sholpo, V., 1975. Identification of Seismically Dangerous Zones in Alpine Folding Zone, *Moscow, Nauka*, 138 p.

[69] Bossolasco, M., Cicconi, G., Eva, C. and Pasquale, V., 1972. La rete Sismica dell Instituto di Genova e Primi Resultati sulla Sismotettonica delle Alpi Maritime ed Occidentali e del Mar Ligure, *Riv. Ital. Geofisica*, **XXI**, 229–247.

[70] Bossolasco, M., Eva, C. and Pasquale, V., 1974 On Seismotectonics of the Alps and Northern Apennines, *Riv. Ital. Geofisica*, **XXIII**, 57–63.

[71] Bousquet, J.C. and Philip, H., 1981. Les Caractéristiques de la Néotectonique en Méditerranée Occidentale, *in* : Sedimentary Basins of Mediterranean Margins, F.C. Wertzel, Ed. C.N.R. Italian project of oceanography, Tecnoprint, Bologna, 389–405.

[72] Braginsky, S.I., 1964. *Transl. Soviet Phys.*, **20**, 726–734.

[73] Braginsky, S.I., and Roberts, P.H., 1987. *Geophys. Astrophys. Fluid Dyn.*, **38** 327–349.

[74] Brandstatter, A., Swift, J., Swinney, H.L., Wolf, A., Farmer, J.D., Jen, E. and Crutchfield, P.J., 1983. Low-dimensional Chaos in a Hydrodynamic System, *Phys. Rev. Letters*, **51** (16), 1442–1445.

[75] Bretherton, F.P., 1969. Momentum Transport by Gravity Waves, *Q. J. R. Meteorol. Soc.*, **95**, 213–243.

[76] Brown, S.R., 1987. A Note on the Description of Surface Roughness using Fractal Dimension, *Geophys. Res. Lett.*, **14**, 1095–1098.

[77] Brun, J.P. and Balé, P., 1990. Cadomian Tectonics in Northern Brittany, *Geol. Soc., Spec. Publ.*, **51**, 95–114.

[78] Bureau Central Sismologique Français, (BCSF), 1983. Observations Sismologiques. Sismicité de la France entre 1971 et 1977. Strasbourg, France.

[79] Bureau Central Sismologique Français, (BCSF), 1983. Observations Sismologiques. Sismicité de la France entre 1978 et 1979. Strasbourg, France.

[80] BRGM (Bureau de Recherches Géologiques et Minières), 1980. Carte Géologique de la France et de la Marge Continentale à 1/1500000, Bureau de Recherches Géologiques et Minières, Orléans, France.

[81] BRGM (Bureau de Recherches Géologiques et Minières), 1981. Carte Sismotectonique de la France, *Mémoire*, **111**, BRGM, Orléans, France.

[82] Burlaga, L.F. and Klein, L.W., 1986, Fractal Structure of the Interplanetary Magnetic Field, *J. Geophys. Res.*, **91**, A1, 347–350.

[83] Burridge, R. and Knopoff, L., 1967. Model and Theoretical Seismicity. *Bull. Seism. Soc. Am.*, **57**, 341–371.

[84] Cande, S.C. and Kent, D.V., 1992. A New Geomagnetic Polarity Time Scale for the Late Cretaceous and Cenosoic, *J. Geophys. Res.*, **97**, 13 917–13 951.

[85] Cande, S.C. and Kent, D.V., 1995. Revised Calibration of the Geomagnetic Polarity Timescale for the Late Cretaceous and Cenozoic, *J. Geophys. Res.* **100**, B4, 6093–6095.

[86] Canny, J.F., 1983. Finding Edges and Lines Images to Edge Detection: *MIT, Cambridge, USA. Artificial Intelligece Laboratory, Master's Thesis*, TR-720.

[87] Canny, J.F., 1986. A Computational Approach to Edge Detection: *IEEE Trans. on Pattern Anal. and Machine Intelligence*, **8**, 679–698.

[88] Caputo, M., 1981. Critical Study of ENEL catalogue of Italian earthquakes from the year 1000 through 1975. *Rassegna Lavori Pubblici*, **2**, 3–16.

[89] Caputo, M., Keilis-Borok, V.I., Oficerova, E., Ranzman, E., Rotwain, I. and Soloviev, A., 1980. Pattern Recognition of Earthquake-prone Areas in Italy, *Phys. Earth Plan. Int.*, **21**, 305–320.

[90] Carlson, J.M., Langer, J.S., Shaw, B. and Tang, C., 1991. Intrinsic Properties of the Burridge-Knopoff Model of a Fault, *Phys. Rev. A*, **44**, 884–897.

[91] Carrozzo, M., DeVisentini, G., Giogetti, F. and Jaccarino, E., 1973. General Catalogue of Italian Earthquakes, CNENRT/PROT (73), **12**.

[92] Carta, S., Figari, R., Sartoris, G., Sassi, E. and Scandone, R., 1981. A Statistical Model for Vesuvius and its Volcanological Implications, *Bull. Volcanol.*, **44**, 2, 129–151.

[93] Cazabat, Ch., 1975. Topologie Hertzienne de la France, *Bull. Soc. Franç. Photogrammétrie*, **60**.

[94] Chaline, J., Notalle, L. et Grou, P., 1999. L'arbre de la Vie a-t-il une Structure Fractale ? *C. R. Acad. Sc. Paris*, **328**, série IIa, 1385–1390.

[95] Chantraine, J., Chauvel, J.J., Balé, P., Denis, E. et Rabu, D., 1988. Le Briovérien (Protérozoïque Supérieur à Terminal) et l'Orogénèse Cadomienne en Bretagne (France), *Bull. Soc. Géol. France*, **IV (5)**, 815–829.

[96] Chantraine, J., Egal, E., Thiéblemont, D., Guerrot, C., Le Goff, E., Ballère, M. and Guennoc, P., 2000. The Cadomian Active Margin, *Tectonophysics*, (in press).

[97] Chapman, S.C., Watkins, N.W., Dendy, R.O., Helander, P. and Rowlands, G., 1998. A Simple Avalanche Model as an Analogue for Magnetospheric Activity, *Geophys. Res. Lett.*, **25**, 13, 2397–2400.

[98] Charles, R.J., 1957. Energy-size Reduction Relationships in Comminution, *Trans. Amer. Inst. Mining Metall. Petrol. Engrs.*, **208**, 80–88.

[99] Choudhury, M., Giese, P. and Visintini, C., 1971. Crustal Structure of the Alps: Some General Features from Explosion Sismology, *Bull. Geofis. Teor. Appl.*, **13**, 211–240.

[100] Cisternas, A., Gvishiani, A., Godefroy, P., Soloviev, A., Gorshkov, A., Kossobokov, V., Lambert, M., Rantsman, E., Sallantin, J., Soldano, A. and Weber, C., 1985. A Dual Approach to Recognition of Earthquakes Prone-areas in the Western Alps, *Annales Geophysicae*, **3 N2**, 249–270.

[101] Cisternas, A., Gvishiani, A., et al., 1989. The Spitak (Armenia) Earthquake of 7 December 1988: Field Observations, Seismology and Tectonics, *Nature*, **339**, 645–679.

[102] Cook, A.E., 1972. Two-disc Dynamos with Viscous Friction and Time Delay, *Proc. Camb. Phil. Soc.*, **71**, 135–153.

[103] Cook, A.E. and Roberts, P.H., 1970. The Rikitake Two-disc Dynamo System, *Proc. Camb. Phil. Soc.*, **68**, 547–569.

[104] Cortini, M. and Barton, C.C., 1993. Nonlinear Forecasting Analysis of Inflation-deflation Patterns of an Active Caldera (Campi Flegrei, Italy), *Geology*, **21**, 239–242.

[105] Cortini, M. and Barton, C.C., 1994. Chaos in Geomagnetic Reversal Records: A Comparison between Earth's Magnetic Field Data and Model Disk Dynamo Data, *J. Geophys. Res.*, **99**, B9, 18,021–18033.

[106] Courtillot, V. and Le Mouel, J.L., 1988. Time Variations of the Earth's Magnetic Field: from Daily to Secular, *Ann. Rev. Earth Planet. Sci.*, **16**, 389–476.

[107] Cox, A., 1968. Lengths of Geomagnetic Polarity Intervals, *J. Geophys. Res.*, **73**, 3247–3260.

[108] Cox, E., 1994. *The Fuzzy Systems Handbook: A Practitioner's Guid to Building, Using and Maintaining Fuzzy Systems*, Boston, A.P. Professional, 624 p.

[109] Cox, C. and Sandstrom, H., 1962. Coupling of Internal and Surface Waves in Water of Variable Depth, *J. Oceanogr. Soc. Jpn.*, **20**, 499–513.

[110] Cuomo, V. Serio, C., Crisciani, F. and Ferraro, A., 1994. Discriminating Randomness from Chaos with Application to Weather Time Series, *Tellus*, **46A**, 299–313.

[111] Croquette, V., 1987. Deux Exemples Mécaniques ayant des Comportements Chaotiques, *Traitement numérique des attracteurs étranges, Éd. CNRS, Paris*, 107–127.

[112] Cundall, P.A., 1971. A Computer Model for Simulating Progressive, Large Scale Movements in Blocky Rock System, *Symp. Int. Soc. Rock Mech., Nancy, 1971*.

[113] Curry, J. and Yorke, J.A., 1977. A Transition from Hopf Bifurcation to Chaos: Computer Experiments with Maps in R2, in *The Structure of Attractors: Dynamical Systems, Lecture Notes in Math.*, **668**, *48, Springer-Verlag*.

[114] Daignières, M., Gallart, J. and Banda, E., 1982. Implications of the Seismic Structure for the Orogenic Evolution of the Pyrenean Range, *Earth Planet. Sci. Lett.*, **57**, 88–100.

[115] Daubechies, I., 1990. The Wavelet Transform, Time-frequency Localization and Signal Analysis, *IEEE Trans. Information Theory*, **36**, 961–1005.

[116] Davy, P., Sornette, A. and Sornette, D., 1990. Some Consequences of a Proposed Fractal Nature of Continental Faulting. *Nature*, **348**, 56–58.

[117] Debelmas, J. and Lemoine, M., 1970. The Western Alps: Paleogeography and Structure, *Earth Sci. Rev.*, **6**, 221.

[118] DeMets, C., Gordon, R.G. and Vogt, P., 1994. Location of the Africa-Australia-India Triple Junction and Motion between the Australian and Indian Plates: Results from an Aeromagnetic Investigation of the Central Indian and Carlsberg Ridges, *Geophysical Journal International*, **119**, 893–930.

[119] Deplus, C., Diament, M., Hébert, H., Bertrand, G., Domingues, S., Dubois, J., Malod, J., Patriat, P., Pontoise, B., and Sibilla, J.J., 1998. Direct Evidence of Active Deformation in the Eastern Indian Oceanic Plate, *Geology*, **26**, 131–134.

References

[120] Deriche, R., 1987. Using Canny's Criteria to Derive Optimal Edge Detector recursuvely Implemented, *Intern. Journ. of Computer Vision*, **1**, 167–187.

[121] Deriche, R. and Faugeras, O., 1990. Tracking Line Segments: *Springer-Verlag, Proceedings of the First European Conference on Computer Vision*, 259–268.

[122] Desmons, J., 1977. Polyphase Methamorphism of the Oceanic and Continental Crust of the Western Alps, Moscow, *J. Geotectonica*, **6**, (in Russian).

[123] De Rubeis, V., Dimitru, P., Papadimitriou, E. and Tosi, P., 1993. Recurrent Patterns in the Spatial Behaviour of Italian Seismicity Revealed by the Fractal Approach, *Geophys. Res. Lett.*, **20**, 1911–1914.

[124] Diodati, P., Marchesoni, F. and Piazza, S., 1991. Acoustic Emission from Volcanic Activity: An Example of Self-Organized Criticality, *Phys. Rev. Lett.*, **67**, 2239–2243.

[125] Diodati, P., Bak, P. and Marchesoni, F., 2000. Acoustic Emission at Stromboli Volcano: Scaling Laws and Seismic Precursors, *Earth Plan. Sci. Lett.*, in press

[126] Ditto, W.L., Rauseo, S.N., and Spano, M.L., 1990. Experimental Control of Chaos, *Phys. Rev. Lett.*, **65**, 26, 3211–3214.

[127] Djibladze, E., 1983. About Deep Sources of Caucasian Earthquakes, *Izvestia AN USSR, Phisica Zemli*, **3**, 22–73.

[128] Dubois, J.E. and Gershon, N., 1996. *Modeling Complex Data for Creating Information*, Data and Knowledge in a Changing World, Springer, Codata, 277 p.

[129] Dubois, J.O., 1995. *La Dynamique Non-Linéaire en Physique du Globe*, Masson, Paris, 292 p.

[130] Dubois, J.O., 1997. *La Gravimétrie en Mer*, Collection Synthèses, Institut Océanographique, Paris, 256 p.

[131] Dubois, J.O., 1998. *Non-Linear Dynamics in Geophysics*, Wiley-Praxis, Series in Geophysics, Chichester, 287 p.

[132] Dubois, J.O. et Cheminée, J.L., 1988. Application d'une Analyse Fractale à l'Étude des Cycles Éruptifs du Piton de la Fournaise (La Réunion) : Modèle d'une Poussière de Cantor, *C. R. Acad. Sci. Paris*, **307**, II, 1723–1729.

[133] Dubois, J.O. and Cheminée, J.L., 1991. Fractal Analysis of Eruptive Activity of Some Basaltic Volcanoes, *J. Volcan. Geotherm. Res.*, **45**, 197–208.

[134] Dubois, J.O., and Gvishiani, 1998. *Dynamic Systems and Dynamic Classification Problem in Geophysical Applications* Springer-Verlag, Berlin, Heidelberg. 275 p.

[135] Dubois, J.O. and Nouaili, L., 1989. Quantification of the Fracturing of the Slab using a Fractal Approach, *Earth Plan. Sci. Lett.*, **94**, 97–108.

[136] Dubois, J.O., Chaline, J. et Brunet-Lecomte, P., 1992. Spéciation, Extinction et Attracteurs Étranges, *C. R. Acad. Sci. Paris*, **315**, 1827–1833.

[137] Dubois, J.O. et Cheminée, J.L., 1993. Les Cycles Éruptifs du Piton de la Fournaise: Analyse Fractale, Attracteurs, Aspects Déterministes, *Bull. Soc. Géol. France*, **164**, 1, 3–16.

[138] Dubois, J.O. et Pambrun, C., 1990. Étude de la Distribution des Inversions du Champ Magntique Terrestre entre -165 Ma et l'Actuel (Échelle de Cox). Recherche d'un Attracteur dans le Systme Dynamique qui les Génère. *C. R. Acad. Sci. Paris*, **311, II**, 643–650.

[139] Dubuc, B., Quiniou, J.F., Roques-Carmes, C., Tricot, C. and Zucker, S.W., 1989. Evaluating the Fractal Dimension of Profiles, *Phys. Rev. A*, **39**, 1500–1512.

[140] Eliutin, A., Morat, P., Pride, S., Le Mouel, J.L. and Gvishiani, A., 1996. Geo-Information Environment for Non-Geographical Data: Processing Spatially Distributed Electrical Signals on Strssed and Deteriorated Rocks Samples, *Cahiers du Cent. Europ. Géodyn. et Séismol.*, **12**,

[141] Ershov, S.V., Malinetskii, G.G. and Ruzmaikin, A.A., 1989. A Generalized Two-Disk Dynamo Model. *Geophys. Astrophys. Fluid Dynamics*, **47**, 251–277.

[142] Essex, C., Lookman, T. and Nerenberg, M.A., 1987. The Climate Attractor over Short Timescales, *Nature*, **326**, 64–66.

[143] Evertsz, C.J.G., and Mandelbrot, B.B., 1992. Multifractal Measures, *in "Chaos and Fractals: New frontiers in Science, Appendix B", H.O Peitgen, H. Jürgen and D. Saupe, Springer*, 921–953.

[144] Fairhead, J.D., Bennett, K.J., Gordon, R.N. and Huang, D., 1994. Euler: Beyond the "Black Box"; *64th Ann? Internat. SEG Meeting, Expanded Abstracts*, 422–424.

[145] Falconer, K., 1990. Fractal Geometry: Mathematical Foundations and Applications, *Ed Jhon Wiley and Sons, Chichester*, 288pp.

[146] Farmer, J.D., Ott, E. and Yorke, J.A., 1983. The Dimension of Chaotic Attractors, *Physica D*, **7**, 153–180.

[147] Fedotov, S., 1965. Distribution of Strong Earthquakes of Kamchatka, Kuril Islands and North-eastern Japan. Questions of Engineering Seismology, *Moscow, Nauka* **10**,

[148] Feigenbaum, M.J., 1978. Quantitative Universality for a Class of Non-linear Transformations, *J. Statist. Phys.* **19**, 1, 25-52.

[149] Feigenbaum, M.J., 1979. The Universal Metric Properties of Non-linear Transformations, *J. Statist. Phys.* **21**, 6, 669-706.

[150] Firebaugh, M.W., 1989. *Artificial Intelligence. A Knowledge-based Approach* Boston: PWS Kent, 740 p.

[151] Fourniguet, J., 1981. Carte Néotectonique de la France à 1/1000000 BRGM

[152] Fourniguet, J., 1977. Mise en Évidence de Mouvements Néotectoniques Actuels, Verticaux dans le Sud-est de la France, par Comparaison de Nivellements Successifs, Rapport BRGM, **77**, S.G.N. 081 GEO.

[153] Fourniguet, J., Vogt, J. and Weber, C., 1981. Seismicity and Recent Crustal Movements in France, *Tectonophysics*, **71**, 195–216.

[154] Fox, C.G. and Hayes, D.E., 1985. Quantitative Methods for Analysing the Roughness of the Sea-floor. *Rev. Geophys.*, **23**, 1–48.

[155] Fraser, A., and Swinney, H., 1986. Independent Coordinates for Strange Attractors from Mutual Information *Phys. Rev. A*, **33**, 1134–1140.

[156] Fréchet, J., 1978. Sismicité du Sud-est de la France et une Nouvelle Méthode du Zonage Sismique, Thèse 3e cycle, Université de Grenoble.

[157] Freeman, J.A. and Skapura, D.M., 1992. *Neural Networks: Algorithms, Applications and Programming Techniques*, New York, etc.: Addison Wesley, 412.

[158] Frisch, U., 1995; *Turbulence*, Cambridge University Press

[159] Frisch, U., Sulem, P.L. and Nelkin, M., 1978. *J. Fluid Mech.*, **87**, 719–736.

[160] Frisch, U. and Parisi, G., 1985. Turbulence and Predictability of Geophysical Fluids Dynamics and Climate Dynamics, *Proc. Int. School of Phys. "Enrico Fermi"*, Ed. M. Ghil, North Holland, Amsterdam, 84.

[161] Froidevaux, C. et Guillaume, A., 1979. Contribution à l'Analyse Structurale des Alpes Liguro-piémontaises par l'Étude du Champ Magnétique Terrestre, *Tectonophysics*, **54**, 139–157.

[162] Frohlich, C. and Willemann, R.J., 1987. Aftershocks of Deep Earthquakes Do not Occur preferentially on Nodal Planes of Focal Mechanisms, *Nature*, **329**, 41–42.

[163] Fujiwara, A., Kamimoto, G. and Tsukamoto, A., 1977. Destruction of Basaltic Bodies by High Velocity Impact. *Icarus*, **31**, 277–288.

[164] Funiciello, R., *et al.*, 1980, Carta Tettonica d'Italia, sc 1/1500000, C.N.R., Progetto Finalizato Geodinamica, publ. n **269**.

[165] Gabrielov, A., Keilis-Borok, V., Levshina, T. and Shaposhnikov, V., 1986. Block Model of the Lithosphere Dynamics, *Vichistelnoiya Seismologia, M., Nauka*, **19**, 168–178.

[166] Galdeano, A., Asfirane, F., Truffert, C., Egal, E. and Debeglia, N., 2000. The Aeromagnetic Map of the French Cadomian Belt, *Tectonophysics*, in press.

[167] Gallet, Y., Besse, J., Krystyn, L., Marcoux, J. and Théveniaud, H., 1992. Magnetostratigraphy of the Late Triassic Bolucektasi Tepe Section (South-Western Turkey): Implication for Changes in Magnetic Reversal Frequency, *Phys. Earth Planet. Inter.*, **93**, 273–282.

[168] Gaudin, A.M., 1926. An Investigation of Crushing Phenomena, *Trans; Amer. Inst. Mining Metall. Petrol. Engrs.*, **73**, 253–316.

[169] Geilikman, M.B., Golubeva, T.V. and Pissarenko, V.F., 1990. Multifractal Patterns of Seismicity, *Earth Plan. Sci. Lett.*, **99**, 253–316.

[170] Gelfand, I., 1971. *Linear Algebra* Moscow, Nauka, 212 p.

[171] Gelfand, I., Guberman, Sh., Izvekova, M., Keilis-Borok, V. and Ranzman, E., 1973. Raspoznavanie Mest Vozmozhnogo Vozniknovenia Silnih Zemletryaseniy. I. Pamir i Tian-Shan. Vishislitielnie i Statisticheskie Metodi Interpretatsii Seismicheskih Dannih. NAUKA. 107–133 (Vichistlit. seismol. ; vip. 6)

[172] Gelfand, I., Guberman, Sh., Zhidkov, M., Kalezkaya, M., Keilis-Borok, V. and Ranzman, E., 1973. Transfer of High Seismicity Criteria from Central Asia to Anatolya and Adjacent Regions, *DAN USSR*, **210, N2**,

[173] Gelfand, I., Guberman, Sh., Zhidkov, M., Kalezkaya, M., Keilis-Borok, V., Ranzman, E. and Rotwain, I., 1974. Recognition of Places Where Strong Earthquake May Occur II. Four Regions of Minor Asia and South-Eastern Europe, *Computational Seismology, Moscow, Nauka*, **7**,

[174] Gelfand, I., Guberman, Sh., Zhidkov, M., Keilis-Borok, V. and Ranzman, E. and Rotwain, I., 1974. Recognition of Places where Strong Earthquake may Occur III. Case, Where the Boundaries of Disjunctive Knots are Unknown, *Computational Seismology, Moscow, Nauka*, **7**, 41–65

[175] Gelfand, I.M., Guberman, Sh.A., Keilis-Borok, V.I., Knopoff, L., Press, F., Rantsman, E.Ya., Rotwain, I.M. and Sadovsky, A.M., 1976. Pattern Recognition Applied to Earthquake Epicenters in California, *J. Phys. Earth and Planet. Int.*, **11**, 227–283.

[176] Geman, S., and Geman, D., Stochastic Relaxation, Gibbs Distributions, and the Bayesian Restoration of Images, *Inst. Elect. Electron. Eng. Trans. on pattern analysis and machine intelligence*, **PAMI-6**, 721-741, 1984.

[177] Giardini, D. and Woodhouse, J.H., 1984. Hirizonta Shear Flow in the Mantle beneath the Tonga Arc, *Nature*, **307**, 505–509.

[178] Giarratano, J. and Riley, G., 1994. *Expert Systems: Principles and Programming, 2nd ed.*, Boston: PWS Publ. 644 p.

[179] Gibert, D. and Courtillot, V., 1987. Seasat Altimetry and the South-Atlantic Geoid, Spectral Analysis *J. Geophys. Res.*, **92**, 6 235–6 248.

[180] Gidon, M., 1977. Carte Géologique Simplifiée des Alpes Occidentales à 1/250000, Ed. Didier-Richard, BRGM.

[181] Giese, P. and Prodehl, C., 1976. Main features of Crustal Structure in the Alps, *in* : Explosion Seismology in Central Europe. P. Giese, C. Prodehl and Stein, Eds., Springer-Verlag.

[182] Gilabert, A., Benayad, M., Sornette, A., Sornette, D. and Vanneste, C., 1990. Conductivity and Rupture in Crack-Deteriorated Systems. *J. Phys. France*, **51**, 247–257.

[183] Gilbert, L.E., 1989. Are Topographic Data Sets Fractal? *Pure Appl. Geophys.*, **131**, 1, 2, 241–254.

[184] Gilbert, L.E. and Malinverno, A., 1988. A Characterization of the Spectral Density of Ocean Floor Topography, *Geophys. Res. Lett.*, **15**, 1401–1404.

[185] Godefroy, P., 1980. Apport des Mécanismes au Foyer à l'Étude Sismotectonique de la France, BRGM, 2ème série, section IV, **2**, 119–128.

[186] Godefroy, P., Dadou, C., Vagneron, J.M., 1983. Esquisse Sismotectonique du Sud-Est de la France et du Nord-Ouest de l'Italie (et schéma a 1/500 000): Évaluation de l'Aléa Sismique dans le Sud-Est de la France, *Documents du BRGM*, **59**.

[187] Goff, J.A. and Jordan, T.H., 1988. Stochastic Modeling of Seafloor Morphology: Inversion of Sea Beam Data for Second-order Statistics, *J. Geophys. Res.*, **93**, 13,589–13,608.

[188] Goff, J.A., 1991. A Global and Regional Stochastic Analysis of Near-ridge Abyssal Hill Morphology, *J. Geophys. Res.*, **96**, B13, 21 713–21 737.

[189] Goff, J.A. and Tucholke, B.E., 1997. Multiscale Spectral Analysis of Bathymetry on the Flank of the Mid-Atlantic Ridge; Modification of the Seafloor by Mass Wasting and Sedimentation, *J. Geophys. Res.*, **102**, B7, 15 447–15 462.

[190] Goldberg, D.E., 1989. *Genetic Algorithm in Search, Optimization and Machine Learning*, Readin Mass. etc.: Addison Wesley, 412 p.

[191] Goodchild, M.F., 1980. Fractals and the Accuracy of Geographical Measures, *Math. Geol.*, **12**, 85–98.

[192] Gordon, R.G., DeMets, C. and Argus, D.F., 1990. Kinematic Constraints on Distributed Lithospheric Deformation in the Equatorial Indian Ocean from Present Motion between the Australian and Indian Plates, *Tectonics*, **9**, 409–422.

[193] Gorskhov, A., Caputo, M., Keilis-Borok, V., Ofizerova, E., Ranzman, E. and Rotwain, I., 1979. Recognition of Places where Strong Earthquakes May Occur. IX. Italy, M ¿ 6,0, *Computational Seismology, Moscow, Nauka*, **12**, 3–18.

[194] Gourvitch, V., 1973. To the Theory of Multi-steps Games, *Computational Mathematics and Mathematical Physics*, **13, N6**,

[195] Grassberger, P. and Procaccia, I., 1983. Characterization of Strange Attractors, *Phys. Rev. Lett.*, **50**, 5, 346–349.

[196] Grasso, J.R. and Bachèlery, P., 1995. Hierarchical Organization as a Diagnostic Approach to Volcano Mechanism: Validation on Piton de la Fournaise, *Geophys. Res. Lett.*, **22**, 21, 2897–2900.

[197] Grasso, J.R. and Sornette, D., 1998. Testing Self-Organized Criticality by Induced Seismicity, *J. Geophys. Res.*, **103**, B12, 29 965–29 987.

[198] Gresta, S. and Patane, G., 1987. Review of Seismological Studies at MountEtna, *Pure Appl. Geophys.*, **125**, 951–970.

[199] Guckenheimer, J., 1986. *Ann. Rev. Fluid Mech.*, **18**, 15–31.

[200] Guckenheimer, J. and Holmes, P., 1983. *Nonlinear Oscillations, Dynamical Systems, and Bifurcations of Vector Fields* Springer ed., New York Berlin Heidelberg and Tokyo.

[201] Guckenheimer, J. and Buzyna, G., 1983. Dimension Measurements for Geostrophic Turbulence, *Phys. Rev. Letters*, **51** (16), 1438–1441.

[202] Gutenberg, B. and Richter, C.F., 1954. *Seismicity of the Earth and Associated Phenomena*, 2nd ed, Princeton University Press.

[203] Gvishiani, A.D., 1982. Time-stability of Strong Earthquake Prone Areas. I. South-Eastern Europe and Minor Asia, *Izvzstia Akademii Nauka USSR; Physics of the Earth*, **8**

[204] Gvishiani, A. and Kossobokov, V., 1979. K obosnovianiu Rezultatov Prognoza Mest Silnih Zemletryasenii Poluchennih Metodani Raspoznavania, *Izv. AN SSR Fisika Zemli*.

[205] Gvishiani, A.D., Zhidkov, M.P., Soloviev, A.A., 1982. Raspoznavanie mest vozmozhnogo vozniknovenia silnih zemletryaseniy. X. Mesta zemletryaseniv magnitudi M 7.75 na Tihookeanskom poberehie Uzhnoi Ameriki. Matematcheskie modeli troenia Zemli i prognoza zemletryasenii. c. 56–67 (Vichislit. seismol. 14).

[206] Gvishiani, A.D., Gorshkov, A. and Kossobokov, V., 1987. Recognition of Seismicaly Active Zones in Pyrénés, *DAN USSR*, **292, N1**, 56–59.

[207] Gvishiani, A.D., Gorshkov, A., Zhidkov, V. Ranzman, E. and Troussov, A., 1987. Recognition of Places where Strong Earthquakes May Occur, XV, Morphostructural Knots of the Great Caucasus $M \geq 5.5$, *Computotianal Seismology Moscow, Nauka*, 136–148.

[208] Gvishiani, A.D., Gorshkov, A., Ranzman, E., Cisternas, A. and Soloviev, A., 1988. *Recognition of Earthquake Prone-areas in Regions of Moderate Seismicity*, Moscow, Nauka, 176 p.

[209] Gvishiani, A.D. and Gurvitch, V.A., 1987. Calculation of Lithosphere Blocks Equilibrium by Convex Programming Methods, *XIX General Assembly IUGG, Vancouver, Canada, Abstracts*, **V1**, 39.

[210] Gvishiani, A.D. and Gurvitch, V.A., 1992. *Dynamical Classification Problems and Convex Programming in Applications*, Moscow, Nauka, 355 p.

References

[211] Gvishiani, A.D. and Gourvitch, V., 1982. Time-stability of Earthquake Prone Areas Recognition II. Eastern Part of Central Asia, *Izvestia AN USSR, Physics of the Earth*, **N9**, 30–38.

[212] Gvishiani, A.D. and Gourvitch, V., 1983. Dual Systems and Sets and its Applications, *Izvestia AN USSR, Technical Cybernetics*, **N4**, 31–39.

[213] Gvishiani, A.D. and Gourvitch, V.A. and Raszvetaev, A., 1985. Dynamical Problems of Pattern Recognition III. Study of Stability of the Recognition of Places of Strongest Earthquakes of Pacific Belt, *Computational Seismology, Moscow, Nauka*, **18**,

[214] Gvishiani, A.D. and Gourvitch, V. and Raszvetaev, A., 1986. Estimation of Seismic Fracturing by Pattern Recognition Technique, *Computational Seismology, Moscow, Nauka*, **19**, .

[215] Gvishiani, A.D., Zelevinsky, A., Keilis-Borok, V. and Kossobokov, V., 1978. Study of the Strongest Earthquake Prone-areas in the Pacific Belt, *Izvestia AN USSR, Physics of the Earth*, **N9**, 31–42.

[216] Gvishiani, A.D. and Kossobokov, V., 1981. On Evaluation of Reliability of Results of Pattern Recognition of Strong Earthquake-Prone Areas, *Izvestia AN USSR, Physics of the Earth*, **N2**, 21–36

[217] Gvishiani, A.D. and Soloviev, A., 1982. Concerning the Solution of the Problem of Recognition of Strong Earthquake Prone-areas on the Pacific Coast of South America, *Izvestia AN USSR, Physics of the Earth*, **N1**, 86–87.

[218] Gvishiani, A., Gorshkov, A., Kossobokov, V., Cisternas, A., Philip, H. and Weber, C., 1987. Identification of Seismicaly Dangerous Zones in the Pyrenees, *Annales Geophysicae*, **5 (6)**, 681–690.

[219] Gvishiani, A.D., Zhizhin, M.N. and Ivanenko, G.I., 1990. Syntactic Analysis of Strong Motion Records, *Vischislitel'naya Seismologiya*, **23**, 235–252.

[220] Gvishiani, A.D., Zhizhin, M.N. and Ivanenko, G.I., 1989. Pattern Recognition of Waveforms of Fore-, Main-, and Aftershocks: Evidence of Different from Strong Motion Data, *Annales Geophysicae, XIV General Assembly EGS, Barcelona 13-17 March 1989, Special Issue*, **13**.

[221] Gvishiani, A.D., Zhizhin, M.N., Mikoyan, A.N., Bonnin, J. and Mohammadioun, B., 1995. Syntactic Analysis of Waveforms from the World-wide Strong Motion Database, *European Seismic Design Practice, Rotterdam, Balkema, Brookfied*, 557–564.

[222] Gwinn, E.G. and Westervelt, R.M., 1987. Scaling Structure of Attractors at the Transition drom Quasiperiodicity to Chaos in Electronic Transport, *Geophys. Rev. Lett.*, **59**, 157–160.

[223] Hall, 1970. Combinatorics, *Moscow, Mir*,

[224] Halsey, T.C. and Jensen, M.H., 1986. Spectra of Scaling Indices for Fractal Measures/ Theory and Experiment, *Physica D*, **23**, 112-117.

[225] Halsey, T.C., Jensens, M.H., Kadanoff, L.P., Procaccia, I. and Shraiman, B.I., 1986. Fractal measures and their Singularties: The Characterization of Strange Sets, *Physical Review A*, **33, 2**, 1141–1151.

[226] Hand, D.J., 1986. Recent Advances in Error Rate Estimation, *Pattern Recognition Letters*, **4**, 335–346.

[227] Hanks, T.C. and Kanamori, H., 1979. A Moment Magnitude Scale, *J. Geophys. Res.*, **84**, 2348–2350.

[228] Haykin, S. and Griffin, J., 1994. *Neural Networks. A Comprehensive Foundation*, New York, etc.: Macmillan Publishing Company, 696 p.

[229] Harris, A.B., Lubensky, T.C., Holcomb, W.K. and Dasgupta, C., 1975. Renormalization Group Approach to Percolation Problems, *Phys. Rev. Lett.*, **35**, 327–330.

[230] Harris, C., Franssen, R. and Loosveld, R., 1991. Fractal Analysis of Fractures in Rocks : the Cantor's Dust Method. - Comments, *Tectonophysics*, **198**, 107–115.

[231] Hartmann, W.K., 1969. Terrestrial, Lunar and Interplanetary Rock Fragmentation, *Icarus*, **10**, 201–213.

[232] Hawkins, G.S., 1960. Asteroidal Fragments, *Astrophys. J.*, **65**, 318–322.

[233] Held, G.A., Solina, D.H., Keane, D.T., Haag, W.J., Horn, P.M. and Grinstein, G., 1990. Experimental Study of Critical-Mass Fluctuations in an Evolving Sandpile, *Phys. Rev. Lett.* **65, 9**, 1120–1123.

[234] Hentschel, H.G.E. and Procaccia, I., 1983. The Infinite Number of Generalized Dimensions of Fractals ansd Strange Attractors, *Physica 8D*, 435–444.

[235] Hertz, J., Krogh, A. and Palmer, R.G., 1994. *Introduction to the Theory of Neural Computation*, Reading Mass. etc.: Addison-Wesley, 327 p.

[236] Hénon, M., 1976. A Two-dimensional Mapping with a Strange Attractor, *Communications in Math. Phys.*, **565**, 29.

[237] Hide, R., 1995. Structural Instability of the Rikitake Disk Dynamo, *Geophys. Res. Lett.*, **22**, 1057–1059.

[238] Hide, R., 1997. The Nonlinear Differential Equations Governing a Hierarchy of Self-exciting Coupled Faraday-disk Homopolar Dynamos, *Phys. Earth Plan. Int.*, **103**, 281–291.

[239] Hide, R., 1998. Nonlinear Quenching of Current Fluctuations in a Self-exciting Homopolar Dynamo, *Nonlinear Proc. Geophys.*, **4**, 201–205.

[240] Hide, R., Skeldon, A.C. and Acheson, D.J., 1996. A Study of Two Novel Self-exciting Single-disk Homopolar Dynamos: Theory, *Proc. R. Soc. Lond. A.*, **452**, 1369–1395.

References

[241] Higuchi, T., 1990, Relationship betwwen the Fractal Dimension and the Power Law Index for a Time Series: a Numerical Investigation, *Physica D*, **46**, 254–264.

[242] Hirita, T. and Imoto, M., 1991. Multifractal Analysis of Spatial Distribution of Micro-earthquakes in the Kanto Region, *Geophys. J. Int.*, **107**, 155–162.

[243] Hirn, A., 1980. Le Cadre Structural Profond d'après les Profils Sismiques, in: Évolution Géologique de la France, colloque C 7, Géologie de la France, 26 CAI, *Mémoire du BRGM*, **107**.

[244] Holland, J.H., 1992. *Adaptation in Natural and Artificial Systems* Cambridge: The MIT Press, 211.

[245] Homeinuk, Yu., Shukin, Yu., Firsova, D. and Filippova, G., 1978. Method of Evaluation of Probability of the Strongest Earthquakes by Geological Geophysical Data Analysis, *Moscow, Nauka*,

[246] Hongre, L., Zhizhin, M. and Dubois, J., 1995. Chaotic Characteristics of the Magnetic Field : Fractal Dimensions and Lyapunov Exponents, *Cahiers du Centre Européen de Géodynamique et de Sismologie*, **9**, 79–94.

[247] Hongre, L., Sailhac, P., Alexandrescu, M. and Dubois, J., 1999. Non-linear and Multifractal Approaches of the Geomagnetic Field, *Phys. Earth Plan. Int.* **110** 157–190.

[248] Hood, P., 1965. Gradient Measurements in Aeromagnetic Surveying, *Geophysics*, **XXX**, 891–902.

[249] Horton, R.E., 1945. Erosional Development of Streams and their Drainage Basins: Hydrophysical Approach to Quantitative Morphology, *Geol. Soc. Am. Bull.*, **56**, 275–370.

[250] Huang, J., and Turcotte, D.L., 1989. Fractal Mapping of Digitized Images: Application to the Topography of Arizona and Comparisons with Synthetic Images. *J. Geophys. Res.*, **94**, 7491–7495.

[251] Huang, J., and Turcotte, D.L., 1990. Are Earthquakes an Example of Deterministic Chaos? *Geophys. Res. Lett.*, **17**, 223–226.

[252] Huang, J., Narkounskaia, G. and Turcotte, D.L., 1992. A Cellular Automatom, Slider-Block Model for Earthquakes 2. Demonstration of Self-Organized Criticality for a Two Dimensional System, *Geophys. J. Int.*, **111**, 259–269.

[253] Hulot, G., 1992. Observations Géomagnétiques et Géodynamo, *PhD Thesis, Univ. Paris VII*, 377 pp.

[254] Hulot, G., Khokhlov, A. and Le Mouel, J.L., 1997. Uniqueness of Mainly Dipolar Magnetic Fields Recovered from Directional Data, *Geophys. J. Int.*, **129**, 347–354.

[255] Hutchinson, A., 1995. *Algorithm Learning* Oxford: Clarendon Press, 436 p.

[256] Hwa, T. and Kardar, M., 1989. Dissipative Transport in an Open System: an Investigation of Self-Organized Criticality, *Phys. Rev. Lett.*, **62, 16**, 1813–1816.

[257] Holland, J.H., Adaptation in Natural and Artificial Systems, University of Michigan Press, 1975.

[258] Ijjasz-Vasquez, E.J., Rodriguez-Iturbe, I. and Bras, R.L., 1992. On the Multifractal Characterization of River Basins, *Geomorphology*, **5**, 297–310.

[259] Illies, J.H., 1975. Recent and Paleo-intraplate Tectonics in Stable Europe and the Rhinegraben Rift System, *Tectonophysics*, **29**, 251.

[260] ISC, 1997. Seismological Observatories. In: *Regional Catalogue of Earthquakes, 1964-1995*, International Seismological Centre, Newbury, United Kingdom.

[261] Ishimoto, M. and Iida, K., 1939. Observations sur les Séismes Enregistrés par le Microséismographe Construit Dernièrement, *I. Bull. Earthq. Res. Inst.*, **17**, 443-478.

[262] Ito, K., 1980. Chaos in the Rikitake Two-disc Dynamo System, *Earth Plan. Sc. Lett.*, **51**, 451–456.

[263] Ito, K. and Matsuzaki, M., 1990. Earthquakes as Self-organized Critical Phenomenon, *J. Geophys. Res.*, **95**, 6853–6860.

[264] Jault, D., 1995. Model Z by Computation and Taylor's Condition, *Geophys. Astrophys. Fluid Dyn.*, **79**, 99–124.

[265] Jones-Cecil, M., Wheeler, R.L. and Dewey, J.W., 1981. Pattern Recognition Program Modified and Applied to Southerneastern United States Seismicity, *U.S. Geol. Survey Open-file Report*, 81–195.

[266] Jensen, M.H., Kadanoff, L.P., Libchaber, A., Procaccia, I. and Stavans, J., 1985. Global Universality at the Onset of Chaos: Results of a Forced Rayleigh-Bénard Experiment, *Phys. Rev. Lett.*, **55**, 2798–2801.

[267] Joswig, M., 1990. Pattern Recognition for Earthquake Detection, *Bull. Seismol. Soc. Am.*, **90**, 170–186.

[268] Kadanoff, L.P., Nagel, S.R., Wu, L. and Zhou, S.M., 1989. Scaling and Universality in Avalanches, *Phys. Rev. A*, **39**, 6524–6532.

[269] Kanamori, H. and Anderson, D.L., 1975. *Bul. Seism. Soc. Am.*, **65**, 1073–1096.

[270] Kaplan, J. and Yorke, J., 1978. Functional Differential Equations and the Approximation of Fixed Points, *Proceedings, Bonn, July 1978, Lectures notes in Math.*, **730**, H.O. Peitgen and H.O. Walther, eds., Springer, Berlin, 1978, 228.

[271] Kapral, R. and Mandel, P., 1985. Bifurcation Structure of Nonautonomous Quadratic Map, *Phys. Rev. A*, **32**, 1076–1080.

[272] Karnik, V., 1969. Seismicity of the European Area, Part I. Reidel Publ. Co, Dordrecht, Holland.

[273] Karnik, V., 1971. Seismicity of the European Area, Part II. Reidel Publ. Co, Dordrecht, Holland.

[274] Kartalopoulos, S.V., 1996. *Understanding Neural Networks and Fuzzy Logic: Basic Concepts and Applications* New York etc.: IEEE Press, 205 p.

[275] Keating, P.B., 1998. Weighted Euler Deconvolution of Gravity Data, *Geophysics*, **63**, 1595–1603.

[276] Keilis-Borok, V.J., 1990. The Lithosphere of the Earth as a Non-linear System with Implications for Earthquake Prediction, *Rev. Geophys.*, **28**, 19–34.

[277] Kew Observatory, 1961. The Geocentric Direction Cosines of Seismological Observatories. Compiled by the Staff of the International Seismological Summary - International Association of Seismology. *Kew Observatory, ISS-IAS*.

[278] Khokhlov, A., Hulot, G. and Le Mouel, J.L., 1997. On the Backus Effect-I, *Geophys. J. Int.*, **130**, 701–703.

[279] Khokhlov, A., Hulot, G. and Le Mouel, J.L., 1999. On the Backus Effect-II, *Geophys. J. Int.*, **137**, 816–820.

[280] King, G., 1983. The Accommodation of Large Strains in the Upper Lithosphere of the Earth and Other Solids by Self-similar Faults Systems: the Geometrical Origin of b-value, *Pure Appl. Geophys.*, **121**, 761–815.

[281] Kirillov, A.A. and Gvishiani, A.D, 1982. *Theorems and Problems in Functional Analysis*, Springer-Verlag, New York Inc., 347 p.

[282] Klein, F.W., 1982. Patterns of Historical Eruptions at Hawaiian Volcanoes, *J. Volcanol. Geoth. Res.*, **12**, 1–35.

[283] Kolmogorov, A.N., 1941; *Dokl. Akad. Nauk. SSSR*, **30**, 9–13 **31**, 538–540.

[284] Kolmogorov, A. and Fomin, C., 1972. *Elements of Theory of Functions "and Functional Analysis"* Nauka, Moscow, 496 p.

[285] Kondorskaya, N.V. and Fedorova, I.V., 1996. Seismic Stations of the "Unified System of Seismic Observations in the USSR". *UIPE RAS*, 36 pp (in Russian).

[286] Kono, M., 1973. Geomagnetic Polarity Changes and the Duration of Volcanism in Successive Lava Flows, *J. Volcanol. Geoth. Res.*, **78**, 5972–5982.

[287] Korcak, J., 1938. Deux Types Fondamentaux de Distribution Statistique, *Bull. Inst. Intern. Stat. III*, 295–299.

[288] Kosko, B., 1992. *Neural Networks and Fuzzy Systems: A Dynamical System Approach to Machine Intelligence*, Englewood Cliffs: Prentice Hall, 452 p.

[289] Kosko, B., 1994. *Fuzzy Thinking. The New Science of Fuzzy Logic*, Glasgow: Harper Collins, 318 p.

[290] Kostyuk, O.P., Rudenskaya, I.M., Moskalenko, T.P. and Pronishin, R.S., 1991. Instrument parameters of Seismic Stations of the Carpathian Experimental-Methodical Party. In: *Seismological Bulletin of Western Zone of the Unified System of Seismic Observations in the USSR for 1988*. Kiev, "Naukova Dumka", 128 pp (in Russian).

[291] Kurthz, J. and Herzel, H., 1987. An Attractor in a Solar Time Series, *Physica D* **25** 165–172.

[292] Lahaie, F., Grasso, J.R., Marcenac, P. et Giroux, S., 1996. Modélisation de la Dynamique Auto-organisée des Éruptions Volcaniques : Application au Comportement du Piton de la Fournaise, Réunion, *C. R. Acad. Sci. Paris*, **323**, ser. II a, 569–574.

[293] Lahaye, Y., Blais, S., Auvray, B. et Ruffert, G., 1995. Le Volcanisme Fissural Paléozoïque du Domaine Nord-armoricain. *Bull. Soc. Géol. France*, **166**, 601–612.

[294] Landau, L. et Lifchitz, E., 1967. *Théorie de l'Élasticité*. Éditions MIR, Moscou.

[295] Landau, L. et Lifchitz, E., 1971. *Mécanique des fluides*. Éditions MIR, Moscou. 669 p.

[296] Lange, M.A., Ahrens, T.J. and Boslough, M.B., 1984. Impact Cratering and Small Failure of Gabbro, *Icarus*, **58**, 383–395.

[297] Laskar, J. and Robutel, P., 1993. The Chaotic Obliquity of the Planets, *Nature*, **361**, 608–612.

[298] Laskar, J., Joutel, F. and Robutel, P., 1993. Stabilization of the Earth's Obliquity by the Moon, *Nature*, **361**, 615–617.

[299] Ledésert, B., Dubois, J., Velde, B., Meunier, A., Genter, A. and Badri, A., 1993. Geometrical and Fractal Analysis of Three-dimensional Hydrothermal Vein Network in a Fractured Granite, *J. Volcanol.Geotherm. Res.*, **56**, 267–280.

[300] Ledésert, B., Dubois, J., Genter, A. and Meunier, A., 1993. Fractal Analysis of Fractures applied to Soultz-sous-Forets Hot Dry Rock Geothermal Program, *J. Volcanol.Geotherm. Res.*, **57**, 1–17.

[301] Legrand, D., Cisternas, A. and Dorbath, L., 1996. Multifractal Analysis of the 1992 Erzincan Aftershock Sequence, *Geophys. Res. Lett.*, **23**, 9 933–936.

[302] Lemoine, M., Trumpy, P. and Carraro, F., 1978. The Alps *in* : Tectonics of Europe and adjacent regions. Moscow, NAUKA (in Russian).

[303] Le Mouël, J.L., Allègre, C.J. and Narteau, C., 1997. Multiple Scale Dynamo, *Proc. Natl. Acad. Sci. USA*, **94**, 5510–5514.

[304] Le Pichon, X., Francheteau, J. and Bonnin, J., 1973. *Plate Tectonics*, Elsevier Scient. Publ. Co., 300 p.

[305] Levenstein, V.I., 1965. Binary Codes with Use of Deletions, Insertions and Substitutions of Symbols, *Dokl. A. N. SSSR*, **163** 845–848.

References

[306] Levy, S., 1993. *Artificial Life. The Quest of a New Creation* London, etc.: Penguin books, 390 p.

[307] Lin, J. and Parmentier, E.M., 1990. A Finite Amplitude Necking Model of Rifting in Brittle Lithosphere, *J. Geophys. Res.*, **95**, 4909–4923.

[308] Linde, A.T. et al., 1996. A Slow Earthquake Sequence on the San Andreas Fault, *Nature*, **383**, 65–68.

[309] Liu, H.H., and Fu, K.S., 1982. A Syntactic Approach to Seismic Pattern Recognition, *IEEE Trans; Pattern Analysis Machine Intelligence*, **PAMI-4**, 136–140.

[310] Liu, C., Curray, J.R., and McDonald, J.M., 1983. New Constraints on the Tectonic Evolution of the Eastern Indian Ocean: *Earth and Planet. Sci. Lett.*, **65**, 331–342.

[311] Loncarevic, B.D. and Parker, R.L., 1971. The Mid-Atlantic Ridge near 45°N, 17 Magnetic Anomalies and Ocean Floor Spreading, *Can. J. Earth Sci.*, **8**, 883–898.

[312] Lorenz, E.N., 1963. Deterministic Non-periodic Flow, *J. Atmosph. Sci.*, **20**, 130.

[313] Lovejoy, S., Schertzer, D. and Ladoy, P., 1986. Fractal Characterization of Inhomogeneous Geophysical Measuring Networks, *Nature*, **319**, 43–44.

[314] Machatschek, F., 1955. Das Relief der Erde. Bd. I, Berlin.

[315] Mackey, M.C., and Glass, L., 1977. Oscillation and Chaos in Physiological Control Systems, *Science*, **197**, 287–289.

[316] Madden, T.R., 1976. Random Networks and Mixing Laws, *Geophysics*, **41**, 1104–1125.

[317] Madden, T.R., 1983. Microcrack Connectivity in Rocks: a Renormalization Group Approach to the Critical Phenomena of Conduction and Failure in Crystalline Rocks, *J. Geophys. Res.*, **88**, B1, 585–592.

[318] Main, I.G., 1987. A Characteristic Earthquake Model for the Seismicity Preceding the Eruption of Mount St Helens on May 1980, *Phys. Earth Planet. Int.*, **49**, 283–293.

[319] Malinverno, A. and Gilbert, L.E., 1989. A Stochastic Model for the Creation of Abyssal Hill Topography at a Slow Spreading Center. *J. Geophys. Res.*, **94**, 1665–1675.

[320] Malinverno, A. and Cowie, P.A., 1993. Normal Faulting and Topographic Roughness of Mid-ocean Ridge Flanks. *J. Geophys. Res.*, **98**, 17 921–17 939.

[321] Malraison, B., Atten, P., Bergé, P. and Dubois, M., 1983. Dimension d'Attracteurs Étranges : une Détermination Expérimentale en Régime Chaotique de deux Systèmes Convectifs, *C. R. Acad. Sci. Paris*, **297**, 209–214.

[322] Malraison, B., Atten, P., Bergé, P. and Dubois, M., 1983. Dimension of Strange Attractors : an Experimental Determination for the Chaotic Regime of two Convective Systems, *J. Phys. Letters*, **44**, 897–902.

[323] Mandelbrot, B.B., 1967. Some Noises with 1/f Spectrum, a Bridge between Direct Current and white Noise, *IEEE Tr. on Information Theory*, **13**, 289–298.

[324] Mandelbrot, B.B., 1967. How long is the coast of Britain? Statistical Self-similarity and Fractional Dimension. *Science*, **155**, 636–638.

[325] Mandelbrot, B.B., 1975. *Les objets fractals*. Flammarion, Paris, 203 p.

[326] Mandelbrot, B.B., 1975. Stochastic Models for Earth Relief, the Shape and the Fractal Dimension of the Coastlines, and the Number-area Rule for Islands, *Proc. Nat. Acad. Sci. USA*, **72**, 3825–3828.

[327] Mandelbrot, B.B., 1977. *Fractal-Form, Chance and Dimension*, Freeman, San Francisco.

[328] Mandelbrot, B.B., 1982. *The Fractal Geometry of Nature*, Freeman, San Francisco.

[329] Mandelbrot, B.B., 1984. Les Images Fractales : un Art pour l'Amour de la Science et ses Applications. *Sciences et Techniques*, 16–19, 34–35.

[330] Mandelbrot, B.B., 1989. Multifractal Measures, Especially for the Geophysicist. *Pure Appl. Geophys.*, **131**, 5–42.

[331] Mandelbrot, B.B. and Wallis, J.R., 1968. Noah, Joseph, and Operational Hydrology, *Water Resour. Res.*, **4**, 909–918.

[332] Mandelbrot, B.B. and Wallis, J.R., 1969. Robustness and Rescaled Range R/S in the Measurement of Noncyclic Long Run Statistical Dependence, *Water Resour. Res.*, **5**, 967–988.

[333] Mandelbrot, B.B., 1977. *Fractal-Form, Chance and Dimension*, Freeman, San Francisco.

[334] Mané, R., 1981. Dynamical Systems and Turbulence, Warwik 1980. *Lecture Notes in Mathematics, Springer, Berlin*, **898**, 230.

[335] Mareschal, J.C., 1989. Fractal Reconstruction of Seafloor Topography, *Pure Appl. Geophys.*, **131**, 1–2, 197–210.

[336] Mark, D.M. and Aronson, B.P., 1984. Scale-dependent Fractal Dimensions of Topographic Surfaces: an Empirical Investigation, with Applications in Geomorphology and Computer Mapping, *Math. Geol.*, **16**, 671–683.

[337] Marks II, R.J., 1994. *Fuzzy Logic Technology and Applications*, New York (IEEE Technology Update Series), 575 p.

[338] Marr, D., 1982. Vision: Freeman.

[339] Marr, D. and Hildreth, E.C., 1980. Theory of Edge Detection: *Proceedings of the Roy. Soc. London B*, **207**, 187–217.

[340] Maruyama, T., 1978. Frequency Distribution of the Sizes of Fractures Generated in the Branching Process: Elementary Analysis, *Bull. Earthqu. Res. Inst.*, **53**, 407–421.

[341] Marzocchi, W., 1997. Missing Reversals in the Geomagnetic Polarity Timescale: Their Influence on the Analysis and in Constraining the Process the Generates Geomagnetic Reversals, *J. Geophys. Res.*, **102**, B3, 5157–5171.

[342] Marzocchi, W. and Mulargia, F., 1992. The Periodicity of Geomagnetic Reversals, *Phys. Earth Planet. Int.*, **73**, 222–228.

[343] Marzocchi, W., Gonzato, G. and Mulargia, F., 1995. Rikitake's Geodynamo Model Analysed in terms of Classical Time Series Statistics, *Phys. Earth Planet. Int.*, **88**, 83–88.

[344] Marzocchi, W., Mulargia, F. and Gonzato, G., 1997. Detecting Low-dimensional Chaos in Geophysical Time Series, *J. Geophys. Res.*, **102**, B2, 3195–3209.

[345] Masek, J.G. and Turcotte, D.L., 1993. A Diffusion-limited Aggregation Model for the Evolution of Drainage Networks, *Earth Planet. Sci. Lett.*, **119**, 379–386.

[346] Mathis, J.S., 1979. The Size Distribution of Interstellar Particles, II, Polarization. *Astrophys. J.*, **232**, 747–753.

[347] Maus, S. and Dimri, V.P., 1994. Scaling Properties of Potential Fields due to Scaling Sources, Geophys. Res. Lett., **21**, 891–894.

[348] Maus, S. and Dimri, V.P., 1995. Potential Field Power Spectrum Inversion for Scaling Geology, *J. Geophys. Res.*, **100** B7, 12,605–12616.

[349] Maus, S. and Dimri, V.P., 1996. Depth Estimation from the Scaling Power Spectrum of Potential Fields. *Geophys. J. Int.*, **124**, 113–120.

[350] McCrosky, R.E., 1968. Distribution of Large Meteoric Bodies, *Spec. Rep. 280*, Smithson. Astrophys. Observ., Washington, D.C.

[351] McFadden, P.L., 1984. Statistical Tools for thr Analysis of Geomagnetic Reversal Analysis, *J. Geophys. Res.*, **89**, 3363–3372.

[352] McFadden, P.L. and Merrill, R.T., 1984. Lower Mantle Convection and Geomagnetism, *J. Geophys. Res.*, **89**, 3354–3362.

[353] McNeil, D. and Freiberger, P., 1994. *Fuzzy Logic. The Revolutionary Computer Technology that is Changing our World*, New York, etc.,: Simon and Schuster, 319 p.

[354] Meakin, P., Stanley, H.E., Coniglio, A. and Witten, 1985. Sufaces, Interfaces and Screening of Fractal Structures, *Phys. Rev. A*, **32**, 2364–2369.

[355] Menard, G., 1979. Relations entre Structures Profondes et Structures Superficielles dans le Sud-est de la France. Essai d'Utilisation des Données Géophysiques, Thèse de doctorat de spécialité. Univ. S. M. Grenoble.

[356] Meneveau, C. and Sreenivasan, K.R., 1989. Measurement of $f(\alpha)$ from Scaling of Histigrams, and Applications to Dynamical Systems and Fully Developed Turbulence, *Phys. Lett. A*, **137**, 3, 103–112.

[357] Mezcua, J. and Martinez, J.M., 1983. Sismicidad del area Ibero-Mogrebi. Instituto Geographico Nacional, Publicacion **206**, Madrid, Spain.

[358] Mikoyan, A., Gvishiani, A. and Zhizhin, M., 1994. EMSC Strong Motion Database: WWW Interface, *EMSC Newsletter*, **6**, 4–7.

[359] Mikoyan, A, Burstev, A., Gvishiani, A. and Zhizhin, M., 1997. EMSC Strong Motion Database: WWW Interface, *EMSC Newsletter*, **11**, 5–6.

[360] Minoux, M., 1983. Programmation Mathématique. Théorie et Algorithmes, Paris.

[361] Mira, C., 1987. Généralités su l'outil de Récurrence. Ed. CNRS, Paris.

[362] Mogi, K., 1962. Study of the Elastic Shocks Caused by the Fracture of Heterogeneous Materials and its Relation to Earthquake Phenomena, *Bull. Earthqu. Res. Inst.*, **40**, 125–173.

[363] Mora, P., 1989a, *Elastic Wavefield Inversion of Reflection and Transmission Data*, **99**, 1211-1233.

[364] Mora, P., 1989b, *Nonlinear Elastic Wavefield Inversiof Multi-offset Seismic Data*, **99**, 1222-2244.

[365] Morat, P. and Le Mouel, J.L., 1992. Signaux Électriques Engendrés par des Variations de Contrainte dans les Roches Poreuses non Saturées, *C. R. Acad. Sci. Paris*, **315**, IIa, 955–963.

[366] Morat, P. and Le Mouel, J.L., Pride, S. et Jaupart, C., 1995. Sur de Remarquables Oscillations de Température, d'Humidité, de Potentiel Électrique Observées dans une Carrière Souterraine, *C. R. Acad. Sci. Paris*, **320**, IIa, 173–180.

[367] Morat, P., Le Mouel, J.L., Poirier, J.P. and Kossobokov, V., 1999. Heat and Water Transport by Oscillatory Convection in an Underground Cave, *C. R. Acad. Sc. Paris*, **328**, 1–8.

[368] Moritz, H., 1980. Advanced physical geodesy, Herbert Wichmann Verlag, Karlsruhe, Abacus Press, Tunbridge Wells, Kent.

[369] Morlet, J., Arens, A., Fourgeau, E., and Giard, D., 1982. Wave Propagation and Sampling Theory-Part II. Sampling Theory and Complex Waves, *Geophysics*, **47**, 222–236.

[370] Morse, P.M., and Feshbach, H., 1953. Methods of Theoretical Physics, McGraw Hill.

[371] Mosegaard, K. and Vestergaard, P.D., A Simulated Annealing approach to ..., *Geophysical Prospecting*, **xx**, yyy-zzz, 1991.

[372] Mulargia, F., Gasperini, P. and Boschi, E., 1987. Identifying Different Regimes in Eruptive Activity: an Application to Etna Volcano, *J. Volcanol. Geotherm. Res.*, **34**, 89–106.

[373] Myers, C., Rabiner, L.R. and Rosenberg, A.E., 1980. Performance Tradeoffs in Dynamic Time Warping Algorithm for Isolated Word Recognition, *IEEE Trans. Acoustics, Speech, Signal Proc., ASSP-* **28**(6), 623–635.

[374] Nagahama, H. and Yoshii, K., 1994. Scaling Laws of Fragmentation, *in* : *Fractals and Dynamic Systems in Geosciences*, 25–36, Kruhl, J. ed., Springer-Verlag, Berlin, 412 p.

[375] Narkounskaia, G. and Turcotte, D.L., 1992. A cellular Automata Slider-Block Model for Earthquakes 1. Demonstration of Chaotic Behavior for a Low Order System, *Geophys. J. Int.*, **111**, 250–258.

[376] Narteau, C., 1995. De l'Application des Groupes de Renormalisation à la Géophysique, *in* : Rapports de Stage, IPGP, **4**, juin 1995, DEA de Géophysique Interne, 21–46.

[377] Needham, H.D., Auffret, G.A., Ballu, V., Beuzart, P., Dauteuil, O., Detrick, R., Dubois, J., Carre, D., Lance, S., Langmuir, C., Le Drezen, E., Normand, A., Renard, V. and Visset, M. (Sigma Scientific Team) and Bougault, H., 1991. The Crest of the Mid-Atlantic Ridge, between 40° N and 15° S: Very Broad Swath Mapping with the EM 12 Echosounding System, *AGU Fall Meeting, EOS*, 470.

[378] NEIC, USGS. Station Book, NEIC, USGS : *http:wwwneic.cr.usgs.gov/neis/stationbook/*.

[379] Nettleton, L.L., 1939. Determination of Density for Reduction of Gravimeter Observations, *Geophysics*, **4**, 176–183.

[380] New Catalogue of Strong Earthquakes on the Territory of USSR from Ancient Time to 1975. *Moscow, Nauka*, 1977 535 p.

[381] Nicolis, C. and Nicolis, G., 1984. Is there a Climatic Attractor?, *Nature*, **311**, 529–532.

[382] Nolet, G., 1985. Solving or Resolving Inadequate and Noisy Tomographic Systems, J. Comp. Phys., 61, 463-482.

[383] Nottale, L., Chaline, J. et Grou, P., 2000. *Les Arbres de l'Evolution, Univers, Vie, Sociétés*, ed. Hachette, Sciences, 379 p.

[384] Nouaili, L., Dubois, J. and Deplus, C., 1987. Analyse Fractale de la Séismicité de la Zone de Subduction des Iles Tonga, *C. R. Acad. Sci. Paris*, II, **305**, 1357–1364.

[385] Nulton, J.D., and Salamon, P., 1988. Statistical Mechanics of Combinatorial Optimization, *Physical Review A*, **37**, 1351-1356.

[386] Ogniben, L. et al., 1972. Modello Structurale d'Italia. 1/1000000. Consiglio Nazionale delle Ricerche. Grafica Editoriale Cartografia Roma.

[387] Ohnaka, M. and Mogi, K., 1981. Frequency Dependence of Acoustic Emission Activity in Rocks under Incremental, Uniaxial Compression, *Bull. Earthqu. Res. Inst.*, **56**, 67–89.

[388] Okubo, P.G. and Aki, K., 1987. Fractal Geometry in the San Andreas Fault System, *J. Geophys. Res.*, **92**, 345–355.

[389] Oono, Y., 1989. Large Deviation and Statistical Physics, *Progress of Theoretical Physics Suplement*, **99**, 165–205.

[390] Orchard, G.A. and Phillips, W.A., 1991. *Neural Computation: A Beginner's Guide*, Hove etc.: Lawrence Erlbaum, 141 p.

[391] Osedelets, V., 1967. Multiplicative Ergotic Theorem. Lyapunov Characteristics Numbers for Dynamic Systems, *Trans. Moscow Mathem. Soc.*, **19**, 197–231.

[392] Otsuka, M., 1972. A Chain-reaction Type Source Model as a Tool to Interpret the Magnitude -frequency Relation of Earthquakes, *J. Phys. Earth*, **20**, 35–45.

[393] Ott, E., Grebogi, C. and York, J.A., 1990. Controlling Chaos, *Phys. Rev. Lett.*, **64**, 11, 1196–1199.

[394] Ouillon, G., 1995. *Application de l'Analyse Multifractale et de la Transformée en Ondelettes Anisotropes à la Caractérisation Géométrique Multi-Échelle des Réseaux de Failles et de Fractures*, Thèse Doct. Univ. Nice-Sophia Antipolis, Éd. BRGM, **246**, 313 p.

[395] Ouchi, T. and Uekawa, T., 1986. Staistical Analysis of the Spatial Distribution of Earthquakes-variation of the Spatial Distribution of Earthquakes before and after Large Earthquakes, *Phys. Earth Planet. Int.*, **44**, 211–225.

[396] Pandey, G., Lovejoy, S. and Schertzer, D., 1998. Multifractal Analysis Including Extremes of Daily River Flow Series for Basins Five to Two Million Square Kilometers, one Day to 75 Years, *J. Hydrol.*,

[397] Panteleyev, A.N., Dubois, J. and Diament, M., 1995. Roughness of the Global Altimetric Geoid, *Cahiers du Centre Eropéen de Géodynamique et de Séismologie*, **9**, 43–77.

[398] Papalashvili, V.G., 1981. List of Seismic Stations of Caucasus on the January 1, 1981 . In: *Seismological Bulletin of Caucasus, 1978.- Tbilisi, "Mecniereba", 1981*, 256–263.

[399] Parsons, B. and Sclater, J.G., 1977. An Analysis of the Variation of Ocean Floor Bathymetry and Heat Flow with Age. *J. Geophys. Res.*, **82**, 803–827.

[400] Patriat, Ph., 1987. Reconstitution de l'Évolution de Systèmes de Dorsales de l'Océan Indien par les Méthodes de la Cinématique des Plaques, *Territoires des Terres Australes et Antartiques Françaises*, Paris, 208 p.

[401] Patriat, P., and Ségoufin, J., 1988. Reconstruction of the Central Indian Ocean: *Tectonophysics*, **155** 211–234.

[402] Patriat, P., Deplus, C., Rommevaux, C., Sloan, H., Hunter, P. and Brown, H., 1990. Evolution of the Segmentation of the Mid-Atlantic Ridge between 28° and 29° N During the Last 10 My: Preliminary Results from the Sara Cruise, *AGU Fall Meeting, EOS*, **71**, 43, p. 1629.

[403] Pavoni, N., 1975. Zur Seismotektonik des Westalpenbogens Fachbl. *Vermess. Photogramm. Kulturtech.*, **3-4**, S. 185–187.

[404] Pavoni, N. and Mayer-Rosa, D., 1978. Seismotektonische Karte der Schweiz 1/750 000, *Eclogae geol. Helv.*, **71**, 293–295.

[405] Peckham, S.D., 1995. New Results for Self-similar Trees with Applications to River Networks, *Water Resour. Res.*, **31**, 1023–1029.

[406] Pecora, L.M. and Caroll, T.L., 1990. Synchronization in Chaotic Systems, *Phys. Rev. Lett.*, **64**, 8, 821–824.

[407] Perroud, H., Auvray, B., Bonhommet, N., Macé, J. and van Voo, R., 1986. Paleomagnetism and K-Ar Dating of Lower Carboniferous Dolerites Dykes from Northen Brittany, *Geophys. J. R. Astr. Soc.*, **87**, 143–153.

[408] Pfeuty, P. and Toulouse, G., 1977. *Introduction to the Renormalization Group and to Critical Phenomena*, John Wiley, New York.

[409] Philip, H., 1980. Tectonique Récente et Sismicité de la France: Caractéristiques Géodynamiques, in: Evolution Géodynamique de la France, coll. C, Géologie de la France, 26 CGI, *Mémoire du BRGM*, **107**.

[410] Philip, H., 1983. Carte de la Tectonique actuelle et Récente du Domaine Méditerranéen et de la Chaîne Alpine. Publication de l'INAG. CNRS. Paris

[411] Philip, H. and Tapponnier, P., 1976. Tectonique Actuelle et Séismicité en Provence; Essai d'Interprétation Cinématique, *in* : A.T.P. Géodynamique de la Méditerranée Occidentale et de ses Abords, colloque final, Montpellier. CNRS *Inag*, 121.

[412] Philip, H., Cisternas, A., Gvishiani, A. and Gorshkov, A., 1989. The Caucasus: an Actual Example of the Initial Stages of Continental Collision, *Tectonophysics*, **161**, 1–21.

[413] Philip, H., Roghozin, E., Cisternas, A., Bousquet, J.C., Borisov, B. and Karakhanian, A., 1992. The Armenian Earthquake of 1988 December 7: Faulting and Folding, Neotectonics and Paleosismicity, *Geophys. J. Int.*, **110** 141–158.

[414] Pilkington, M. and Todoeschuck, J.P., 1993. Fractal Magnetization of Continental Crust. *Geophys. Res. Lett.*, **20**, 7, 627–630.

[415] Pilkington, M., Gregotski, M.E. and Todoeschuck, J.P., 1994. Using Fractal Crustal Magnetization Models in Magnetic Interpretation. *Geophysical Prospecting* **42**, 677–692.

[416] Pinsker, I., 1973. Evaluation of the Method of Learning and Learning Set. Modeling and Automated Analysis of Electrocardiograms, *Moscow, Nauka*.

[417] Pisarenko, V.F. and Pisarenko, D.V., 1991. *Phys. Lett. A*, **153** 169–173.

[418] Pisarenko, D.V.. and Pisarenko, V.F., 1995. Statistical Estimation of the Correlation Dimension *Phys. Lett. A*, **197** 31–39.

[419] Presgrave, B.W., Needham, R.E. and Minsch, J.H., 1985. Seismograph Station Codes and Coordinates, 1985 Edition. *Open-File Report, National Earthquake Information Center, U.S.G.S.* 85-714.

[420] Prewitt, J.M.S., 1970. Object Enhancement and Extraction: *Academic Press, Picture Processing and Psychopictorics*.

[421] Purcaru, W., and Berkhemer, H., 1978. A Magnitude Scale for very Large Earthquakes. *Tectonophysics*, **49**, 189–198.

[422] Queiros-Condé, D., 1999. Géométrie de l'Intermittence en Turbulence Développée, *C. R. Acad. Sc. Paris*, **327**, Série IIb, 1385–1390.

[423] Queiros-Condé, D., 2000. Le Modèle des Peaux Entropiques en Turbulence Développée, *C. R. Acad. Sc. Paris*, **328**, Série IIb, 541–546.

[424] Queiros-Condé, D., 2000. Principe de Conservation du Flux d'Entropie pour l'Évolution des Espèces, *C. R. Acad. Sc. Paris*, **330**, 445–449.

[425] Queiros-Condé, D., 2001. Internal Symmetry in a Multifractal Spectrum of a Fully Developed Turbulence, *Phys. Rev. E*, **64**.

[426] Ranzman, E.Ia., 1979. Places of Earthquakes and Morphostructures of Mountain Countries, *Moscow, Nauka*, 172 p.

[427] Raszvetaev, A., 1984. Local Stability of Seismic Prediction and Cluster Analysis, *Computational Seismology, Moscow, Nauka*, **17**, 67–69.

[428] Regional Catalogue of Earthquakes. International Seismological Center, *Newberry, Edinbourg, 1964-1982*, **V. 1-18**,

[429] Réhault, J.P., Olivet, J.L. and Auzende, J.M., 1974. Le Bassin Nord-occidental Méditerranéen: Structure et Évolution, *Bul. Soc. Geol. France* **XV**, 3, 281–294.

[430] Reid, A.B., Allsop, J.M., Grancer, H., Millet, A.J. and Somerton, I.W.,1990. Magnetic Interpretation in Three Dimensions using Euler Deconvolution, *Geophysics*, **55**, 80–91.

[431] Reynolds, P.S., Klein, W. and Stanley, H.E., 1977. A Real Space Renormalization Group for Site and Band Percolation, *J. Phys. C*, **10**, 1167–1172.

[432] Rice, J.R., 1979. Theory of Precursory Processes in the Inception of Earthquake Rupture, *Gerlands Butrage zur Geophysik*, **88**, **N2**,

[433] Richardson, L.F., 1961. The Problem of Contiguity: an Appendix of Statistics of Deadly Quarrels. *General Systems Yearbook*, **6**, 139–187.

[434] Rigon, R., Rinaldo, A., Rodriguez-Iturbe, I., Bras, R.L. and Ijjasz-Vasquez, E., 1993. Optimal Channel Networks: A Framework for Study of River Basin Morphology, *Water Resources Res.*, **29**, n 6, 1635–1646.

[435] Rikitake, T., 1958. Oscillations of a System of Disk Dynamos, *Proc. Camb. Phil. Soc.*, **54**, 89–105.

[436] Rinaldo, A., Rodriguez-Iturbe, I., Rigon, R., Ijjasz-Vasquez, E. and Bras, R.L., 1993. Self-Organized Fractal River Networks, *Phys. Rev. Lett.*, **70**, n 6, 822–825.

[437] Rinaldo, A., Dietrich, W.E., Rigon, R., Vogel, G.K. and Rodriguez-Iturbe, I., 1995. Geomorphological Signatures of Varying Climate, *Nature*, **374**, 632–635.

[438] Rössler, O.E., An Equation for Continuous Chaos, 1976. *Phys. Lett. A* **57**, 397–398.

[439] Rossler, O.E., 1978. Continuous chaos, in Sygernetics - A Workshop, H. Haken, ed., Springer, New York, 184 p.

[440] Roux, S. and Hansen, A., 1990. Introduction to Multifractality, *in Disorder and Fracture*, J.C. Charmet, S. Roux and E. Guyon ed. Plenum Press, New York.

[441] Royer, J.Y., and Sandwell, D.T., 1989. Evolution of the Eastern Indian Ocean since Late Cretaceous from Geostat Altimetry: *J. Geophys. Res.*, **94**, 13 755–13 782.

[442] Ruelle, D. and Takens, F., 1971. On the Nature of the Turbulence, *Com. Mathem. Phys.*, **20**, 167–192, **23**, 343–344.

[443] Ruelle, D., 1987. *Chaotic Evolution of Strange Attractor*, Cambridge Universty Press.

[444] Ruthen, M.G., 1969. *The Geology of Western Europe*, Elsevier P.C., Amsterdam.

[445] Rybach, L., 1979. The Swiss Geotraverse from Basel to Chiasso, *Schweitz. Mineral. Petrol. Mitt.*, **59**, 199.

[446] Sailhac, P., 1997. Introduction to Nonlinear Geophysics, Chaos and Fractals, *Cahiers Cent. Europ. Géodyn. Séismol.*, **14**, 121–161.

[447] Sallantin, J., 1983. Méthodologie de l'Apprentissage pour des Variables Binaires, Publication Structures de l'Information.

[448] Sammis, C.G. and Biegel, R.L., 1989. Fractals, Fault-gouge and Friction, *Pure Appl. Geophys.*, **131**, 255–271.

[449] Sammis, C.G., King, G.C.P. and Biegel, R.L., 1987. The Kinematics of Gouge Deformation, *Pure Appl. Geophys.*, **125**, 777–812.

[450] Schalkoff, R.J., 1992. *Pattern Recognition: Statistical, Structural and Neural Approches*, New York: John Wiley, 364 p.

[451] Sayles, R.S. and Thomas, T.R., 1978. Surface Topography as a non Stationary Random Process. *Nature*, **271**, 431–434.

[452] Scheidegger, A.E., 1967. A stochastic Pattern for Drainage Patterns into an Intermontane Trench, *Int. Assoc. Sci. Hydrol. Bull.*, **12**, 15–20.

[453] Scholz, C.H., 1991. Earthquakes and Faulting: Self-organized Critical Phenomena with a characteristic dimension, In: Riste T. and Sherrington D. (eds), *Spontaneous Formation of Space-Time Structures and Criticality*, Kluwer Dordrecht, pp. 41–56.

[454] Sempere, J.C., Purdy, G.M. and Schouten, H., 1990. Segmentation of the Mid-Atlantic Ridge between 24° and 30°40′N, *Nature*, **334**, 427–431.

[455] Shapiro, S.,C., 1990. *Encyclopedia of Artificial Intelligence, in 2 vol.*, New York etc.: Wiley, vol.1, 679 p.

[456] Shapiro, S.,C., 1990. *Encyclopedia of Artificial Intelligence, in 2 vol.*, New York etc.: Wiley, vol.2, 680–1219.

[457] Shaw, H.R. and Chouet, B., 1989. Singularity Spectrum of Intermittent Seismic Tremor at Kilauea Volcano, Hawaii, *Geophys. Res. Lett.*, **16**, 2, 195–198.

[458] Shaw, P.R. and Smith, D.K., 1990. Robust Description of Statistically Heterogeneous Seafloor Topography through its Slope Distribution, *J. Geophys. Res.*, **95**, B6, 8705–8722.

[459] Shebalin, P., Girardin, N., Rotwain, I., Keilis-Borok, V. and Dubois, J., 1996. Local Overturn of Active and Non-active Seismic Zones as a Precursor of Large earthquakes in the Lesser Antillean Arc, *Physics of the Earth and Planetary Interiors*, **97**, 163–175.

[460] Shilov, G., 1952. *Introduction to Linear Spaces Theory* GITTL, Moscow, 312 p.

[461] Shinbrot, T., Grebogi, C., Ott, E. and Yorke, J.A., 1993. Using Small Perturbations to Control Chaos, *Nature*, **363**, 411–417.

[462] Scholtz, C.H., 1968. The Frequency-magnitude Relation of Microfracturing in Rock and its Relation to Earthquakes, *Bull. Seimol. Soc. Am.*, **58** 399–416.

[463] Schoutens, J.E., 1979. Empirical Analysis of Nuclear and High-explosive Cratering and Ejecta, *Nuclear Geophysics Sourcebook*, vol. **55**, part 2, section 4, Rep. DNA 65 01H-4-2, Def. Nucl. Agency, Bethesda.

[464] Schuhmann, R., Jr., 1960. Energy Input and Size Distribution in Comminution, *Trans. Amer. Inst. Mining Metall. Petrol. Engers.*, **217**, 22–25.

[465] She, Z.S. and Lévêque, E., 1994. *Phys. Res. Lett.*, **72**, 336–339.

[466] Shinbrot, T., Grebogi, C., Ott, E. and Yorke, J.A., 1993. Using Small Perturbations to Control Chaos, *Nature*, **363**, 411–417.

[467] Simpson, P.K., 1996. *Neural Networks Theory, Technology and Applications*, New York: (IEEE Technology Update Series) 943 p.

[468] Slack, H.A., Lynch, V.M. and Langan,L., 1967. The Geomagnetic Gradiometer, *Geophysics*, **XXII**, 877–892.

[469] Sloan, H. and Patriat, P., 1993. Kinematics of the North-American-African Plate Boundary between 28° and 29° during the last 10 Ma; Evolution of the Axial Geometry and Spreading Rate and Direction, *Geophys. Res. Lett.*, **20** 2 139–2 141.

[470] Smalley, R.F., Turcotte, D.L. and Solla, S.A., 1985. A Renormalization Group Approach to the Stick-slip Behavior of Faults, *J. Geophys. Res.*, **90**, 1894–1900.

[471] Smalley, R.F., Chatelain, J.L., Turcotte, D.L. and Prévot, L., 1987. A Fractal Approach to the Clustering of Earthquakes: Application to the Seismicity of the New Hebrides, *Bull. Seismol. Soc. Am.*, **77**, 4, 1368–1381.

[472] Smith, W.D., 1986. Evidence for Precursory Changes in Frequency-magnitude b-value, *Geophys. J. Roy. Astron. Soc.*, **86**, 815–838.

[473] Smith, D.K. and Jordan, T.H., 1988. Seamount Statistics in the Pacific Ocean, *J. Geophys. Res.*, **93**, B4, 2899–2918.

[474] Smith, D.K. and Shaw, P.R., 1990. Seafloor Topography: A Record of a Chaotic Dynamical System?, *Geophys. Res. Lett.*, **17**, 1541–1544.

[475] Sobel, I., 1990. An Isotropic 3 × 3 Image Gradient Operator: *Academic Press, Machine Vision for Three-Dimensional Scenes*, 376–379.

[476] Sobolev, G., 1981. Modeling of the Process of Preparation and Precursors of Earthquakes. Physical Properties of the Mountains Rocks, *Tashkent*,

[477] Sornette, D., 1993. Physique Statistique, les Phénomènes Critiques Auto-Organisés, in: *Images de la Physique 1993*, 9–17.

[478] Sornette, A. and Sornette, D., 1989. Self-Organised Criticality and Earthquakes, *Europhys. Lett.*, **9**, 197–202.

[479] Sornette, D., Davy, P. and Sornette, A., 1990. Structuration of the Lithosphere in Plate Tectonics as a Self-Organized Criticality Phenomenon, *J. Geophys. Res.*, **95**, 17353–17361.

[480] Sornette, A. and Sornette, D., 1999. Renormalization of Earthquake Aftershocks, *Geophys. Res. Lett.*, **26**, 13, 1981–1984.

[481] Sornette, A., Dubois, J., Cheminée, J.L. and Sornette, D., 1991. Are Sequences of Volcanic Eruptions Deterministically Chaotic ? *J. Geophys. Res.*, **96**, 11,931-11,945.

[482] Spicher, A., 1972. Carte géologique de la Suisse à 1/500 000. *Commission géologique suisse*

[483] Spicher, A., 1980. Carte Tectonique de la Suisse à 1/500 000. *Commission géologique suisse*

[484] Starovoit, O.Ye. and Chernobai, I.P. (eds.), 1996. Parameters, Amplitude and Phase Characteristics of the Instruments of the Strond-Motion Seismological Stations, 1993-1994.- Experimental-Methodical Expedition, United Institute of Physics of the Earth, Russian Academy of Sciences, Obninsk, 1996.

[485] Steenland, N.C., 1968. Discussion on: The Geomagnetic Gradiometer by Slack et al. (1967), *Geophysics*, **XXIII**, 680–693.

[486] Strahler, A.N., 1957. Quantitative Analysis of Watershed Geomorphology, *Trans. Am. Geophys. Un.*, **38**, 913–920.

[487] Stinchcombe, R.B. and Watson, B.P., 1976. Renormalization Group Approach for Percolation Conductivity, *J. Phys. C*, **9**, 3221–3247.

[488] Sugihara, G. and May, R.M., 1990. Non-linear Forecasting as a Way of Distinguishing Chaos from Measurement Error in Times Series, *Nature*, **344**, 734-741.

[489] Sukmono, S., Zen, M.T., Kadir, W.G.A., Hendrajaya, L., Santoso, D. and Dubois, J., 1996. Fractal Geometry of the Sumatra Active Fault System and its Geodynamical Implications, *J. Geodynamics*, **22 1/2**, 1–9.

[490] Scales, L. E., 1985. Introduction to Non-linear Optimization, Macmillan.

[491] Tait, S.C., Jaupart, C. and Vergniolle, S., 1989, Pressure, Gas Content and Eruption Periodicity of a Shallow Crystallising Magma Chamber, *Earth Planet. Sc. Lett.*, **92**, 107–123.

[492] Tapponnier, P. and Francheteau, J., 1978. Necking of the Lithosphere and the Mechanics of Slow Accreting Plate Boundaries, *J. Geophys. Res.*, **83**, 3955–3970.

[493] Tarantola, A., 1986. A Strategy for Nonlinear Elastic Inversion of Seismic Reflection Data, Geophysics, 51, 1893-1903.

[494] Tarantola, A., 1987. Inverse Problem Theory; Methods for Data Fitting and Model Parameter Estimation, Elsevier.

[495] Tarantola, A., and Valette, B., 1982a. Inverse Problems = Quest for Information, J. Geophys., 50, 159-170.

[496] Tarantola, A., and Valette, B., 1982b. Generalized Nonlinear Inverse Problems Solved Using the Least-squares Criterion, Rev. Geophys. Space Phys., 20, No. 2, 219-232.

[497] Taylor, A.E., and Lay, D.C., 1980. Introduction to Functional Analysis, Wiley.

[498] Takens, F., 1981. Detecting Strange Attractors in Turbulence, *Lecture Notes in Mathematics, Springer, Berlin*, **898**, 366.

[499] Telesca, L., Cuomo, V., Lapenna, V., Macchiato, M. and Serio, C., 1998, Detecting Scaling Laws and Stochastic Behaviours in Daily Geomagnetic Time Series, *Phys. Earth Planet. Int.*, in press

[500] Tarboton, D.G., Bras, R.L. and Rodriguez-Iturbe, I., 1988. The fractal Nature of River Networks, *Water Resour. Res.*, **24**, 1317–1322.

[501] Tartaron, F.X., 1963. A general Theory of Comminution, *Trans. Amer. Inst. Mining Metall. Petrol. Engrs.*, **226**, 183–190.

[502] Tauxe, L. and Wu, G., 1990. Normalized Remanence in Sediments of the Western Equatorial Pacific: Relative Paleointensity of the Geomagnetic Field? *J. Geophys. Res.*, **95**, B8, 12,337–12,350.

[503] Terano, T., Asai, K. and Sugeno, M., 1987. *Fuzzy Systems Theory and its Applications*, Boston, etc.: Academic Press, 268.

[504] Thatcher, W., and Hanks, T.C., 1973. Source Parameters of Southern California Earthquakes, *J. Geophys. Res.*, **78**, 8,547–8,576.

[505] Thompson, D.T., 1082. EULDPH: A new Technique for Making Computer-assisted Depth Estimates for Magnetic Data, *Geophysics*, **47**, 31–37.

[506] Thorarinsson, F. and Magnusson, S.G., 1990. Bouguer Density Determination by Fractal Analysis, *Geophysics*, **55**, 7, 932–935.

[507] Thouveny, N., 1991. Variations du Champ Magnétique Terrestre au Cours du Dernier Cycle Climatique (depuis 120 000 ans); *CERLAT Memoirs*, **3**, 349 pp.

[508] Tokunaga, E., 1978. Consideration on the Composition of Drainage Networks and their Evolution, *Geographical Rep. Tokyo Metro. Univ.* **13**, 1–27.

[509] Tokunaga, E., 1984. Ordering of Divide Segments and Law of Divide Segment Numbers, *Jap. Geomorph. Un.*, **5**, 71–77.

[510] Tsallis, C., 1988. Possible Generalization of Boltzmann-Gibbs Statistics, *J. Stat. Phys.*, **52**, 479–487.

[511] Tsonis, A.A. and Elsner, J.B., 1988. The Weather Attractor over very Short Timescales, *Nature*, **333**, 545–547.

[512] Tu, D. and Gonzales, P., 1978. Principals of Pattern Recognition, *Moscow, Mir*

[513] Tucholke, B.E., Lin, J. and Kleinrock, M.C., 1992. Crustal Structure of Spreading Segments on the Western Flank of the Mid-Atlantic Ridge at 25°25′ N to 27°10′ N, *AGU Fall Meeting, EOS*, 537.

[514] Turcotte, D.L., 1986. Fractals and Fragmentation, *J. Geophys. Res.* **91**, 1921–1926.

[515] Turcotte, D.L., 1989. Fractals in Geology and Geophysics, *Pure Appl. Geophys.*, **131**, 171–196.

[516] Turcotte, D.L., 1992. *Fractals and Chaos in Geology and Geophysics*, Cambridge University Press, 221 p.

[517] Turcotte, D.L., 1994. Crustal Deformation and Fractals, a Review, in *Fractals and Dynamic Systems in Geoscience*, J.H. Kruhl editor, Springer-Verlag, Berlin, 7–23.

[518] Turcotte, D.L., 1997. *Fractals and Chaos in Geology and Geophysics*, Second ed. Cambridge University Press, 398 p.

[519] Van Damme, H., Obrecht, F., Levitz, P., Gatineau, L. and Laroche, C., 1986. Fractal Viscous Fingering in Clay Slurries, *Nature*, **320**, 731–733.

[520] Vainzvaig, M., 1973. Algorithms of Pattern Recognition with Learning "Cora" Algorithms of a Pattern Recognition Learning, *Moscow, Sov. Radio*,

[521] Velde, B. and Dubois, J., 1991. Fractal Analysis of Fractures in Rocks: the Cantor's Dust Method - Reply, *Tectonophysics*, **198**, 112–115.

[522] Velde, B., Dubois, J., Moore, D. and Touchard, G., 1991. Fractal Patterns of Fractures in Granites, *Earth Planet. Sci. Lett.*, **104**, 25–35.

[523] Velde, B., Dubois, J., Touchard, G. and Badri, A., 1990. Fractal Analysis of Fractures in Rocks: the Cantor's Dust Method, *Tectonophysics*, **179**, 345–352

[524] Velde, B., Moore, D., Badri, A. and Ledesert, B., 1993. Fractal and Length Analysis of Fractures During Brittle and Ductile Changes, *J. Geophys. Res.*, **98**, 11 935–11 940.

[525] Vere-Jones, D., 1976. A Branching Madel of Crack Propagation, *Pure and Appl.Geophys.*, **114**, 711–725.

[526] Vidal, P., 1980. L'Évolution Polyorogénique du Massif Armoricain: Rapport de la Géochronologie et de la Géochimie Isotopique du Strontium, *Mém. Soc. Géol. Minér. Bretagne*, **21**, 162.

[527] Vincent, A. and Meneguzzi, M., 1991. *J. Fluid Mech.*, **225**, 1–20.

[528] Vine F.J. and Matthews, D.H., 1963. Magnetic Anomalies over Oceanic Ridge, *Nature*, **199**, 947–949.

[529] Vogt, J., 1979. Les tremblements de Terre en France, *Mémoire du BRGM*, **96**, Orléans, France.

[530] Vogt, J. et al., 1981. Carte Sismotectonique de la France à 1/1 000 000, *Mémoire du BRGM*, **111**, Orléans, France.

[531] Volant, Ph., 1993. Mécanisme des Déformations et Aspect Fractal de la Sismicité Induite par l'Exploitation d'un Gisement d'Hydrocarbures (Lacq, France), Thèse de Doctorat, Univ. Joseph Fourier, Grenoble, France 158 pp.

[532] Voss, R.F., 1985a. Random fractals : Characterization and Measurement, in *Scaling Phenomena in Disordered Systems*, R. Pynn & A. Skejeltorp, eds., Plenum Press, New York, PP 1–11.

[533] Voss, R.F., 1985b. Random Fractal Forgeries, in *Fundamental Algorithms for Computer Graphics*. Ed. R.A. Earnshaw, NATO ASI Ser. F13, Springer, New York, **13-16**, 805–835.

[534] Warren, B.A., 1973. Transpacific Hydrographic Sections at Lats 43° S and 28° S: The SCORPIO Expedition, II. Deep Water. *Deep Sea Res.*, **20**, 9–38.

[535] Weber, C., Gvishiani, A., Godefroy, P., Gorshkov, A., Kossobokov, V., Lambert, J., Ranzman, E., Sallantin, J., Soldano, A., Cisternas, A. and Soloviev, A., 1986. Recognition of Places where Strong Earthquakes May Occur XII. Two Approaches to Recognition of Strong Earthquakes in Western Alps, *Computational Seismology, Moscow, Nauka*, **19**, 132–154

[536] Weber, C., Gvishiani, A., Godefroy, P., Lambert, J., Ranzman, Soloviev, A. and Trusov, A., 1986. Recognition of Places where Strong Earthquakes May Occur XIII. Neotectonic Scheme of Western Alps; $M \geq 5,0$, *Computational Seismology, Moscow, Nauka*, **19**, 82–94.

[537] Wickman, F.E., 1966. Repose Period Pattern of Volcano, *Arkiv. Mineral. Geol.*, **4**, 7, 291–364.

[538] Wickman, F.E., 1976. Markov Models of Repose-period Patterns of Volcanoes. In: *Random Processes in Geology, Springer Verlag*, 135–161.

[539] Wiens, D.A., DeMets, C., Gordon, R.G., Stein, S., Argus, D., Engeln, J.F., Lundgren, P., Quible, D., Stein, C., Weinstein, S. and Woods, D.F., 1985. A Diffuse Plate Boundary Model for Indian Ocean Tectonics, *Geophys. Res. Letters*, **12**, 429–432.

[540] Willgoose, G., Bras, R.L. and Rodriguez-Iturbe, I., 1991. A Coupled Channel Network Growth and Hillslope Evolution Model, I: Theory, *Water Resour. Res.*, **27**, 1671–1684.

[541] Wolf, A., Swift, J.B., Swinney, H.L. and Vastano, J.A., 1985. Determining Lyapunov Exponents from a Time Series, *Physica D*, **16**, 285–317.

[542] Yaglom, I., 1980. Boolean Structure and its Models, *Moscow, Sov. Radio*,

[543] Young, A.P. and Stinchcombe, R.B., 1975. A Renormalization Group Theory for Percolation Problems. *J. Phys. C*, **8**, 1535–1540.

[544] Yuan, A.T., McNutt, S.R., and Harrow, D.H., 1984. Seismicity and Eruptive Activity at Fuego Volcano, Guatemala: February 1975 January 1977. *J. Volcanol. Geotherm. Res.*, **21**, 277–296.

[545] Zax, Sh., 1975. Theory of Statistical Conclusions, *Moscow Mir*,

[546] Zhang, C., Mushayandebvu, M.F., Reid, A.B., Fairhead, J.D. and Odegard, M.E., 2000. Euler Deconvolution of Gravity Tensor Gradient Data, *Geophysics*, **65**, 512–520.

[547] Zhizhin, M., Gvishiani, A., Bottard, S., Mohammadioun, B. and Bonnin, J., 1992. Clasification of Strong Motion Waveforms from Different Geological Regions using Syntactic Pattern Recognition Scheme, *Cahiers Centre Europ. Géodyn. Séism.*, **6**, 33–42.

[548] Zhizhin, M., Gvishiani, A. and Mikoyan, A., 1994. European-Mediterranean Seismological Center, *Newsletter, dec.1994*, **6**.

[549] Zhizhin, M., Gvishiani, A., Bonnin, J., Madariaga, R., Mohammadioun, B. and Rouland, D., 1995. Syntactic Pattern Recognition Scheme (SPARS) Applied to Seismological Waveform Analysis, *Cahiers Centre Europ. Géodyn. Séism.*, **9**, 17–26.

[550] Zhizhin, M., Gvishiani, A., Mikoyan, A.N., Bonnin, J. and Mohammadioun, B., 1996. Syntactic Pattern Recognition and Strong Motion Data Bank, *Cahiers Centre Europ. Géodyn. Séism.*, **12**, 271–281.

[551] Zhizhin, M., Bataglia, J., Dubois, J. and Gvishiani, A., 1996. Application of Dynamic Programming for the Reconstruction of Isochrons along the Mid-Atlantic Ridge, *Cahiers du Centre Europ. de Géodyn. et Séismol.*, **12**, 151–159.

[552] Zhizhin, M., Battaglia, J., Dubois, J. and Gvishiani, A., 1997. Syntactic Recognition of Magnetic Anomalies along the Mid-Atlantic Ridge, *C. R. Acad. Sci. Paris*, **325**, 983–990.

Part IV

Index

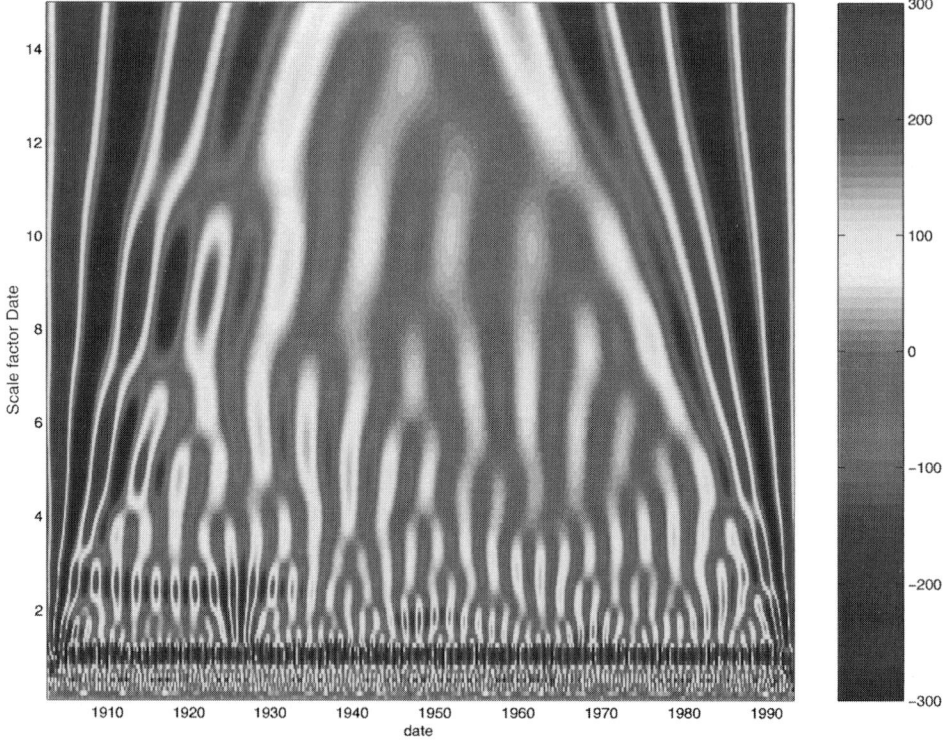

Wavelet transform - Scalogram of Amazon river discharge, the wavelet is a Gaussian thirtieth derivative (after David Aubert, 2001).

Index

A

Acoustic emission, 256,
Aeromagnetic map, 109,
    aeromagnetic measurements, 110,
Accretion rate, 88, 182, 183
Aftershocks, 148, 211, 212, 213, 229,
    aftershock sequence, 227,
Agregation, 32,
AI Artificial intelligence, 1-293,
    AI approach, 285, 286,
    AI tools, 287,
Algorithms
    ATTILA algorithm, 120,
    circle recognition algorithm, **64**,
    class of CORA-i algorithm, 128, 131, 137, 139, 140,
    classification algorithms, **128**, 286,
    Clustering algorithm, **34**,
    expert communication algorithm, 123, 128, 129, 140,
    FSPARS algorithm, 23, **24**,
    linear recognition algorithm, **64**,
    pattern recognition algorithm, 123,
    Rodin algorithm, 40-64, 286, 287,
    SOFT algorithm, 199,
    SPARS algorithm, **7**, 18,
    syntactic algorithm, 1,
    Voting by a Set of Features (VSF) algorithm, 123, 128,
Alignment path, 90, 91, 92,
Altimetric geoid, 192,
Anak Krakatau, 253,
Anisotropy, 218,
Annual Cycle, 239, 241,
Anomalies, 106,
    Magnetic anomalies, 106,
Appeerence, disappearence (cyclones), 279,
Arabic plate, 225, 226,

Artificial intelligence, 1-193
Ascending plumes, 282,
Asthenosphere, 234,
Asperities, 209,
Attractor, 220, 250, 252, 262-266, 272,
    chaotic attractor, 270,
    Feigenbaum attractor, 202,
    Lorenz attractor, 270,
    Rossler attractor, 252,
    Strange attractor, 271,
    system attractor, 293,
Automated expert system, 286,

B

Backward pass, 93,
Ball, 175,
Bathymetric data, 114,
    bathymetric profile, 180, 181, 182,
Bathymetry, 115, 179,
    bathymetry image, 121,
    multibeam bathymetry, 115,
Bayesian *a posteriori* probability, 156,
Bell shaped curve, 229,
Bifurcation, 273,
    bifurcation cascade, 282, 282,
Bi-logarithmic graph, 177, 178, 192, 194, 259,
    semi log graph, 259,
Binary space, 128, 132,
Binary descriptors, 136
Blocks, 146, 148, 210, 211,
    block dynamics, 210,
    moving blocks, 212,
Bodies, 108,
    causative bodies, 105, 108, 113, 114,
    magnetic bodies, 110,
Bouguer anomalies, 138, 141, 151, 168,

193,
  Bouguer anomaly variogram, 194, 195,
  Bouguer anomaly surface, 195,
Boundaries of silence, 136,
Box(es) 187, 189, 223, 227,
  box counting algorithm, 179, 215, 220, 290,
  box size, 225, 227,
Broad asperities, 75,
BRGM, Bureau de recherches géologiques et minières, 133, 152,

C

Cantor dust method **215**, 216, 220, 234, 243, 248, 249, 251,
Cantor dust set, 216, 218, 290,
Cascades, 209,
  direct, inverse cascade, 278, 281,
Catastrophe Theory
Cave, 282,
Central field oscillator
Chambon la Forêt observatory, 262,
CGDS Center of Geophysical Data Studies, 71, 72,
CGI Common Gateway Interface, 73,
CGMN Critical Golden Mean Nonlinearity, 256, 257,
Chaos, 247, 259, 292,
  chaotic chaos, 247,
  controlling chaos, 292,
  deterministic chaos, 247, 259, 265,
  low dimensional chaos, 261, 263,
  mastering chaos, 292,
  route to chaos, 283,
Chaotic 273,
  chaotic behaviour, 270, 271, 275,
  chaotic circuit, 293,
  chaotic obliquity, 288,
  chaotic region, 288,
  chaotic trajectories, 292,
  chaotic variations, 292,
  non linear chaotic system, 269,
Cell(s), 188, 200, 201, 204, 205, 208, 209,
Circular feature, 65,
Cl-generation, 48,
Classification, 74, 85, 87, 130, 137, 139, 150, 161, 166, 167, 170,
  classification algorithms, 128, 286,
  main classification of recognition, 153-156,
  random classification, 156,
  main classification of recognition, 166, 169,
Clusters, 41, 42, 44, 47, 48, 50, 51, 52, 54, 58, 60, 99, 106, 108, 111, 149, 211,
  cluster analysis, 114,
  linear clusters, 111,
Clustering, 39, 56, 101, 164, 245, 285,
  clustering algorithm, **34**,
  clustering analysis, 111,
  clustering solutions, 112,
  clustering threshold, 243,
Cocoon shape, 267,
Coil, 276,
Coimbra magnetic observatory, 268,
Co-image, 20, 24, 25,
Compaction, 44,
Complexity, 223, 247, 288,
Composition, 32,
Configuration, 200, 206,
  topological configuration, 206, 207,
Connectivity, 55,
Conjunction, 30, 32,
  probabilistic conjunction, 30,
  Zadeh conjunction, 30,
Construction, 39,
Continental lithosphere, 214,
Continental tectonics, 214,
Control Experiments, 132, 133, **142**, 143, 154, 160, 166, 169,
Control of Chaos, 292, 293,
Control parameters, 295,
Convolution 66, 67, 68, 83,
  Convolution integral **66**
Cora-i algorithms, **131**, 137, 139, 140, 141, 143, 144, 150, 154, 156, 160, 164, 168,
Correlation, 247,
Correlation function method, 179, 182, 239, 250, 251, 259,
Correlation integral, 181,
Covering circles, 222,
Cox scale, 258, 261,

Critical value,
    critical phenomenon, 211,
    critical pressure, 255,
Cube, 204, 208,
Cyclone(s), 277-280,
    cyclonic turbulence, 277,
    neighbouring cyclone, 278,

D

Daily variations, 262,
Dangerous area, 152,
Dangerous class, 149, 164,
    seismically dangerous, 168, 169,
Dangerous (Non-dangerous) zones, 142, 167,
Database, 16, 22, 24, 71, 72, 78, 285,
Declination, 107, 264, 265,
Deconvolution, 99,
    Euler deconvolution, **99**, 101, 102, 111, 112,
    Euler solutions, 99, 100, 101, 102, 107, 109, 111, 112, 113,
    Euler equation, 100, 103,
Deletion, 4, 5, 6, 83, 84,
Degrees of freedom, 247, 251, 253, 262, 265, 288, 291,
Density, 44, 49, 52, 53, 58,
Derivative, 117,
    first partial derivative, 117,
Descriptors, 135, 136,
Detectability problem, 192, 193,
Detectors,
    Gaussian detector, 115, 120,
    zero crossing detector, **67**
    gradient edge detector, **67**, 115, 118, 120,
    Laplacian edge detector, **68**, 115, 120,
    Canny edge detector, **68**, 115, 116, 119, 120,
    LoG edge detector, 115, 118, 120,
Deterministic Chaos, 247, 253,
    deterministic behaviour, 291,
    deterministic systems, 247, 253,
    deterministic relation, 289,
Differentiation, 69,
Dilation, 83,

Dimension, 174,
    attractor dimension, 250, 252, 253, 255, 259, 262, 267, 268, 269, 273, 291,
    correlation dimension, 180-183, 227, 290, 291,
    embedding dimension, 180, 183, 241, 250, 262, 263, 266,
    Euclidian dimension, 217,
    fractal dimension, 175, 177, 178, 181-189, 193-198, 203, 214, 215, 217, 221, 224, 244,
    fractal roughness dimension, 237, 238,
    generalized dimension, 189, 223, 224, 227, 228, 229, 240,
    Hausdorf dimension, 179,
    information dimension, 290,
    similarity dimension, 180, 181, 182, 234, 290,
    topological dimension, 233,
Dipole, 99, 100, 108,
Disjunction, 22, 32,
Discrete dynamic system,
Disk, 270, 272-276,
Distribution, 202, 209, 223, 228, 243, 247, 258,
    distribution of probability, 27,
    determinist distribution, 252,
    cumulated distribution, 229, 231, 232,
    exponential distribution, 204, 232,
    fractal distribution, 202, 204,
    Γ distribution, 259,
    normal distribution, 229,
    Poisson type distribution, 232, 243, 259, 261,
    power law distribution, 202, 203,
    random distribution, 252,
    Weibull distribution, 202, 203, 204, 209,
D$_{LA}$, Diffusion-limited aggregation, **186**, 187, 188,
D$_{LA}$ drainage network, 187, 188, 224,
D$_{MBS}$ Database Management System, 81,
Drainage basins, 188, 189,
    drainage network, 186, 188,
Dust, 216,
Dual Systems of Sets,
Dyke, 100, 101, 110,

thin dyke, 99,
dyke thickness, 254,
Dynamic Pattern Recognition, 285,
Dynamic Systems, 179, 181, 182, 183, 220, 232, **239**-283, 288,
natural dynamic systems, 282, 283,
instable dynamic systems, 270,
Dynamo
chaotic dynamo action, 276,
Faraday disk homopolar dynamo, 273,
Rikitake dynamo, 273,
self-exciting homopolar dynamo, 273, 274,
steady dynamo action, 276,

E

Earthquakes, 29, 71, 86, 134, 161, 164, 210, 211, 212, 220, 227, 233, 235, 236, 238, 243, 244, 292,
earthquake clusters, 234,
Earthquake-prone Areas, **123**, 143, 144, 153, 158, 164, 168,
E.C. (expert experiments), **128**, 129, 130, 135, 136, 137, 138, 142, 143, 144,
EDD Energy density distribution, 83,
Edge, 65,
edge points, 65,
edge detection, **64**,
edge pixel, 64, 65, 68, 116,
linking edge pixel, 69,
ramp edges, 65,
Electric currents, 270, 276, 278,
Electrodes, 282, 283,
Elements, 200, 201, 204, 205,
fragile elements, 206,
sound elements, 206,
Elimination,
quality-based elimination, 60,
elimination in cluster, 62,
Embedding space, 193,
Empty set,
empty element, 5,
EMSC European Mediterranean Seismological Center, **72**, 78, 80,
Energetic balance, 211,

Energy distribution, 76,
energy dissipation, 210, 289,
Enlargement factor, 240,
Entropy jumps, 289,
Epicenters, 86, 126, 128, 133, 134, 135, 148, 149, 159, 162, 169,
Eruption(s), 248-256
eruptionj distribution, 255,
eruption duration, 254,
eruption dynamics, 255,
eruption volume, 254,
pycritic eruption, 248,
ESK Eskalemuir observatory, 262, 264,
Etna volcano, 248,
Erzincan aftershocks sequence, 227, 228,
Error rate estimation, 75,
Euclidean norm, 6,
Euclidean distance, 17,
Euler deconvolution, **99**, 101, 102, 111, 112,
Euler solutions, 99, 100, 101, 102, 107, 109, 111, 112, 113,
Euler equation, 100, 103,
Events, 72, 229, 243, 247,
Experts, 130, 136,

F

Faults, 124, 126, 146, 147, 162, 164, 209, 212, 213, 237, 292,
fault dimension, 233,
fault fields, 214, 215, 223, 226,
fault length, 233,
fault network, 212, 213, 215,
fault plane, 233,
fault segments, 213,
fault zone, 212,
surface faults, 220,
Features, 24, 132, 141, 146, 147, 152, 156, 166,
Feigenbaum attractor, 201,
Fixed points,
unstable fixed point, 270,
First return map(ping), 207, 209, 243, 251, 253-255,
Flood(s), 240,
Floquet Matrix,

Index 339

Flow(s), 242, 255,
Foreshocks, 212, 213,
Forward pass, 94,
Fourier transform, 1, 83,
  Fourier power spectrum, 263,
Fractals, **173**-238, 289, 290,
  fractal analysis, 173, 174, 192, 193, 213-216, 249, 288,
  fractal Brownian motion, 262,
  fractal distribution, 245,
  fractal geometry, 178, 182, 193, 223, 237,
  fractal law, 243,
  fractal object(s), 174, 175, 233,
  fractal properties, 196,
  fractal sets, 213, 214, 218,
  fractal skin of turbulrnce, 289,
  fractal structure, 181, 197, 198, 228,
  fractal support, 174,
  fractal surface, 194,
  fractal value, 207,
  FSA fractal singularity analysis, 256, 257,
  self affine fractals 177, 179, 181,
Fractional Brownian walks, 178,
  fractional Brownian surface, 193,
  fractional Brownian motion, 240,
Fracturation, **202**, 214,
  fracturation processes, 236,
Fracture fields, 213, 215, 216, 219,
Fracture network, 224,
  length of fracture, 213,
Fracture tectonics, 211, 212,
Fragment(s), 202, 237,
Fragmentation, 199, **202**, 205, 208, 217, 236, 237,
  fragmentation law, 213,
  fragmentation model, 204,
  catastrophic fragmentation, 208, 209,
  lithospheric fragmentation, 238,
Fracturization cascade, 225,
Free air anomalies, 194,
  free air anomaly variogram, 194,
Frequency domain analysis, 198,
FSPARS algorithm, 23, **24**
Fuji volcano, 248,
Functions,
  closed function, 20,
  continuous function, 90,
  correlation integral function, 240, 241, 243, 247, 290,
  covariance function, 179,
  channel initiation function, 188,
  density function, 44,
  distribution function, 64,
  fractional Brownian function, 196,
  Gaussian scalar function, 196,
  Heavyside function, 181, 228,
  homogeneous function, 103,
  image function, 66, 68,
  monotonic function, 37, 90,
  membership function, 18,
  non-linear function, 152,
  probability function, 200,
  space function, 277,
  step function, 33,
  time function, 277,
Fuzzy Logic, **1**, **7**, 15, 285, 287,
  fuzzy binary relation, 14, 15,
  fuzzy clustering, 1, **56**, 286,
  fuzzy conjunction, 23, 25, 53,
  fuzzy disjunction, 23, 26, 27, 53,
  fuzzy distribution, 27,
  fuzzy extension, 22, 23,
  fuzzy mapping, 18, 22, 23,
  fuzzy measure, 32,
  fuzzy metrics, 56, 57,
  fuzzy modeling, 286,
  fuzzy SPARS, 33,
  fuzzy structure, 21, 23, 27, 56, 57, 58, 62,
  fuzzy technique, 25,
Fuzzy sets, 7, 8, 9, 13, 14, 15, 18, 22, 58, 285,
  Fuzzy set intersection and union,
  Fuzzy set operations,
  Fuzzy set products, 21,
  Fuzzy set "weakly", "moderately", "strongly", 9, 10,
Fuzzy subsets, 8, 59,
Fuzzy Rodin, **60**, 62,
Fuzzyness, 56, 57,
  Fuzzy to non-fuzzy transformation, 56,
Fuzzy space, 56,

G

Gaussian kernel, 115,
Generalized mean-value, 14,
Generators,
    additive generators, 14,
Geodynamo reversals, 270,
Geomagnetism, **196**, **258**-281,
Geomagnetic field, 197, 198,
    geomagnetic time series, 197,
    geomagnetic reversals,**258**-261,
Geomorphology, **177**, 184, 191, **239**-242,
Geomorphological features, 213,
GEOSCOPE, 81, 86,
Geophysical analysis, 287,
    geophysical data acquisition, 285,
    geophysical management and study, 285,
Geotectonic regions, 74,
GIS Geographical Information System, **64**, 120,
Goal, 54,
Gradient, 103,
Granitic batholith, 218, 220,
Grassberger and Procaccia method, 220, 264, 266,
Gravity anomaly, **192**, 193,
Gravity field, 100,
Gravity gradient tensor, 99,
Great Caucasus, **163**-170, 287,
Gulf of Saint Malo, 109,
Gutenberg-Richter law, 213, **226**, 227, 235, 238,
    Gutenberg-Richter constant, 238,
Grid, 91, 93, 94, 97, 106, 186, 187, 189,

H

Hawaii (volcano), 248-251, 253, 255,
    Kilaua, 248, 251, 253,
    Mauna Loa, 248, 251,
Heat and water transport, **282**-283,
Hierarchical organisation, 255,
    hierarchical construction, 281,
Helicity, 278, 279,
Helical motion, 279,
Hereditary principle, 51,

Hilbert transform, 1, 2,
Histograms, 229, 230,
Hölder exponent, 173,
Horizontal magnetic components, 198,
Horton's law, 185,
Hydrology 177, **184**, **239**-242,
    hydrographic network, 184,
Hypersphere, 181,
HTML Hypertext mark-up language, **73**, 81,

I

Illumination, 35, 36, 53, 57, 59,
    ball type illumination, 35,
    Gauss law illumination, 35,
    illumination density, 59,
    illumination intensity, **37**
    inner illumination, 42, 57,
    level of illumination, 43,
    outer illumination, 36, 57, 58,
    outer fuzzy illumination, 59, 60,
    smooth descending illumination, 35,
Image, 65, 67,
    gradient image, 117,
Inclination, 107, 264, 266, 269,
Inclusion, 9, 44,
    degree of inclusion, 10,
    stict inclusion, 16,
Initialisation, 6, 47, 93,
Insertion, 4, 5, 6, 84, 85,
Intensity, 37, 38, 39,
Interflow periods, 255,
Intermission order, 243,
Intermittency, 289,
Intersection, 14, 219,
Inversions, 258, 261, 274, 275,
Irreductibility, 44,
IRIS, 81,
Isochrons, 98,
Isostatic compensation, 193,
    isostatically compensated topography, 194,
ISS International Seismological Summary, 78,
ISC International Seismological Centre, 79,

Index 341

Iterate, Iteration, Iterative, 216,

J

Jack Knife estimation, 85, 136, 154, 162,
  Jack Knife experiment, 154, 160, 169
Java technology, 74,

K

K-mean distances **7**,
Kanto region, 228,
Kaplan-Yorke formula, 271,
Kernel, 18, 21, 22, 66, 67,
  convolution kernel, 67,
  Gaussian kernel, 115, 116,
Knot(s), 163-167, 169,
Knowledge base, 75, 85, 86,
Koenigs-Lemeray construction,
Kolmogorov spectrum, 196, 197,
Kolmogorov-Smirnov statistics, 247,
Kolmogorov theory, 289,
Krakatau volcano, 248,

L

Lacustrine sediments, 264,
Lake du Bouchet, 264-268,
Lamps, **25**, 30, 37, 44, 45,
  Lamp function, 26, 28, 30,
Laplacian, 68, 115,
Largest Common Divisor, LCD,
L<small>DG</small> Laboratoire de Détection Géophysique (C<small>EA</small>), 150,
Lava flows, 246,
Learning Materials, 128, 129, 143, 148, 166, 169,
  learning sets **133**, 135, 136, 154, 156, 170,
Legendre transform, 224,
  Legendre polynomials, 267,
Level of trust, 30,
Levenstein distance, 4, 5, 6, 17, 23, 56, 85,
Light, 26, 27, 32, 49, 50, 57, 59,
  lighting, 34,

light dispersion laws, 35,
Lineaments, 124, 126, 136, 137, 138, 141, 144, 146, 147, 150, 161, 163, 164, 168,
Linear clustering, 114,
Linear features, 65,
Linearity, 42,
Lithosphere, 234, 236,
Logical
  logical stability, 131,
  logical justification, 130,
  logical contestation, 130,
  logical consistency, 143,
  logical silence, 130,
Logistic block, 48,
Logistic map(ping), 202,
Log-log graph, 218, 220, 231, 232, 243,
Long time series, **239**-283,
Lyapunov,
  F<small>SLE</small> Finite Size Lyapunov exponents, 292,
  Lyapunov exponents, 292,

M

Magnetic anomalies, 88, 94, 95, 97, 105, 110, 114,
Magnetic data clustering, 105,
Magnetic field, 88, 99, 103, 196, 264, 267, 272,
  earth magnetic field, 258, 264, 265, 266,
  large scale magnetic field, 277,
  interplanetary magnetic field, 196, 197,
  secondary magnetic field, 279, 280,
Magnetic inversion, 88,
Magnetic sources, 103,
Magnetization, 106, 107,
Magnitude, 86, 164, 165, 226, 227, 232, 239,
  optical magnitude, 173, 174,
Mass, 57,
Mapping, 17, 18, 24, 39,
Mathematical microscope, 240,
Matrix, 83, 93, 132, 274,
  matrix of pixels, 66,

dissimilarity matrix, 82,
Maximum likelyhood criteria, 291,
Measure, 22, 36, 37, 44, 174, 189, 264,
    conditionnal measure, 21, 31,
    dissimilarity measure, 85,
    inner measure, 58,
    outer measure, 58,
Mechanical friction, 275,
Megablocks, 126, 142, 146, 147, 148,
Mériel quarry, 282,
Mesosphere, 236,
Metric Spaces, 20,
    finite metric spaces, 34,
    non-Archimedial metric spaces, 40,
Metrics, 20, 42,
Micro-fissures, 233,
Mid Atlantic Ridge, 88, 94,
Minimal information criteria, 129,
Models, 202, 205,
    Blanter et al. model, **279**-281,
    block-spring model, 245,
    branching model, 233, 234,
    disk dynamo model, 261,
    Erschov et al. model, 272,
    experimental model, 214,
    fragmentation model, 204,
    Goischi model, 233,
    Hide's models, **273**-276,
    hierarchical model of blocks, 212, 213, 277,
    Le Mouel model, **276**-279,
    Lorenz model, 270, 292,
    magnetic model, 89,
    multiple scale dynamo model, 276, 290,
    permeability model, 199,
    Rikitake model, 259, 270-272,
    scale invariant hierarchical model, 209,
    seismicity model, 234, 292,
    seismological model, 29,
    slider-blocks model, 211,
    SOFT model, 210, 211, 290,
    sound pillar model, 206, 208, 209,
    spreading rate model, 95,
    two asperities model, 211,
    two-block model, 246,
    two-disk dynamo model, 270,
    up and down cascade model, **279**-281,
    updating global model, 269,
    weakness plane model, 206, 207, 208, 210,
    theoretical modeling, **270**-281,
Module, 262,
Moho, 124, 226,
Moment, 223,
    angulae moment, 271,
Monotonous function, 49,
Monolithic, 43, 44,
Mount Pelée volcano, 248,
Mount St Helen volcano, 248,
Morphostructural zoning scheme, 126, **127**, 135, 137, 139, 144, 146, 147, 148, 163,
Multibeam bathymetry, 114, 115,
Multifractal, 173, 187, 213, 223,
    multifractal approach, 223,
    multifractal analysis, 173, 184, 187, 191, 226, 227, 228, 256, 257,
    multifractal distribution, 173,
    multifractal object, 174,
    multifractal spectrum, 187, 189, 190, 191, 223, 263,
Multiscale characterization, 224,

N

NEIC National earthquake information center, 79, 81,
Neighbours, 45, 130,
    nearest neighbour approach, 24, 82,
    nearest neighbour decision, 87,
Neotectonic scheme, 124, 125, 133, 138,
Network, 214, 255,
    homogeneous network, 193,
    network grid size, 193,
New Hebrides, 243, 244,
Non-Archimediarity, 40,
Non-linear, 90, 288,
    non-linear alignment, 96, 98,
    non-linear analysis, 179,
    non-linear differential equations, 273, 274, 276,
    non-linear forecasting approach, 197,

Index

non-linear chaotic system, 198,
non-linear quenching, 275,
Non-maximum suppression, 69,
Non-randomness of classification, 156, 158, 170,
Norm, 12-14,
    Archimedian t-norm, 14,
    Dombi's t-norm, 12,
    Lukasiewicz's t-norm, 12,
    probability t-norm, 12,
    triangle norm (t-norm), 12, 13, 14,
    triangle co-norm (t-co-norm), 12, 13, 14,
    Zadeh's t-norm, 12,
Null element, 5, 6,

O

Objects, 124, 126, 128, 132, 133, 137, 138, 139, 143, 144, 149, 151, 154, 159, 160, 164, 166, 169, 203, 207, 209, 237,
    dangerous objects, 135, 140, 141, 152, 153, 159-162,
    non-dangerous objects, 141, 152,
Ohmic dissipation, 270,
    ohmic coefficient, 271,
    ohmic heating, 275,
Omori law, 212, 213,
Ophiolit belt, 124, 125,
Orogenic belt, 146,
Orbit, 274,
    orbit jumps, 274,
    fixed orbit, 293,
    stable orbit, 293,
ORFEUS, 81,
ORSTOM Office de la Recherche Scientifique et Technique Outre-Mer, 86,
Oscillatory convection, 282,
    oscillatory convective process, 282,
Oubangui river discharge, 239-241,
Overthrusting, 124, 125,

P

Paleointensity, 264,

Paleomagnetic analysis, 261,
    paleomagnetic data, 264,
Paradigm, 288,
Parametrisation, 2, 3,
Path, 84, 88,
    path grid, 223,
    space of paths, 92,
Pattern recognition, 1, 123, 146, 150,
    pattern recognition operator, 129,
    dynamic pattern recognition, 158,
Percolation, 199, 200,
Period
    period-doubling, 273, 282, 283,
    direct, inverse period, 259,
Periodic behaviour, 198,
Periodicity
    multiple periodicity, 276,
Permeability, 199, 201,
Permeable, impermeable, 200, 201,
P-generator, 47,
Phase space, 239, 251, 266-268, 272, 274, 291,
    phase space orbits, 263,
    pseudo phase space, 239, 259, 263,
Phase trajectory, 181, 267, 271, 272,
Piton de la Fournaise volcano, 230, 232, 248-252, 254,
Plate boundary dynamics, 233,
Poincaré,
    Poincaré cross section, 252, 293,
Poisson law, 229,
Pole, 99,
Potential, 44, 50, 51,
    potential product, **30**,
    anomalous potential field, 99,
Power law, 175, 181, 197, 214, 232, 236, 253,
    power law distribution, 227, 229, 243, 255,
    power law exponent, 223, 243, 256,
    power index, 197,
    power law spectrum, 196,
Power spectrum, 196, 197, 198, 240, 241,
Precursor, 234,
Predictibility, 293,
Principle,
    hereditary principle, 51,
    S-principle, 57, 59, 60, 63,

Probability, 10, 20, 26, 169, 170, 176, 204, 205, 206, 210, 227, 243, 277,
   probability law, 200,
   conditional probability, 243,
   critical probability, 207, 209, 210, 211,
   uniform probability, 247,
Prism, 106, 107, 108, 109,
   prism body dimension, 102,
PTD person taking the decision, 56,
Pyrénées, **146**-158,

Q

Quadratic mapping, 201,
Quality, 47, 62,
Quenching, 275,
   partial quenching, 276,

R

Ramp edge, 65,
Random, 268,
Random, Non-random, 229,
   random behaviour, 246,
   random data sets, 285,
   random event, 246,
   random numbers, 230, 231,
   random problems, 156, 158, 162,
   random learning sets, 156,
   Random path, 186,
   random process, 247,
   randomness test, 248,
Ranks, 146, 148, 163, 200, 201, 206,
RAS Russian Academy of Sciences, **71**, 80,
Rayleigh number, 282,
Recognition, 88, **123**, 124, 132, 150, 286,
   recognition objects, 124, 151, 153, 160, 164, 166, 169,
Recording instruments, 72,
Recurrence time, 256,
Recursion, 6, 93, 94,
Renormalized groups method, 199, 200, 202, 205, 209, 210,
   renormalization group theory, **199**,
207, 212, 276,
Repose periods, 246,
Rescaling, 199,
Reversal time series, 261,
   polarity reversals, 281,
Richardson method, 177, 178, 183, 220, 221,
Ridge segments, 183,
River basins, 184, 190, 191,
   river discharge (record) 142,
   river flow, 184, 239,
   fractal river network, 191,
RODIN algorithm **44**, 47, 52, 55, 57, 61, 102, 106, 108,
   block initialisation, 47,
   block quality, 47,
   block P-generation, 47, 50, 62,
   block cluster, 47, 63,
   block cl-generation, 48,50, 63,
   logic of the block, 48,
   free parameters, 48, 105,
Roughness (degree of), 192, 193, 220, 236, 238,
   seafloor roughness, 179, 181,

S

S-principle, 57,
San Andreas fault system, **220**-223,
(evolving) Sandpile experiment, 245,
Saturation (level) 241, 250, 251, 259, 266,
Saudi Arabia, 224,
Scale(s) 185, 189, 199, 200, 213, 214, 218, 220, 223, 224,
   scale invariant geometry, 253, 292,
   scale range, 255,
   scale transfer, 279,
   time scale, 196,
   worldwide scale, 286,
Scaling properties, 185, 190, 270, 280,
   scaling behaviour, 190,
   scaling exponent, 198,
   scaling index, 198,
   scaling law, 255, 276,
   scaling organization, 211, 212,
   scaling technique, 212,
   spatial scaling, 253,

Index

temporal scaling, 253,
universal scaling, 289,
Seafloor imagery, 114,
Secular variations, 264,
Segments, 90, 124, 215, 216, 217, 220, 223,
    elementary segments, 90, 92,
Seismic signal, 15,
Seismic observation network
    seismic belt, 152,
    seismic break zone, 238,
    seismic crisis, 215,
    seismic cycle, 210, 212, 213,
    seismic fault, 233,
    seismic future experiment, 154, 156, 157, 166,
    seismic instabilities, 245,
    seismic hazard assessment, **123**,
    seismic moment, 232, 233,
    seismic records, 78,
    seismic sources, 238,
    seismic vulnerability, 75,
    seismic zone, 254,
Seismically dangerous zone, 146,
Seismicity, 133, 136, 150, 158, 159, **199**, 212, 213, 223, 226, 228, 236, 243-245, 287,
Seismograms, 33, 34, 56, 86, 87,
    synthetic seismograms, 85,
Seismological application (FSPARS),
Seismological data classification, 78,
Seismology, **243**-246,
    engineering seismology, 71, 76,
Self-affine, 196,
    statistically self-affine curves, 197,
Self invariance, 227,
Self-potential, 282, 283,
Self similar, 196,
    self similarity, 213, 214, 218, 220, 255,
    self similar fault system, 234,
Sets, 189, 193,
    dataset, 116, 260, 261, 268, 285, 290,
    triadic Cantor set, 216, 232,
Singularity (ies), 175,
    singularity spectrum, 175,
Strength, 175,
Signal, 15, 16, 17, 18, 24,

signal origin definition (SOD), **31**,
signal-to-noise ratio, 69,
Sloping contact, 99,
Smallest Common Multiplior, SCM,
SMDB Strong motion database, **71**, 72, 80, 286,
Smoothed first derivative, 97,
Smoothing, 69,
SOC, Self Organized Criticality, 210, 245, 292,
SOFT (process), 211, 212, 213, 276,
Soufrière de la Guadeloupe volcano, 248,
Sources,
    source detectability, 192,
    causative source, 112, 114,
SPARS algorithm, 18, 19, 20, 24, 32, 75, 82, 85, 86, 87, 286,
SPARS classification, 25,
Spatial domain, 215, 220,
Spectral analysis, 181,
    spectral power law, 198,
Spectrum energy, 181,
Spin-axis orientation, 288,
Spreading rate, 88, 91, 97,
SQL Structural querry language, 73,
Stable (fixed) point, 207,
    stable regime, 293,
Stations, 72,
Statistics, 176,
    statistical analysis, 217, 291,
Step, 41, 106,
    step discontinuities, 65,
Stick-slip behaviour, 209,
Stochastic, 56, 209,
    stochastic behaviour, 198, 291,
    stochastic dynamics, 239,
    stochastic process, 240, 246,
Storm, 262,
Stream ordering system, 185, 186,
    stream flow, 240,
    stream ordering parameter, 186,
Stress, 209, 234, 236, 245, 292,
    critical stress, 209,
Stress distribution, 75,
Stromboli volcano, 256,
Strong ground motion, **71**, 75, 76, 77,
Subduction zone, 234,
Sub-series, 252, 254,

Subset, 175, 218, 229,
Substitution, 4, 5, 84, 85,
Sucheno's negation, 12,
Symmetry breaking, 279, 280, 281,
    internal symmetry, 290,
Syntactic classification, **71**, 81,
    syntactic distance, 87,
    syntactic pattern recognition, 88, 287,

T

Tectonics, **199**, 213, 220, 236,
    tectonic field, 238,
Termination, 93, 94,
Terrestrial planets, 288,
Threshold, 69, 129, 130, 136, 150,
Threshold classification algorithms,
Thresholding, 69, 120,
Time domain, 215, 220, 244,
    time embedding, 291,
    time-frequency domain, 76,
    time interval, 258,
    time series, 198, 215, 229, 241, 242, 246, 247, 262, 291,
    time scale, 262,
    filtered time series, 239,
    short time series, 290,
Topography, 177,
Topographic features, 114,
Translation, 83,
Transference, 159, 160,
Tremors, 256,
    Hawaiian tremors, 256, 257,
Torque, 271, 276,
Triadic Cantor series, 229, 230, 231, 232,
    triadic set, 232,
Turbulence, 196, 197, 289,
    fully developped turbulence, 289, 290,
Turbulent motion, 279,

U

Unicity,
Union, 14, 49,
    set union, 8,
Unstability, 293,
Useful zone, 218,

Usso Unified system of seismic observation, 79, 80, 81,

V

Vanuatu, 244,
Variance, 196,
Variograms, 241,
Vectors, 262,
    vector edges, 121,
    vector of descriptors, 143,
    binary vectors, 128,
    magnetic field vector, 262,
Vesuvius volcano, 250, 251,
Vibrating strip, 293,
Viscous friction, 270, 271,
    viscous coefficient, 271,
Volcanic eruptions, 215, 246,
    volcanic eruption time series, 248,
Volcanology, **199**, **246**-257,
Vortices, 278, 279, 280,
    helical vortices, 279, 280,
Vote, 136, 138, 155, 157,
Voting, 150, 155, 157,
    Voting by elementary features VEF-i, 131,
    Voting by a set of features VSF-i, 131,

W

Waveform, 1, 72, 75, 82,
Wave length,
Wavelet, 82, 223, **240**,
    wavelet scalogram, 242,
    wavelet transform, 82, 83, 240, 242,
    Gaussian wavelet, 83, 240,
    Morlet wavelet, **82**,
    mexican hat wavelet transform, 240, 242,
Western Alps, **123**-145, 287,
Wharton Basin, 114,
Wharton fossil ridge, 115,
Window, 101, 102, 104,
    window point, 101,
    window size, 101, 102, 111, 114,

Index

moving window, 180, 182, 227, 260, 261, 262, 268,
running window, 106, 109,
Wolf number, 263,
Worlswide meteorological network, 192,
Wwssn World wide standard seismograph network, 78,

Z

Zadeh's t-norm, 12,
conjunction, 26,
Zero crossing detector, 67,

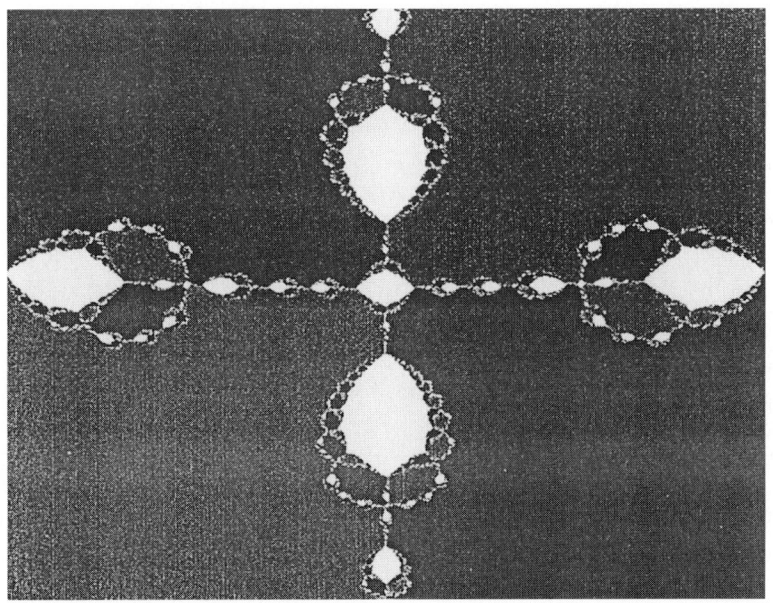

Julia sets associated with one-parameter family of dynamical systems (after Barnsley, 1988).

Druck: Strauss Offsetdruck, Mörlenbach
Verarbeitung: Schäffer, Grünstadt